Bibliography of the Philosophy of Technology

Bibliography of the Philosophy of Technology

Compiled by
Carl Mitcham
and
Robert Mackey

The University of Chicago Press
Chicago and London

This work initially appeared as volume 14, number 2, part 2 (April 1973) of Technology and Culture under the editorship of Carl Mitcham and Robert Mackey and published by the University of Chicago Press. Publication of the Bibliography has been made possible by generous grants from the John W. Hill Foundation of New York City and the Xerox Corporation of Stamford, Connecticut.

The University of Chicago Press, Chicago 60637
The University of Chicago Press, Ltd., London

Published 1973 by The University of Chicago Press for
The Society for the History of Technology
Printed in the United States of America

International Standard Book Number: 0-226-53195-3
Library of Congress Catalog Card Number: 72-96343

Contents

Bibliography of
the Philosophy of Technology

FOREWORD

Slightly more than a half-dozen years ago the Society for the History of Technology devoted a program session at its annual meeting and an issue of *Technology and Culture* (vol. 7, no. 3 [Summer 1966]) to the topic, "Toward a Philosophy of Technology." The great efflorescence of literature bearing on this topic during the past half-dozen years is a measure of our increasing awareness of the significant role played by technology in our daily lives, our social institutions, and our value schemes.

How are we to account for the previous neglect of the philosophy of technology and the recent growth of interest? Looking back into the intellectual tradition of Western man, we find that most systematic philosophic thought — including scientific thought — from the time of the Greeks onward was part of a speculative tradition far removed from the workaday concerns of the technical world. Then, at the beginning of the 17th century, Sir Francis Bacon argued that "The true and lawful goal of science is that human life be endowed with new powers and inventions." Fortified by the material conquests of the Industrial Revolution and by the growth of utilitarian modes of thought, Western man has, until recently, accepted this Baconian vision of knowledge as leading man to power over nature.

While the popular mind still holds to the idea that technology is "a good thing," the fear is being increasingly voiced that it might be too much of a good thing. Questioning of the beneficence of technology has become part of the questioning of the totality of institutions and value systems, as we recognize that technology is imbedded in the larger matrix of society. This *Bibliography* reveals how philosophic investigations into technology cover nearly every aspect of technology's involvement with man, society, and nature.

Readers of this *Bibliography* will note two facts whose explanation might be the concern of future intellectual — and technological — historians. First is the prevalence of European works in the early period. Second is the relative absence of professional philosophers among the authors cited; instead, most are historians, anthropologists, political scientists, sociologists, scientists, and "generalists"—and even some engineers, a group little noted for literary articulation of speculative thoughts.

Contemporary philosophers have concerned themselves primarily with problems of language systems and logical analysis. If they have dealt with the philosophy of technology at all, it has largely been through investigations of artificial intelligences, which involve language and logical systems. Now, however, Messrs. Mitcham and Mackey, by this meaningful compilation of publications in the philosophy of technology, have provided a foundation for other philosophers to apply the analytical and synthetical tools for their field to the philosophic problems revolving around technology and its meaning to man and society.

Finally, a word must be said about the interests of the Society of the History of Technology in encouraging philosophical discussions of technology. The interests of the philosopher and the historian converge on many central issues involving the meaning and role of technology in human affairs: How has technology served man and been served by man? What is the impact of technology on culture and societies, and vice versa? How does man achieve knowledge of and expertise in technology? How have technological changes altered our concepts of beauty, truth, justice? How has industrialized society transformed man's relations to his neighbor, the state, the society, and the natural environment? Above all, historians and philosophers are interested in how technological developments have helped change man's views of the meaning and purpose of life—in the past, present, and future. Because all of us should be concerned with such problems, our Society is vitally interested in the philosophy as well as history of technology.

MELVIN KRANZBERG
Editor-in-Chief, *Technology and Culture*
Callaway Professor
of the History of Technology,
Georgia Institute of Technology

Bibliography of
the Philosophy of Technology

INTRODUCTION

Awareness of the philosophical significance of modern technology has produced a body of literature for which a comprehensive bibliographical survey in English has been needed for some time. Although there exist bibliographies of the history of technology, technology and public affairs, philosophy of science, and the philosophy of artificial intelligence — all of which are relevant to the study of the philosophy of technology — none is specifically directed to this end. As a result, existing bibliographies fail, singly and in combination, as general references to the extensive philosophical literature surrounding technology.

As for historical and cultural bibliographies, Eugene S. Ferguson's *Bibliography of the History of Technology* (Cambridge, Mass.: M.I.T. Press, 1968) contains only a brief, two-item section on the philosophy of technology, although it does contain a larger, related section on technology and culture. (This latter section can, of course, be supplemented by the yearly bibliographies in *Isis* and *Technology and Culture*. Note, as well, Magda Whitrow, ed., *ISIS Cumulative Bibliography,* 2 vols. [London: Mansell, 1972].) Also of help with historical and cultural studies of technology are Maxwell H. Goldberg's *Needles, Burrs, and Bibliographies* (University Park: Pennsylvania State Univ. Center for Continuing Liberal Education, 1969), a resource manual on technological change, and the *Checklist for the Study of Science in Human Affairs,* edited by Elisabeth Weis and Geoffrey Dutton (New York: Columbia Univ. Institute for the Study of Science in Human Affairs, 1969), with its *Supplement* (October 1970).

In addition to these general works, there are a number of more specialized bibliographies available. "Social Implications of Technical Advance," prepared by S. C. Gilfillan and A. B. Stafford, with introductions by W. F. Ogburn and Gilfillan, is a valuable annotated bibliography on the social causes and social effects of invention, concentrating on the period 1945-53. It is published in *Current Sociology; La sociologie contemporaine* 1, no. 4 (1953-54):

187–266. Other, less directly relevant bibliographies in this series are "Social Implications of Technical Advance in Underdeveloped Countries: A Trend Report and Bibliography" by G. Balandier, *Current Sociology; La sociologie contemporaine* 3, no. 1 (1954–55): 5–75; "Sociology of Science; A Trend Report and Bibliography" by Bernard Barber, *Current Sociology; La sociologie contemporaine* 5, no. 2 (1956): 91–153; and "Industrial Sociology 1951–62; A Trend Report and Bibliography" by Jean-René Tréanton and Jean-Daniel Reynaud, *Current Sociology; La sociologie contemporaine* 12, no. 2 (1963–64): 123–245.

Of more immediate service are eight specialized *Research Reviews* published under the direction of the Harvard Program on Technology and Society (Cambridge, Mass.: Harvard Univ. Press, 1968–71): I. Taviss and J. Koivumaki, eds., *Implications of Biomedical Technology* (1968); I. Taviss and W. Gerber, eds., *Technology and Work* (1969); I. Taviss and L. Silvermann, eds., *Technology and Values* (1969); I. Taviss and J. Purbank, eds., *Technology and the Polity* (1969); I. Taviss, J. Purbank, and J. Rothschile, eds., *Technology and the City* (1970); I. Taviss, ed., *Technology and the Individual* (1970); I. Taviss and J. Purbank, eds., *Implications of Computer Technology* (1971); J. H. Weiss, ed., *Technology and Social History* (1971). Each *Research Review* is about fifty pages in length and contains a brief "state of the art" essay followed by substantial abstracts of a number of the relevant articles and books. Together with the Program's 35 "Assession Lists," these *Reviews* have proved extremely useful in preparing the present bibliography. But although all these bibliographies of cultural and historical analyses provide a good survey of relevant background material and some specialized philosophical works, they do not deal adequately with the larger body of philosophical literature on the subject of technology.

The same is true with bibliographies of technology and public affairs. *Science, Technology, and Public Policy,* a two-volume annotated bibliography edited by Lynton K. Caldwell (Bloomington, Ind.: Department of Government, Indiana Univ.; and Washington, D.C.: National Science Foundation, 1969) is undoubtedly the most complete in this area. Yet while it, too, contains a section on the philosophy of technology, the selection of works is limited and the annotations are sketchy. *Science Policy Reviews,* published quarterly by the Battelle Memorial Institute of Columbus, Ohio, also contains good information in the area of technology and public affairs and is thus a good supplement to the Caldwell bibliography.

Bibliographies in the standard works and anthologies in the philosophy of science might be expected to be of more relevance to the philosophy of technology. Among these bibliographies are those in Philipp Frank's *Philosophy of Science* (Englewood Cliffs, N.J.: Prentice-Hall, 1957), Israel Sheffler's *Anatonomy of Inquiry* (New York: Knopf, 1963), Carl Hempel's *Aspects of Scientific Explanation* (New York: Free Press, 1965), H. Feigl and M. Brodbeck's *Readings in the Philosophy of Science* (New York: Appleton-Century-Crofts, 1953), A. Danto and S. Morgenbesser's *Philosophy of Science* (Cleveland: World, 1960), and Arthur Pap's *An Introduction to the Philosophy of Science* (New York: Free Press, 1962). But because the standard philosophies of science tend to ignore technology as an independent subject, here the results are even less rewarding than with the two types of bibliographies already mentioned.

With regard to literature on artificial intelligence, the most extensive general and specialized bibliography—including at least two sections of philosophically relevant material—is Marvin Minsky's "A Selected Descriptor-indexed Bibliography to the Literature on Artificial Intelligence," *IRE Transactions on Human Factors in Electronics*, HFE-2 (March 1961), pp. 39–55. This has been revised and is easily available in Edward Feigenbaum and Julian Feldman, eds., *Computers and Thought* (New York: McGraw-Hill, 1963), pp. 453–523. Despite the descriptor-index, however, it suffers from the absence of annotations. The same is true with the bibliographies in Alan R. Anderson, ed., *Minds and Machines* (Englewood Cliffs, N.J.: Prentice-Hall, 1964); and in Kenneth W. Sayre and Frederick J. Crosson, eds., *The Modeling of Mind* (Notre Dame, Ind.: Univ. Notre Dame Press, 1963). There are some very helpful notes on current research and attitudes in Zenon W. Pylyshyn, ed., *Perspectives on the Computer Revolution* (Englewood Cliffs, N.J.: Prentice-Hall, 1970). Nevertheless, there is no substitute for a good working bibliography. The further limitation, of course, is that intelligent artifacts are only one element of modern technology, and none of these bibliographies makes any real attempt to relate its special subject to the larger field of the philosophy of technology.

Outside the brief, unannotated section on "Philosophy and Technology" in George F. McLean's *A Bibliography of Christian Philosophy and Contemporary Issues* (New York: Ungar, 1967)—which uncritically picks up a number of works from the short section on "Technique et cultura" in G. A. DeBrie's *Bibliographia philosophica, 1934–1945*, 2 vols. (Editiones Spectrum, 1950–54)—the only bibliographies to deal explicitly with the philosophy of technology are in foreign languages; and of these the best are in German. The two most important German bibliographies are by Friedrich Dessauer in his book *Streit um die Technik* (Frankfurt: Knecht, 1956) and by Erwin

Herlitzius entitled "Technik und Philosophie" in *Informationsdienst Geschichte der Technik* (Dresden) 5, no. 5 (1965): 1–36.

The bibliography by Dessauer is arranged chronologically, beginning with 1807, to reflect "the awakening of self-consciousness concerning technology and the independence of interest in it from contemporary circumstances" and contains entries on over seven hundred books and articles. Nevertheless, Dessauer readily admits that "Completeness has not been attempted, as it seemed to me impossible. The theme of the book determines the selection. Not technology purely and simply, but the basis of technology and the controversy concerning that basis as well as the illumination of technology as a unity in social consciousness—these are the problems." Dessauer's bibliography is thus oriented around his own attempt to give a Kantian critique of technology. Although not intended to be limited to German sources, Dessauer admits that "literature in the German language predominates" since "foreign literature was not accessible to me to the same degree." At the same time, he may well be right when he suggests that "German literature is the richest in our theme and probably contains almost all the important subjects." Nevertheless, there are a number of important omissions, and naturally there have been significant additions to the literature since his survey was completed.

The bibliography by Herlitzius is explicitly limited to German works. It is divided into three sections, each arranged alphabetically by author. The three major divisions are: "Independent Publications up to 1945"; "Independent Publications from 1945–1964"; and "Selected Works from Twenty Periodicals of the German Democratic Republic." Although containing good references to East German Marxist literature during the periods covered, it is otherwise limited. It is much less complete than Dessauer. A more complete and analytic bibliography of this same material has been announced, but, so far as is known, has not yet appeared. It should also be noted that neither the Dessauer nor Herlitzius bibliographies is annotated.

There are, of course, still other relevant bibliographies—not only in English and German, but also in French, Dutch, Italian, Spanish, and Russian—but usually only in specialized areas or as appended to books. These have been cited or mentioned in annotations where they are known. It is not unlikely that there are unknown bibliographies, especially in the Scandinavian and East European languages which we have not been able to consult adequately. But the major fact remains: there is none readily available to the English-speaking reader.

The present bibliography, by concentrating on the period 1925-72, is, then, intended to remedy a deficiency which cannot help but become more acute with the inevitable growth of the philosophy of technology. Yet in so doing no attempt has been made to duplicate the Dessauer and Herlitzius bibliographies already cited; instead, we have consciously chosen to build upon them, especially insofar as articles are concerned. When an article in German pre-1956 is cited by Dessauer, it is rarely included in the present work; and the same is true for East German periodical literature cited by Herlitzius. Our bibliography concentrates, as far as German periodical literature is concerned, on the periods following those surveyed by Dessauer's and Herlitzuis's works. We have tried to include annotations when these bibliographies ought to be consulted further. This policy points up the general fact that we have been more selective in citing foreign than English works.

For purposes of organization, we have adopted a topical as opposed to a simple alphabetical or chronological framework. Thus the present bibliography is divided into five major categories in the following manner:

I. COMPREHENSIVE PHILOSOPHICAL WORKS. — Books which take technology as a primary theme for reflection and deal with it in terms of two or more of the categories listed below. Includes important collections of articles, individual articles by the major philosophers of technology, and virtually all essays with titles like "philosophy of technology" or "philosophy and technology."

II. ETHICAL AND POLITICAL CRITIQUES
    A. PRIMARY. — Basic analyses of the ethical and political implications and problems of modern technology.
    B. SECONDARY. — Less philosophically valuable analyses on this same theme, along with works on man, society, and civilization in relation to technology. Includes marginal topics such as: technology and economics, automation and labor, futurology, technology transfer, technology assessment, technology and law, technology and education, environment and ecology. Not exhaustive.
    C. APPENDIX: SOVIET AND EAST EUROPEAN MATERIALS. — Lists mainly some Soviet and East German items, many of which could not be adequately checked. Many might be of value, but are suspected of being highly ideological.

xiv

III. RELIGIOUS CRITIQUES
   A. PRIMARY. — Basic analyses of the religious implications
      and problems of modern technology.
   B. SECONDARY. — Less philosophically valuable analyses on
      this same theme. Includes some materials on the emerging
      theology of ecology.

IV. METAPHYSICAL AND EPISTEMOLOGICAL STUDIES
   Primarily studies from within the West European traditions of
   philosophy — especially existentialism and phenomenology
   — on the metaphysics of technology, along with more
   epistemological studies on cybernetics and artificial intelligence from within the Anglo-American analytic tradition.

V. APPENDIX
   A. CLASSICAL DOCUMENTS. — Important philosophical works
      and related materials mostly from the period prior to
      1925.
   B. BACKGROUND MATERIALS. — Includes the basic histories
      of technology, essays on the relation between technology
      and history, and historico-philosophical studies which are
      an important background for philosophical reflection on
      technology. Some literary-philosophical works as well.
      Not exhaustive.

   This outline is, it is hoped, largely self-justifying and
self-explanatory. Still, a few clarifying remarks are no doubt in order.
   As all students of this field will readily admit, the subject is not well
defined. What is the philosophy of technology? It is easier to say what
it is not — that is, it is not the philosophical examination of science, or
of some restricted aspect of technology such as cybernetics or automation, although reflections on these specialties often have more
general implications. The general problem of the relation between
science and technology presents, however, a special difficulty. While
most philosophies of science identify technology with applied science,
some philosophies of technology argue that modern science is essentially theoretical technology. Both tendencies blur our commonsense
distinctions, so that sometimes a thinker will use the term "science"
to refer to what might more popularly be designated "technology"
and vice versa. This is especially true with works on the "social
implications of science." Confronted with this ambiguity, we have
simply tried to cite important works which, while using the term
"science," mean to refer to technology as well.

But what is technology? This, it turns out, is one of the central questions for the philosophy of technology. The philosophy of technology attempts to grasp the ultimate nature of technology — to answer the basic questions "What is it?" and "What does it mean?" The difficulty is that different thinkers answer such questions in radically different ways, reflecting, no doubt, their own philosophic predelictions. Some consider the basic philosophical issues raised by technology to be metaphysical and/or epistemological in character, others take them to be ethical-political, and still others religious. It is because of this, and in an attempt to avoid undue restrictions, that this bibliography adopts the categories it does. Nevertheless, it is well to mention that the actual classification of material was not always easy. Also, as a consequence of organizing the bibliography according to subject, works by one author do not always appear together, and works in one category are sometimes relevant to others as well. Still, given our limitations of space, there are few cross-references or double entries.

By entitling some categories "critiques," we do not refer only to negative evaluations; the word is used in a stricter sense to mean "a careful analysis." At the same time, the careful reader will discover works of unequal philosophic merit included within a single subdivision, so that questions may arise about the intentions behind the primary-secondary distinction in Parts II and III. We have tried to include all works of high philosophic merit within the primary subdivision, but philosophical acumen has not been the sole criterion. Our primary intention has been to list representatives of the main alternative positions or arguments relevant to that category — and the truth is that some alternatives have had more thoughtful spokesmen than others. Furthermore, if one piece by an author is listed in the primary subdivision of some category, we have often included related works by the same author within this same subdivision. Sometimes this has been done simply by collapsing a number of separate entries into the annotation. (Indeed, this technique has been employed extensively throughout the bibliography.) And for a number of influential works a list of some important reviews and/or commentaries are also cited.

Instead of the standard practice of listing all works by a single author alphabetically by title, they are ordered here chronologically by date of publication. In our opinion this gives a better immediate indication of the development and scope of a particular author's work.

Finally, a word about the principles of citation. No unexamined foreign-language books have been included unless they have been cited either in the *National Union Catalog* or in two independent sources. Where it seemed advisable and possible, titles have been translated and placed in brackets. For foreign-language works available in English, the translation is cited, usually with a reference to the original. (For English-language books we have preferred to cite American editions where there was a choice.) In principle, we have meant to annotate important and/or hard-to-secure sources more extensively than others. And only when they have been annotated or deserve special notice, are articles cited independently of general collections.

As for the bibliography as a whole, although it is obviously unreasonable to expect to have included every work which rightfully belongs, we do hope to have listed the majority of works which are truly important and a good deal of those which (without proper annotation) might otherwise be thought to be important – since one main function of a bibliography of this type is to steer others away from deadwood encountered at one's own expense. This last-mentioned principle accounts for perhaps more material than it ought to, especially in Part II, Section B. But when in doubt we tended to err on the side of inclusiveness. Be that as it may, we would welcome comments or suggestions from readers, both with regard to omissions and inclusions – as well as annotations in those cases where we have been unjust or inadequate – for we hope to publish supplements to this work periodically. Readers may also wish to consult the *Philosopher's Index* under the subject "Technology," and the new section on "Philosophie de la culture et de la technique" which was added in 1971 to the *Repertoire bibliographique de la philosophie*, for supplementary bibliographical information.

As always with a work of this sort there has been more than a little help from our friends. The perfunctory character of acknowledgments should not, however, obscure our great debts. First, where others have contributed annotations to this bibiliography, their entries are initialed; the names behind these initials (roughly in order of the amount of contribution they made) are: Richard J. Rundell, Mary Ann Stanley, James Vincent, and Forest Williams. Special thanks should also be rendered to William Carroll, Paula Nye Ellis, Gerald Janacek, and Kris Kogerma for a great deal of otherwise unacknowledged translation work – as well as to Merlyn Baird, Irene Taviss of the Harvard Program on Technology and Society, and Virginia Boetcher and the staff of the Interlibrary Loan Department of Norlin Library at the University of Colorado for research assistance. Andrew Baskin, Sue Weddington, and especially Yvonne Williams all deserve thanks for typing the manuscript. Miss Williams was also of great assistance in reading proofs. A grant from the Committee on

Professional Growth of Berea College has aided in the preparation of this bibliography.

CARL MITCHAM

ROBERT MACKEY

I. COMPREHENSIVE PHILOSOPHICAL WORKS

ASSUNTO, ROSARIO. *L'integrazione estetica, studi e ricerche* [Aesthetic integration, studies and research]. Milan: Edizioni di Comunità, 1959. Pp. 102. Philosophical discussion centering on the relation between art and technology.

AUZIAS, JEAN MARIE. *La philosophie et les techniques.* Paris: Presses Universitaires de France, 1965. Pp. 127. A brief historical survey of the different attitudes toward technology from the classics to the present.

_____. *Clefs pour la technique* [Keys for technology]. Paris: Seghers, 1966. Pp. 192. A more analytic approach to technology. Topics discussed are: technology and machines; problems of technology and society; cultural problems. The conclusion contains a brief discussion of the relation between *technique* and *poesis.*

BECK, HEINRICH. *Philosophie der Technik: Perspektiven zu Technik, Menschheit, Zukunft* [Philosophy of technology; perspectives on technology, mankind, future]. Trier: Spee-Verlag, 1969. Pp. 226. A negative assessment undertaken from a Heideggerian perspective. Technology is conceived as the "encounter of the human spirit with the world, in which man further molds and alters . . . nature . . . to his purposes according to the measure of the known laws of nature" (p. 31). These purposes are generally defined as the "mobilization and masking of nature" (p. 41).

_____. "Die philosophische Frage mach der Zukunft und das Ereignis der Technik" [The philosophical problem facing the future and the event of technology], *Wissenschaft und Weltbild* 22 (1969): 122-32.

BENSE, MAX. *Technische Existenz, Essays.* Stuttgart: Deutsche Verlags-Anstalt, 1949. Pp. 250.

_____. "Oswald Spengler, Die Technik als Taktik des Lebens." Pp. 425-32 in *Die Philosophie.* Frankfurt: Suhrkamp, 1951.

_____. "Die spirituelle Reinheit der Technik" [The spiritual purity of technology]. In *Plakatwelt; vier Essays.* Stuttgart: Deutsche Verlags-Anstalt, 1952. A comprehensive essay delineating the nature of technical existence and thought. – R. J. R.

_____. "Das technische Bewusstsein des modernen Menschen" [The technological consciousness in modern man], *Vortragsreihe des Deutschen Industrieinstituts* (Cologne), vol. 26 (1953). Pp. 5.

_____. "Philosophie der Technik," *Physikalische Blätter* 10, no. 11 (1954): 481-85.

_____. "Technik und Aesthetik," *Das Kunstwerk; eine Monatsschrift für alle Gebiete der bildenden Kunst* 11, no. 3 (1957): 3-10.

_____. *Die Idee der Politik in der technischen Welt.* Dortmund: Kulturamt der Stadt Dortmund, 1960. Pp. 8.

_____. "Der Mensch im Technischen Zeitalter; Theorie der Meta-Technik," *Vorträge anlässlich der Hessischen Hochschulwochen für staatswissenschaftliche Fortbildung* (Bad Homburg V.D. Höhe, Berlin) 34 (1962): 192-203.

_____. *Ungehorsam der Ideen; Abschliessender Traktat über Intelligenz und technische Welt* [Insubordination of ideas; concluding tract on intelligence in the technological world]. Cologne-Berlin: Kiepenheuer & Witsch, 1966. Pp. 96. Technology is perceived as an integral development in the history of philosophy. – R. J. R.

BERNHART, JOSEPH. *Der technisierte Mensch.* Abendländische Reihe, vol. 4. Augsburg: Naumann, 1946. Pp. 47. Study by a German Catholic theologian.

_____. "Technik und Menschenseele" [Technology and human soul], *Theologische Quartalschrift* 135 (1955): 1-27. Included in *Joseph Bernhart: Gestalten und Gewalten* (Würzburg, 1962).

BORANETSKII, P. *Filosofia tekhniki; tekhnika i novoe mirosozertsanie* [Philosophy of technology; technology and the new outlook]. Paris, 1947. Pp. 222. A comprehensive study by a Russian expatriot. Chapter titles: "Introduction" (in which technology is outlined as an object of contemplation), "Understanding Technology," "Overcoming Law and Surmounting Necessity," "Victory over Space," "Ancient and Modern Wonders," "Revelation and Invention," "Ideal and Real Value," "Material Liberation," "Technology and Spiritual Liberation," "Victory over Illness, Old Age, and Death," "Technology and the Basic Attitudes toward the World," "Objections to Technology," and "Conclusion." Despite a recognition of problems, the author remains basically optimistic, as is indicated by the last section on the "Necessity of a New, Higher Spirituality." – M.A.S.

BRANDENSTEIN, BELA VON. "Philosophie und Technik," *Wissenschaft und Weltbild* 23 (1970): 44–50.

BRINKMANN, DONALD. *Mensch und Technik; Grundzüge einer Philosophie der Technik* [Man and technology; the main features of a philosophy of technology]. Bern: Franke, 1945. Pp. 167. Table of contents: "Philosophy and Technology," "Technological Elements in Philosophical Thinking," "The Essential Nature of Technology," "Technological Man," "Concluding Considerations."

———. "Geistige Grundlagen der modernen Technik" [Spiritual foundations of modern technology], *Universitas; Zeitschrift für Wissenschaft, Kunst und Literatur* (Stuttgart) 8 (1953): 289–94.

———. "Technik und Naturwissenschaft," *Schweizerische Bauzeitung* 72, no. 1 (1954): 1–3.

———. "Technik und Naturwissenschaft im Zeitalter der Wasserstoffbombe" [Technology and natural science in the age of the hydrogen bomb], *Carinthia Una* 146 (1956): 178–89.

———. "Der Mensch im Zeitalter der Automation." Pp. 96–113 in *Festschrift H. J. de Vleeschauwer.* Pretoria, South Africa: Publications Committee, Univ. South Africa, 1960.

———. "Mensch und Technik," *Schicksalsfragen der Gegenwart; Handbuch politische-historischer Bildung* (Tubingen) 5 (1960): 60–82.

———. "Technology as Philosophic Problem," *Philosophy Today* 15, no. 2 (Summer 1971): 122–28. Brief examination of interrelations between technology and philosophy, followed by a critical discussion of four theories: technology as applied science, economic means, neutral means, and aspiration for power. Concluding argument: technology is modern man's attempt to take salvation into his own hands; it is an active, this-worldly religiosity. Translation of "Die Technik als philosophisches Problem," *Herders Zeitbericht* (Freiburg: Herder, 1963), cols. 1235–46. For a number of other articles pre-1956, see Dessauer's bibliography.

BROWNHILL, R. J. "Towards a Philosophy of Technology," *Scientia* 104, no. 7 (November–December 1969): 602–14. A critique of both excessively realistic and neo-Kantian theories of science and technology (in the persons of M. Polanyi and M. Oakeshott, respectively). Technology equals applied science. But it has more than just a utilitarian value; it has intrinsic value as increasing our knowledge of appearances. "Metaphysics lays the ground for a further apprehension of reality, and the task of the scientist is to stablize this extremely uncertain knowledge. . . . In this sense it is closely allied to technology which ideally is an attempt at stabilizing to an even greater extent our theoretical knowledge by making it practical." At the same time, metaphysics, science, and technology are all "involved in the task of comprehending nature for the sake of mastering our environment."

BRUIN, P. DE. "Philosophie der Technik," *Studia catholica* 13 (1937): 436–64.

BULLE, GERHARD. *Die Technik als philosophisches Problem; zur Kritik der Gegenwartsphilosophie* [Technology as a philosophical problem; on a critique of contemporary philosophy]. Erfurt: Koenig, 1934. Pp. 147. A dissertation.

Civilization technique et humanisme. Bibliothèque des archives de philosophie, n.s. 6. Paris: Beauchesne, 1968. Pp. 292. A colloquium of the International Academy of the Philosophy of

Sciences. Section (1) on technology and culture contains M. Barzin's "Valeurs et technique," S. Watanabe's "La simulation mutuelle de l'homme et la machine," and F. P. A. Tellegen's "Einige Betrachtungen über Technik und Kultur"; Section (2) on technology and sociology contains A. Catemario's "Technique sociale et reconstruction." V. Tonini's "La rationalité technologique dans la sociologie moderne de la connaissance," and J.-L. Destouches's "Développement de la technique et structuration sociale"; Section (3) on technology and philosophy contains S. Breton's "Réflexion philosophique et humanisme technique," J. Hommes's "Die humanistiche Bewältigung des wissenschaftlich-technischen Zeitalters," J. Hollak's "Technik und Dialektik," M. Bunge's "Towards a Philosophy of Technology," J. Ladrière's "Technique et eschatologie terrestre," S. S. Acquaviva's "Technique et désacralisation de l'homme," and F. Gonseth's "Valeur et défense de la personne dans une civilisation technicienne."

COHEN, JOSEPH W. "Technology and Philosophy," *Colorado Quarterly* 3, no. 4 (Spring 1955): 409-20. "Dewey's instrumental theory of values taken together with his philosophy of experience . . . reveals that his philosophical aim is an integration of science, technology, democracy, ethics, and art." Thus pragmatism is the philosophy which most readily takes account of the technological reality and faces up to the problems it poses. Weakness of the article is that this argument tends to be left at the level of mere assertion.

CONZE, EDWARD. "Philosophers and Techniques," *Hibbert Journal* 55, no. 1 (October 1956): 14-19. Different philosophical theories are constructed on top of different techniques. "Far from being . . . universal, the validity of scientific propositions is strictly relative to scientific technology, and circumscribed by the range of its effectiveness. . . . Those facets of reality which respond to more loving and considerate methods, such as yogic, astrological, and so on, must remain hidden from them" (p. 19).

DESSAUER, FRIEDRICH. *Technische Kultur? Sechs Essays* [Technical culture? Six essays]. Kempten – Munich: Kosel, 1908. Pp. 57.

_____. *Leben, Natur, Religion; das Problem der transzendenten Wirklichkeit* [Life, nature, religion; the problem of transcendent reality]. Bonn: F. Cohen, 1924. Pp. 140.

_____. *Philosophie der Technik; das Problem der Realisierung* [Philosophy of technology; the problem of its realization]. Bonn: F. Cohen, 1927. Pp. 180. First complete statement of the main themes of Dessauer's philosophy of technology. Reviewed by J. Perthel, "Philosophie der Technik – eine Auseinandersetzung mit dem Buch von Friedrich Dessauer," *Christengemeinschaft* 7, no. 2 (1930): 35-40. Part II, "Technology in Its Proper Sphere," translated in C. Mitcham and R. Mackey, eds., *Philosophy and Technology* (New York: Free Press, 1972).

_____. *Technik als Sinn und Bestimmung* [Technology as meaning and destiny]. Einsiedel: Benzinger, 1944. A short pamphlet of an article first published in *Schweizerische Rundschau* 44 (1944): 154-77.

_____. *Mensch und Kosmos* [Man and cosmos]. Frankfurt: Knecht-Carolusdruck, 1949. Pp. 110. A historico-philosophical examination of modern scientific cosmology.

_____. "Technik, Wirtschaft und Gesellschaft," *Theologische Quartalschrift* 133 (1953): 257-77.

_____. *Kultur, Technik und Gesellschaft* [Culture, technology and society]. Wolfshagen-Scharbeutz: F. Westphal, 1954. Pp. 16.

_____. *Streit um die Technik* [The controversy about technology]. Frankfurt: Knecht, 1956. Pp. 471. 2d ed., 1958, pp. 480; abridged ed. (Freiburg: Herder, 1959), pp. 205. This is Dessauer's *magnum opus*, a survey of the history of ideas and arguments concerning technology, with a restatement of his own position. Reviewed by H. Muckermann, *Humanismus und Technik* 4, no. 3 (June 15, 1957): 199.

――――. *Naturwissenschaftliches Erkennen; Beiträge zur Naturphilosophie* [Natural scientific knowledge; contribution to a philosophy of nature]. Frankfurt: Knecht, 1958. Pp. 445. Contains a philosophy of science to complement his philosophy of technology.

――――. *Prometheus und die Weltübel* [Prometheus and world-calamities]. Frankfurt: Knecht, 1959. Pp. 204. Christian man battles against moral evil, Promethean man against physical evils (pain, sickness, natural disasters, limitation). Thanks to technology, man can deal successfully with things, but can he deal with himself? To save himself, Prometheus requires Christ. – R. J. R.

――――. *Durch die Tore der neuen Zeit, ein Ausblick* [Through the gate of the new era; an outlook]. Göttingen: Sachse & Phol., 1961. Pp. 63. An autobiographical memoir. – R. J. R.

――――. "Technik und Gesellschaft," *Studium generale* 15 (1962): 490-93.

DESSAUER, FRIEDRICH, and KARL AUGUST MEISSINGER. *Befreiung der Technik* [Liberation of technology]. Stuttgart-Berlin: Cotta, 1931. Pp. 120. Concerned with "liberating" technology from various misunderstandings.

DESSAUER, FRIEDRICH, and XAVIER VON HORNSTEIN. *Seele im Bannkreis der Technik* [Man's soul under the influence of technology]. Olten: Otto Walter, 1945. Pp. 307. Major parts of the book: "Meaning and Mission of Technology" (Dessauer), "Technology Impresses Itself upon Men" (Dessauer), "The Soul of the Technologist" (Dessauer), "Theologico-Philosophical Access to Technology" (Hornstein), "A Timely Ideal: the Economic Saint" (Dessauer and Hornstein).

[NOTE: The Dessauer corpus contains the most sustained study of technology available in German. Its ultimate aim is to place beside the three Kantian critiques a fourth critique of the technological realm. Dessauer defines technology as "real being from ideas through purposeful construction and active working out which uses resources given in nature." This definition is first developed in *Philosophie der Technik* and further elaborated in *Streit um die Technik*, his two most important books. The best bibliography of his work through 1956 is in *Streit*. The best study of Dessauer is Klaus Tuchel's *Die Philosophie der Technik bei Friedrich Dessauer* (Frankfurt: Knecht, 1964), which also contains supplementary bibliography references. But Dessauer's work extends from travel letters, scientific biography, and jurisprudence, to philosophy of science, and theology, and there does not yet exist a complete bibliography of all this material. For two critical studies from a Marxist perspective, see G. Bohring, "Der dialektische Materialismus und der 'Streit um die Technik,'" *Technik* 15, no. 6 (1960): 389-92; and Horst Jacob, "Die idealistische Technikphilosophie Friedrich Dessauer – eine religiös verbrämte Apologetik des westdeutschen Imperialismus und Militarismus," *Deutsche Zeitschrift für Philosophie* 8, no. 8 (1960): 974-86.]

DIEMER, ALWIN. "Philosophie der Technik." Pp. 540-49 in *Grundriss der Philosophie*, Vol. 2: *Die Philosophischen Sonderdisziplinen*. Meisenheim: Anton Hain, 1964. A brief sketch which forms part of a larger systematic study of philosophy, After examining the concept and phenomenon of technology, distinguishes between two types of philosophical theories of technology – ontological (i.e., anthropological) and metaphysical theories. Concludes by outlining three problems: the nature of technology itself, the relation between technology and man, and the relation between technology and culture.

DUCASSÉ, PIERRE. "Les techniques et la philosophie contemporaine." Pp. 1048-50 in *Proceedings of the Xth International Congress of Philosophy; Amsterdam, 1948*. Amsterdam: North-Holland, 1949.

———. "Science, Technology, and Leisure," *Impact of Science on Society* 3 (1952): 26-42.

———. "Technocratie ou sagesse?" *Revue de synthèse* 74 (1953): 69-74.

———. "Intelligence technique et culture ouvert," *Structure et évolution des techniques,* vol. 6, nos. 41-42 (February-August 1955).

———. *Les techniques et le philosophe.* Paris: Presses Universitaires de France, 1958. Pp. 176. An uneven work. Primarily concerned with how philosophy ought to face technology. Begins by noting the historical tendency of philosophy to either dissociate itself from technology (ancients) or be subsumed within it (moderns). Tries to indicate a path between these two extremes. Although it contains some interesting arguments, the book is written in an oracular style that undercuts much potential usefulness. But see also the same author's *Historie des techniques* (Paris: Presses Universitaires de France, 1958).

ELLUL, JACQUES. *The Technological Society.* Translated by J. Wilkinson. New York: Knopf, 1964. Pp. 449. From *La technique, ou l'en jeu du siecle* (Paris: Colin, 1954). Ellul's major work in which modern technology is analyzed as a social phenomenon. Technique is defined as "the totality of methods rationally arrived at and having absolute efficiency (for a given stage of development) in every field of human activity."

———. *Propaganda; the Formation of Men's Attitudes.* Translated by K. Kellen and J. Lerner. New York: Knopf, 1965. Pp. 320. From *Propagandes* (Paris: Colin, 1962). Some reviews: C. Lasch, *Nation* (April 4, 1966); C. Mitcham, *Cross Currents,* vol. 17, no. 1 (Winter 1967); M. McLuhan, *Book Week* (November 28, 1965). See also the author's "Information and Propaganda," *Diogenes* 18 (Summer 1957): 61-77.

———. *The Political Illusion.* Translated by K. Kellen. New York: Knopf, 1967. Pp. 258. From *L'illusion politique, essai* (Paris: Laffont, 1964).

[NOTE: The above three works form a kind of tripartate study of technology. But for a full appreciation of the perspective from which Ellul writes, they should be read in conjunction with his *The Presence of the Kingdom* (New York: Seabury, 1951; reprinted 1967). Other works are also available in English. But see especially the following articles.]

———. "Modern Myths," *Diogenes* 23 (Fall 1958): 23-40.

———. "Western Man in 1970." Pp. 27-64 in B. de Jouvenel, ed., *Futuribles: Studies in Conjecture.* Geneva: Droz, 1963.

———. "The Biology of Technique," *Nation* (May 24, 1965), pp. 567-69. This is part of a dialogue with Robert Theobald elicited by the latter's review of *The Technological Society* in *Nation* (October 19, 1964). Other reviews: R. A. Hall, *Scientific American* (February 1965); C. E. Silberman, *Fortune* (February 1966); H. Wheeler, "Means, Ends and Human Institutions," *Nation* (January 2, 1967); C. G. Benello, *Our Generation,* vol. 4, no. 4 (1967); R. A. Nisbet, "The Grand Illusion: an Appreciation of Jacques Ellul," *Commentary* (August 1970). See also J. Y. Holloway, ed., *Introducing Jacques Ellul* (Grand Rapids, Mich.: Eerdmans, 1970); and C. Mitcham and R. Mackey, "Jacques Ellul and the Technological Society," *Philosophy Today,* vol. 15, no. 2 (Summer 1971).

———. "The Technological Revolution and Its Moral and Political Consequences." Pp. 97-107 in J. Metz, ed., *The Evolving World and Theology.* New York: Paulist Press, 1967.

———. "Between Chaos and Paralysis," *Christian Century* (June 5, 1968), pp. 747-50.

———. *Critique of the New Common Places.* New York: Knopf, 1968. Pp. 303. Forceful analyses of a number of contemporary clichés. See especially the chapters on "The Machine Is a Neutral Object and Man Is Its Master" and "It Is Fashionable to Criticize Technology."

———. "Techniques, Institutions, and Awareness," *American Behavioral Scientist* 11, no. 6 (July-August 1968): 38-42.

———. "Technological Morality." Pp. 185–98 in *To Will and to Do*, translated by C. Edward Hopkin. Philadelphia: United Church Press, 1969.

———. "Cain, Theologian of 1969," *Katallagete* (Winter 1969), pp. 4–7.

ENGELMEYER, P. K. VON. "Philosophie der Technik." Pp. 587–96 in *Atti del 4. Congresso internazionale di filosofia; Bologna*. Vol. 3. Genoa, 1911. See also three other pieces by this same author: "Grundriss der Philosophie der Technik," *Köln Zeitung*, no. 608 (1894); "Allgemeinen Fragen der Technik," *Dinglers Polytechnishes Journal* (Berlin-Stuttgart), vol. 80 (1899); and "Vorarbeit zur Philosophie der Technik," *Technik und Kultur* (Verband deutscher diplom-ingenieure, Berlin) 18 (1927): 85.

FREYER, HANS. "Philosophie der Technik," *Blätter für deutsche Philosophie* 3, no. 2 (1929): 192–201.

———. *Theorie des gegenwärtigen Zeitalters* [Theory of the present age]. Stuttgart: Deutsche Verlags-Anstalt, 1955. Pp. 259.

———. *Das soziale Ganze und die Freiheit der einselnen unter den Bedingungen des industriellen Zeitalters* [The social whole and the freedom of the individual under the conditions of the industrial age]. Göttinger: Musterschmidt, 1957. Pp. 34. From a pro-capitalist point of view, Freyer gives essentially a socioeconomic description of individual freedom in the modern industrialized state. Technology is equated with emancipation from natural limitations.—R. J. R.

———. *Über das Dominantwerden technischer Kategorien in der Lebenswelt der industriellen Gesellschaft* [Concerning the domination of technical categories in the life of industrial society]. Mainz: Akademie der Wissenschaften und der Literatur, 1960. Pp. 15.

FREYER, HANS, JOHANNES C. PAPALEKAS, and GEORG WEIPPERT, eds. *Technik im technischen Zeitalter; Stellungnahmen zur geschichtlichen Situation* [Technology in the technological era; attitudes toward the historical situation]. Düsseldorf: Schilling, 1965. Pp. 414. A major anthology. Contents: G. Weippert's "Einführung," M. Rassem's "Bemerkungen zur Entstehung der modernen Technik," W. Gerlach's "Naturwissenschaft im technischen Zeitalter," H. Freyer's "Der Ernst des Fortschritts," A. Gehlen's "Anthropologische Ansicht der Technik," F. Jonas's "Technik als Ideologie," H. Klages's "Marxismus und Technik," H. Schomerus's "Schöpfung und Nichtung," W. Schöllgen's "Die Begegnung von moderner Technik und christlichem Daseinsverständnis," P. R. Hofstätter's "Das Stereotyp der Technik," E. Forsthoff's "Technisch bedingte Strukturwandlungen des modernen Staates," J. C. Papalekas's "Herrschaft—technisch herausgefordert," K. Oftinger's "Konfrontation der Technik mit dem Recht," H. Kesting's "Der neue Leviathan," F. Ronneberger's "Die emanzipierte Verwaltung," E. Schneider's "Technik und Wehrwesen," H. Ritschl's "Die Stellung der Technik in den modernen Wirtschaftsordnungen," E. Leitherer's "Technik und Konsum," H. Bausinger's "Technik und Volkstum," and A. Seifert's "Technik in der Landschaft."

FRIEDT, HEINZ. "Technik und Philosophie," *Wissenschaftliche Zeitschrift der Martin-Luther-Universität Halle-Wittenberg; Gesellschafts- und sprachwissenschaftliche Reihe* 11, no. 10 (1962): 1315–28. Other works by this author listed under Ethical and Political Critiques, Section C.

GLOCKNER, HERMANN. *Philosophie und Technik. Fragen der Zeit*, vol. 1. Krefeld: Agis-Verlag, 1953. Pp. 28.

GOLDBECK, GUSTAV. "Die Philosophie der Technik und die Realität" [The philosophy of technology and reality], *Humanismus und Technik* 10, no. 1 (1965): 43–47.

HALDER, ALOIS. "Technology." Pp. 205–10 in Karl Rahner, S.J., ed., *Sacramentum mundi*. Vol. 6. New York: Herder & Herder, 1970. A good introductory article on the general concept of technology followed by a discussion of how a number of European philosophers have assessed modern technology as a social problem.

HERLITZIUS, ERWIN. "Technik und Philosophie," *Bergakademie* 13, no. 6 (1961): 330-48. Other work by this author listed under Ethical and Political Critiques, Section C.

HILDESHEIMER, A. "Philosophie über Technik," *Philosophie und Leben* (Leipzig) 5 (1929): 218-26.

HUNGER, EDGAR. "Zur Frage einer Philosophie der Technik," *Pädagogische Provinz* 17 (1963): 243-49.

KAPP, ERNST. *Grundlinien einer Philosophie der Technik; Zur entstehungsgeschichte der Cultur aus neuen gesichtspunkten* [Fundamentals of a philosophy of technology; the genesis of culture from a new viewpoint]. Braunschweig: Westermann, 1877. Pp. 351. The first book to bear the title "philosophy of technology." The thesis is that weapons and tools are extensions of human organs and senses. The argument is documented with extensive comparisons between the human body and man's technological inventions. — R. J. R.

KLIBANSKY, RAYMOND, ed. *Contemporary Philosophy; a Survey*, Vol. 2: *Philosophy of Science*. Florence: La Nuova Italia Editrice, 1968. Pp. 521. Primarily survey articles on the philosophy of science, but also a number relevant to the philosophy of technology: J. Wilkinson's "Methodology of Science," M. Bunge's "Scientific Laws and Rules," R. Ruyer's "Cybernetique et information," J. Zeman's "Cybernetics and Philosophy in Eastern Europe," K. Gunderson's "Minds and Machines: a Survey," H. Skolimowski's "Technology and Philosophy," and T. Kotarbiński's "L'évolution de la praxéologie en Pologne."

KOESSLER, PAUL. "Technik und Ethik," *Lebendiges Zeugnis; Warteheft der Akademie Bonifatius-Einigung Paderborn* 2-3 (1949): 10-20.

――――. "Natur und Technik," *Studium generale* 7 (1954): 308-11.

――――. "Ursprung, Weg und Grenzen der Technik" [Origin, path and limits of technology], *VFDB-Zeitschrift; Forschung und Technik im Brandschutz* (Stuttgart) 3, no. 3 (1954): 77-82.

――――. "Wesen und Grenzen der Technik" [Essence and limits of technology], *Mineralöl; eine Zeitschrift der Mineralölwirtschaft* (Munich) 3, no. 9 (1958): 1-3. Also in *Vortragsreihe des Deutschen Industrieinstituts* (Cologne) 49 (1957): 4 ff.

――――. *Christentum und Technik* [Christianity and technology]. Aschaffenburg: Pattloch, 1959. 2d ed., 1966. Pp. 107. A popular, rather optimistic discussion by a German engineer which relies heavily on the work of Friedrich Dessauer.

――――. "Wo steht die Technik [Where is technology standing], *Österreichische Ingenieur-Zeitschrift; Zeitschrift des Österreichischen Ingenieur- und Architekten-Vereines* (Vienna) 2 (1959): 67-71.

――――. "Technik und menschliche Verpflichtung" [Technology and human obligation], *Zeitschrift für praktische Psychologie* (Paderborn) 1 (1961): 129-42.

――――. "Die technische Denkweise und der Mensch" [Technological ways of thinking and man], *VDI-Zeitschrift* 104 (1962): 1331-35. Technology may alter thinking processes, but such processes are "far more influenced by individual characteristics than by surroundings."

――――. "Bildungswerte der Technik" [Educational value of technology], *VDI-Zeitschrift* 110, no. 5 (1968): 161-66. An education in the engineering sciences promotes the development of logical thinking and the subordination of imagination to scientific reality. For a number of other articles pre-1956, see Dessauer's bibliography.

KOHLEN, WILHELM. "Technik als Ausgang zur Philosophie" [Technology as point of departure to philosophy], *Lebendiges Zeugnis; Warteheft der Akademie Bonifatius-Einigung Paderborn* 2-3 (1949): 21-47.

KRANZBERG, MELVIN, and WILLIAM H. DAVENPORT, eds. *Technology and Culture; an Anthology*. New York: Schocken, 1972. A collection of important articles from the journal of the Society for the History of Technology.

KUHNS, WILLIAM. *The Post-industrial Prophets; Interpretations of Technology.* New York: Weybright & Talley, 1971.

Pp. 280. Popular but valuable introduction to the theories of L. Mumford, S. Giedion, J. Ellul, H. W. Innis, M. McLuhan, N. Wiener, and R. B. Fuller.

"Die Künste im technischen Zeitalter," *Gestalt und Gedanke; Jahrbuch,* special issue, vol. 3 (Munich: Oldenbourg, 1954). This special issue was edited by Clemens G. Podervils and includes articles by W. Heisenberg, M. Heidegger, Praetorius, F. G. Jünger, and W. Riezler, plus R. Guardini's "Die Situation des Menschen" and M. Schröter's "Bilanz der Technik; Schlusswort."

LAFITTE, JACQUES. *Réflexions sur la science des machines.* Supplément: P. Archambautt, R. Aigrain, J. Soulairol, G. Rabeau: Les idées et les livres. Paris: Bloud & Gay, 1932. Pp. 162. After a survey of different kinds of machines, the author seeks to develop a science of machines, first, through a system of classification and, second, by means of an analysis of the principal characteristics of machines. Views man and machine as intimately related. "Without man, no machine; no man without machine. The machine took its birth also by an *élan* of primordial life, like us, whom it made men, at the same time as we made it. As earth and water form rivers, the one always conforming itself to the other since primitive times, so mechanical structures and social structures have made up, without cease, across the ages, the course of our destiny, have woven the web of our human life. Both see the same splendors and virtues and vices, unnamed abysses, identical for both" (p. 119).

LENK, HANS. *Philosophie im technologischen Zeitalter* [Philosophy in a technological age]. Stuttgart: Kohlhammer, 1971. Pp. 174. Popular. Does not present a philosophic analysis of technique, technology, or technical systems but outlines "what opportunities offer themselves to one who philosophizes rationally in the face of 'technological challenges,' what tasks and responsibilities philosophy must take on in the technological age of the future."

LEY, HERMANN. *Information zum Thema: Zum Verhältnis von Philosophie und Technik* [Information on a theme: on the relation of philosophy and technology]. Gesellschaft zur Verbreitung wissenschaflicher Kenntnisse, Sektion Technik, vol. 1. Berlin: Gesellschaft zur Verbreitung wissenschaflicher Kenntnisse, 1959. Pp. 15.

_____. "Philosophie und Technik," *Bergakademie* (Leibniz) 14, no. 10 (1962): 671–77.

_____. "Naturwissenschaft, Philosophie und Technik," *Technische Gemeinschaft* (Berlin) 12, no. 7 (1964): 320–23. Other works by this author listed under Ethical and Political Critiques, Section C.

LILJE, HANNS. *Das technischen Zeitalter; Grundlinien einer christlichen Deutung* [The technical age; outline of a Christian interpretation]. Berlin: Furche-Verlag, 1928. Pp. 198. A comprehensive and important work divided into four parts: I, "The Intellectual Presuppositions of the Technical Age"; II, "The New Realism"; III, "The Feel of Life in the Technical Age"; IV, "The Gospel in the Technical Age." For a brief English statement of Lilje's general philosophical position, see his *Atheism, Humanism, and Christianity* (Minneapolis: Augsburg, 1964).

_____. "Ethos und Dämonie der Technik." Pp. 27–45 in *Menschenwürde, Wirtschaftsordnung und Technik: Vorträge auf der Bergbautagung in Essen 1948.* Essen-Kettwig: Gluckauf Verlag, 1948.

_____. *Der Christ im planetarischen Zeitalter* [Christ in a planetary age]. Hamburg: Furche-Verlag, 1960.

LUCIA, P. J. "Bosquejo de une filosofía de la técnica" [Sketch of a philosophy of technology], *Revista de Occidente* 40 (April–May–June 1933): 38–57. Begins with a consideration of the question "What is technology?" and the nature of man as *Homo faber,* and moves through a discussion of knowledge and technology, and economics and technology to a concluding statement on the grandeur and misery of technology.

MCLEAN, GEORGE F., ed. *Philosophy in a Technological Culture.* Washington D.C.: Catholic Univ. America Press, 1964. Pp. 438. Proceedings of the workshop on Philosophy in a Technological Culture held at the Catholic University of America in June 1963. The topics discussed range from "Technology and the Philosophy of Science" to "Philosophy's Contribution to the Formations of Sisters Professionally Engaged in a Technological Culture"; the papers are very uneven in quality. The more important articles are: G. F. McLean's "The Contemporary Philosopher and His Technological Culture," D. E. Marlowe's "Opportunities and Dangers of Technology," E. McMullin's "The Nature of Scientific Knowledge: What Makes It Science?" and "Medieval and Modern Science: Continuity and Discontinuity," J. A. Weisheipl's "The Continuity of Ancient and Modern Science," V. E. Smith's "Technology and the Image of Man," W. A. Wallace's "Cybernetics and the Christian Philosophy of Man," and L. A. Foley's "God and Man in a Technological Culture." There are also some discussions of "Technology and Culture," and "Technology and Ethics," along with seminars on various subsidiary themes.

"Die marxistisch-leninistische Philosophie und die technische Revolution," *Deutsche Zeitschrift für Philosophie,* special issue (1965). Pp. 369. Contains. the proceedings of a philosophical congress held in East Berlin, April 22-24, 1965. Theses and main reports: "Die marxistisch-leninistische Philosophie und die technische Revolution (Thesen)," G. Heyden's "Die marxistische-leninistische Philosophie und die technische Revolution," A. Kosing's "Der dialektische Materialismus als Methodologie der modernen Wissenschaft," K. Fuchs's "Moderne Physik und marxistische-leninistische Philosophie." Discussions on (1) "Essence and Historical Position of the Technological Revolution" by K. Tessman, K. D. Wüstneck, W. Müller,

H. Vogel, L. Kuntz, H. Ullrich, G. Hoppe, E. Stüber, H. Ley, G. Bohring, and H. Quitzsch; (2) "The Role of Science and of Men in the Technological Revolution" by L. Agoston, D. Pasemann, J. Bernhardt, K. Fuchs, M. Döbler, W. Heise, F. Kohlsdorf, E. Herlitzius, H. Kallabis, G. Hoppe, R. Weidig, O. Eisenblätter, and W. Hähnlein; (3) "Socialist Image of Man and the Technological Revolution" by J. Schmollack, H. G. Eschke, M. Zivotic, R. Löther, R. Miller, H. Kulak, H. Engelstädter, R. Kramer, H. Bober, F. Staufenbiel, D. Mühlberg, G. Neuner, E. J. Giessman, E. Drefenstedt, A. Bendmann, and B. Bittighöfer; (4) "Problems of Guidance and Planning in the Technological Revolution" by J. Vorholzer, H. Edeling, J. Tripotzky, H. Taubert, H. Metzler, J. Gustmann, G. Schnauss and W. Maltusch, G. Pawelzig, G. Domin, D. Schulze, and A. Kahenitz; (5) "Methodological Problems of Modern Science" by G. Mende, E. Herlitzius, H. Metzler, H. Seidel, D. Wittich, K. D. Wüstneck, E. Schlegel, W. Strauss, K. Berka, E. Albrecht, H. Skala, K.-H. Müller, G. Gruse, H. Korch, H. Hörz, H. Parthey, and J. Müller; (6) "Philosophical Problems of Industrial Science" by B. Wenzlaff, H. Laitko, G. Fuchs, H. Felke, D. Bernhardt, I. T. Frolow, K. H. Kannegiesser, H. Meyer, and M. Guntau. See also *Deutsche Zeitschrift für Philosophie,* vol. 13, no. 9 (1965) for reflections and comments on the congress: A. Kosing's "Über die Funktionen der marxistischen Philosophie," J. Schmollack's "Technische Revolution und sozialistisches Menschenbild," H. Schaler's "Philosophische Probleme der Gesellschaftswissenschaften," M. Strauss's "Gedanken und Vorschläge zu Philosophie, Logik und moderne Physik," and J. Müller's "Den philosophischen Fragen des technischen Fortschritts mehr Beachtung schenken!" See also the main articles in this last issue: K. Tessmann's "Mensch, Produktion und Technik in der Revolution," and E. Lesciewitz's "Technische Revolution und sozialistisches Arbeitskollektiv," K. Fuchs-Kittowski's "Kybernetik in der molekularen Biologie — zum Determinismus-problem und den Beziehungen zwischen technischem Automaten und lebendem Organismus," A. Uhlmann's "Neuere Vorstellungen aus dem Bereich der Elementarteilchenphysik," H. Lesser's "Das zeitliche Verhältnis von Ursache und Wirkung," and S. N. Braines and W. B. Swetschinski's "Elemente einer allgemeinen Theorie der Steuerung des Organismus." For information about planning for the congress, see *Deutsche Zeitschrift für Philosophie*, vol. 13, no. 8, (1965) which also contains these further articles relevant to this theme: E. Pracht's "Über den Zusammenhang on Bildung und Kultur," H. Liebscher's "Kybernetik und philosophische Forschung," R. Rochhausen's "Zur weiteren Arbeit auf dem Gebiet philosophischer Probleme der modernen Biologie," M. Guntau's "Philosophische Probleme der geologischen Wissenschaften," and E. Albrecht's "Philosophische Probleme der Sprachwissenschaft und die wissenschaftlich-technische Revolution." As the titles of these articles make clear, since Marxist doctrine regards science as theoretical technology, the philosophy of science and the philosophy of technology are discussed as one subject.

MELSEN, ANDREW G. VAN. *Science and Technology*. Pittsburgh: Duquesne Univ. Press, 1961. Pp. 373. Unifies a neo-Thomist philosophy of science with a comprehensive philosophical analysis of technology. Technology is conceived as the primary influence of physical science upon culture. Surveys the major criticisms of this influence and arrives at optimistic but well-argued conclusions. See also the author's *Physical Science and Ethics* (Pittsburgh: Duquesne Univ. Press, 1968) and *Science and Responsibility* (Pittsburgh: Duquesne Univ. Press, 1970).

―――. "De wijsgerige implicaties van de wisselwerking tussen wetenschap en techniek," *Algemeen Nederlands Tijdschrift voor Wisjsbegeerte en Psychologie* 57, no. 3 (October 1965): 172–81.

―――. "Filosofie van de techniek," *Algemeen Nederlands Tijdschrift voor Wijsbegeerte en Psychologie* 60, nos. 3–4 (October 1968): 245–58.

MITCHAM, CARL, and ROBERT MACKEY, eds. *Philosophy and Technology; Readings in the Philosophical Problems of Technology*. New York: Free Press, 1972. Pp. 399. A major anthology in five parts: "Conceptual Issues," "Ethical and Political Critiques," "Religious Critiques," "Existentialist Critiques," and "Metaphysical Studies." Includes: Mitcham and Mackey's "Technology as a Philosophical Problem" (original), J. K. Feibleman's "Pure Science, Applied Science and Technology: an Attempt at Definitions," H. Skolimowski's "The Structure of Thinking in Technology," I. C. Jarvie's "The Social Character of Technological Problems: Comments on Skolimowski's Paper" and "Technology and the Structure of Knowledge," M. Bunge's "Toward a Philosophy of Technology," L. Mumford's "Technics and the Nature of Man," J. Ellul's "The Technological Order," E. G. Mesthene's "Technology and Wisdom" and "How Technology Will Shape the Future," G. Ander's "Commandments in the Atomic Age," R. M. Weaver's "Humanism in an Age of Science and Technology," C. S. Lewis's "The Abolition of Man," N. Rotenstreich's "Technology and Politics," C. B. Macpherson's "Democratic Theory: Ontology and Technology," Y. R. Simon's "Pursuit of Happiness and Lust for Power in Technological Society," G. Grant's "Technology and Empire," N. Berdyaev's "Man and Machine," E. Gill's "Christianity and the Machine Age," R. A. Buchanan's "The Churches in a Changing World," W. N. Clarke's "Technology and Man: a Christian Vision," L. White's "The Historical Roots of Our Ecological Crisis," E. Jünger's "Technology as the Mobilization of the World through the Gestalt of the Worker" (original trans.), J. Ortega y Gasset's

"Thoughts on Technology" (revised trans.), F. Dessauer's "Technology in Its Proper Sphere" (original trans.), H. Jonas's "The Practical Uses of Theory," and W. F. Hood's "The Aristotelian versus the Heideggerian Approach to the Problem of Technology" (original).

MOSER, SIMON. *Philosophie und Gegenwart; Vorträge.* Meisenheim-Glan: Hain, 1960. Pp. 205.

———. "Zur philosophischen Diskussion der Kybernetik in der Gegenwart," *Zeitschrift für Philosophische Forschung* 21, no. 1 (January–March 1967): 64–77. See also the author's "Philosophische Bemerkungen zum nachrichtentechnischen und psychologischen Begriff des Lernens und der Intelligenz," *Studium generale* 15, no. 1 (1962): 71–80.

———. "Toward a Metaphysics of Technology," *Philosophy Today* 15, no. 2 (Summer 1971): 129–56. A critical analysis of: relations between philosophy of science and philosophy of technology; technology as applied science, neutral means, and this-worldly religiosity (D. Brinkmann); the experiment in science and technology (J. Conant and C. F. von Weisäcker); technology as real being from ideas and participation in divine creation (F. Dessauer); Plato's concept of *techne;* the control loop as the essence of technology (H. Schmidt); modern technology as a provoking, setting-up disclosure of nature (M. Heidegger). Translation of "Zur Metaphysik der Technik," in *Metaphysik einst und jetzt* (Berlin: DeGruyter, 1958). See also the author's "Der Begriff der Natur in aristotelischer und moderner Sicht," *Philosophia naturalis* 6, no. 3 (1961): 261–87.

MÜLLER, MAX. "Philosophie — Wissenschaft — Technik," *Philosophisches Jahrbuch der Görres-Gesellschaft* 68 (1960): 309–29. See also the author's "Person und Funktion," ibid., 69 (1961): 371–404.

MUMFORD, LEWIS. *Technics and Civilization.* New York: Harcourt Brace, 1934. Pp. 495. Classic study of the influence of the machine upon civilization through three phases of the machine age — eotechnic, paleotechnic, and neotechnic — in an attempt to determine the rightful place of the machine in the life of man.

———. *Art and Technics.* New York: Columbia Univ. Press, 1951. Paperback reprint, 1960. Pp. 162. Traces the historical relations between the subjective artistic impulse and the objective technical impulse.

———. *The Myth of the Machine.* 2 vols. New York: Harcourt Brace Jovanovich, 1967–70. This is Mumford's *magnum opus.* Vol. 1, *Technics and Human Development* (pp. 342), is an extended critique of the definition of man as a tool-using animal; vol. 2, *The Pentagon of Power* (pp. 496), is a detailed analysis of the closed circle of technological rationality. Vol. 1 reviewed by P. F. Drucker, *Technology and Culture* 9, no. 1 (January 1968): 94–98; vol. 2 reviewed by E. J. Hobsbawn, *New York Review of Books,* vol. 15, no. 9 (November 19, 1970). Extensive bibliographies in both volumes.

[NOTE: Mumford's work, virtually all of it dealing in one way or another with technology, is so large that (like Dessauer's) it really deserves a separate bibliography. Yet his major theses can be found in the three books cited above — especially when some secondary use is made of such works as *Sticks and Stones* (New York: Boni, 1924; 2d rev. ed., New York: Dover, 1955), *The Culture of Cities* (New York: Harcourt Brace, 1938), and *The City in History* (New York: Harcourt, Brace & World, 1961). *In the Name of Sanity* (New York: Harcourt, Brace & World, 1954) collects a number of essays and lectures from 1947 to 1954. A number of his more important articles since that time relevant to the topics of this bibliography are cited below.]

———. "Science as Technology," *Proceedings of the American Philosophical Society* 105, no. 5 (October 1961): 506–11. Although Bacon tied the value of modern science to technology, we should not do so.

———. "Authoritarian and Democratic Technics," *Technology and Culture* 5, no. 1 (Winter 1964): 1–8. "From late neolithic times down to our own day, two technologies have recurrently existed side by side: one authoritarian, the other democratic, the first system-centered, immensely powerful, but inherently unstable, the other man-centered, relatively weak, but resourceful and durable." Slightly revised from a lecture first published as "Technology and Democracy" in E. Reed, ed., *Challenges to Democracy: the Next Ten Years* (New York: George Praeger, 1963).

———. "Man the Finder," *Technology and Culture* 6, no. 3 (Summer 1965): 375-81. The first stage of man's technological development was finding, not making; collecting, not hunting.

———. "Technics and the Nature of Man," *Technology and Culture* 7, no. 3 (Summer 1966): 303-17. Questions the assumption that man is basically a tool- using animal, a theory which has been the basis of our commitment to the present form of technical and scientific progress as an end in itself. Subsequently included in P. H. Oehser, ed., *Knowledge among Men* (New York: Simon & Schuster, 1966), and in C. Mitcham and R. Mackey, eds., *Philosophy and Technology* (New York: Free Press, 1972).

———. "Speculations on Prehistory," *American Scholar* 36 (Winter 1966-67): 43-53. Too much emphasis has been placed on tool using in the attempt to understand man's development.

———. "Closing Statement." Pp. 91-102 in R. Disch, ed., *The Ecological Conscience*. Englewood Cliffs, N.J.: Prentice-Hall, 1970.

[NOTE: Since Mumford's theory of the nature of man is crucial to his evaluation of technology, one might also examine *The Condition of Man* (New York: Harcourt Brace, 1944) and *The Conduct of Life* (New York: Harcourt Brace, 1951). In both books Mumford puts forward a polycentric theory of human nature in search of self-realization. See also *The Transformations of Man* (New York: Harper & Row, 1956). Finally, for a good critical analysis of some of Mumford's key ideas, see J. Carey and J. Quick, "The

Mythos of the Electronic Revolution," *American Scholar* 39, nos. 2-3 (1970): 219-41 and 395-424.]

PARIS, CARLOS. *Mundo técnico y existencia auténtica*. Madrid: Guadarrama, 1959.

———. "Téchnica y filosofía," *Crisis*, vol. 11 (1964).

Proceedings of the XIth International Congress of Philosophy; Brussels, Aug. 20-26, 1953. 14 vols. Amsterdam:' North-Holland, 1953. The section on "Philosophy and Culture," vol. 8, contains D. Brinkmann's "L'homme et la technique."

Proceedings of the XIIth International Congress of Philosophy; Venice and Padua, Sept. 12-18, 1958. 12 vols. Florence: Sansoni, 1958-61. Vol. 8 contains the following articles: R. M. del Castillo's "Para una gnoseología de la automacion" (pp. 153-59), R. D. McKinney's "Science and the Humanities in a Space Age" (pp. 161-67), E. Obert's "Natura, tecnica, umanesimo" (pp. 169-76), H. Pohl's "Das Wesen der Tecknik" (pp. 195-200), M. Ralea's "Dialectique du donné et du construit" (pp. 209-13), H. Scheler's "Über das Verhältnis von Mensch und Technik" (pp. 231-38), P. A. Schilpp's "Does Philosophy Have Anything to Say to Our Atomic Age?" (pp. 239-45), J. A. M. Van Moll's "Philosophy and the Progress of Cultural Endeavor" (pp. 269-76).

Proceedings of the XIIIth International Congress of Philosophy; Mexico, D.F., 1964. Memorias del XIII Congreso Internacional de Filosofia. Univ. Nacional Autónoma de Mexico, 1964. Vol. 6 contains the following articles: J. L. Fischer's "Science, Philosophy and the Future of Culture" (pp. 263-78), Nolan Jacobson's "The Problem: the Fatal Mistake of Modern Man; the Solution: the Faith of the Scientist" (pp. 303-15), Irinen Strenger's "Problematica humana de tecnica" (pp. 403-37), and Satosi Watanabe's "Civiliation and Science, Man and Machine" (pp. 477-91). See also Vernon J. Bourke's "Man in the Space Age" (vol. 4, pp. 23-29).

Proceedings of the XIVth International Congress of Philosophy; Vienna, Sept. 2-9, 1968. Vienna: Herder, 1968. Vol. 2, pp. 477–614, contains Colloquium VI on "Cybernetics and the Philosophy of Technical Science," which includes the following papers: A. Adam's "Philosophie der Technik," J. Agassi's "The Logic of Technological Development," D. A. Cardone's "La technique et l'anarchisme intérieur," H. L. Dreyfus's "Cybernetics as the Last Stage of Metaphysics," G. G. Granger's "D'une vraie et d'une fausse technicité dans les sciences humaines," H. Greniewski's "Notions fondamentales de la cybernetique," Z. Kleyff's "Science, Technology and Economics as Integral Parts of Production Processes," T. Kotarbiński's "L'attitude active et la passivité apparente," W. Mays's "The Use and Misuse of Logical Principles in Cybernetic Discussions," M. Mazur's "Concept of Autonomous System and Problem of Equivalence of Machine to Man," J. J. Ostrowski's "Essai d'une typologie méta-praxéologique," M. Serres's "Information et sensation," H. Skolimowski's "On the Concept of Truth in Science and in Technology," H. Titze's "Das Problem des Bewusstseins und die kybernetische Maschine," L. Tondl's "Der Janiskopf der Technik," K. Tuchel's "Zum Verhältnis von Kybernetik, Wissenschaft und Technik," B. Walentynowicz's "On Methodology of Engineering Design," Z. Wasiutynski's "Sur le postulat du plus grand effort d'activité intellectuelle et technique," S. Watanabe's "Epistemological Implications of Cybernetics," J. Zieleniewski's "Why, 'Cybernetics and the Philosophy of Technical Science' Only? Some Comments," and A. Zvorykine's "Technology and the Laws of Its Development."

[NOTE: These successive volumes of *Proceedings* indicate the progressive recognition of technology as a philosophical issue. (See also in this series the work by Engelmeyer from the IVth International Congress.) The XVth International Congress on Philosophy, to be in Varna, Bulgaria, in September 1973, will feature as a main theme, "Science, Technology and Man."]

PRUCHA, MILAN. "Technik und Philosophie," *Marxismusstudien* (Tubingen) 6 (1969): 172–86. First printed (in Czech) in *Filosofický časopis*, vol. 16, no. 3 (1968).

PYLYSHYN, ZENON W., ed. *Perspectives on the Computer Revolution.* Englewood Cliffs, N.J.: Prentice-Hall, 1970. Pp. 540. Chap. 1, "Some Landmarks in the History of Computers," contains T. M. Smith's "Some Perspectives on the Early History of Computers," C. Babbage's "Of the Analytical Engine," H. Aiken's "Proposed Automatic Calculating Machine," A. W. Burks, H. H. Goldstine, and J. von Neumann's "Preliminary Discussion of the Logical Design of an Electronic Computing Instrument," V. Bush's "As We May Think." Chap. 2, "Theoretical Ideas: Algorithms, Automata and Cybernetics," contains B. A. Trakhtenbrot's "Algorithms," J. von Neumann's "The General and Logical Theory of Automata," C. E. Shannon's "Computers and Automata," J. O. Wisdom's "The Hypothesis of Cybernetics," S. Toulmin's "The Importance of Norbert Wiener." Chap. 3, "Man-Machine Confrontation," contains S. Butler's "The Destruction of Machines in Erewhon," R. Taylor's "The Age of the Androids," W. G. Walter's "Totems, Toys, and Tools," B. Mazlish's "The Fourth Discontinuity," P. Armer's "Attitudes toward Intelligent Machines." Chap. 4, "Machine Intelligence," contains A. M. Turing's "Computing Machinery and Intelligence," H. Block and H. Ginsburg's "The Psychology of Robots," H. A. Simon and A. Newell's "Information-processing in Computer and Man," J. R. Pierce's "What Computers Should Be Doing." Chap. 5, "Man-Machine Partnership," contains G. Herbkersman's "Talking to a Monster," J. C. R. Licklider's "Man-Computer Symbiosis," D. D. Bushnell's "Applications of Computer Technology to the Improvement of Learning," J. Weizenbaum's "Contextual Understanding by Computers," A. M. Noll's "The Digital Computer as a Creative Medium." Chap. 6, "The Impact of the Computer and

Information Revolution," contains R. W. Hamming's "Intellectual Implications of the Computer Revolution," G. E. Forsythe's "Educational Implications of the Computer Revolution," M. Greenberger's "The Computers of Tomorrow," M. Mead's "The Information Explosion." Chap. 7, "Automation, Technology, and Social Issues," contains H. A. Simon's "The Shape of Automation," D. N. Michael's "Cybernation: the Silent Conquest," J. Ellul's "The Technological Society." Chap. 8, "Ethical and Moral Issues," contains E. C. Berkeley's "The Social Responsibilities of Computer People," A. F. Westin's "Legal Safeguards to Insure Privacy in a Computer Society," A. M. Hilton's "An Ethos for the Age of Cyberculture." Bibliography.

RIESSEN, I. H. VAN. "Philosophie der Technik," *Philosophia reformata* 3 (1938): 202-24.

———. *Filosofie en techniek.* Kampen: J. H. Kok, 1949. Pp. 715. A major work divided into three parts. Part I contains analyses of the relation between philosophy and technology in the thought of Comte, Spencer, Ostwald, Mach, Dilthey, Rickert, James, Nietzsche, Sorel, and Jaspers. Part II analyzes the philosophies of technology of Ure, Reuleaux, Kapp, Wendt, Eyth, Zschimmer, Engelhardt, Dessauer, Hardensett, Korevaar, von Hanffstengel, Spengler, Lilie, and Bangerter. Part III contains van Riessen's own analysis of the technological object, technological production, and technological design.

———. "De strutuur der technik," *Philosophia reformata* 26, nos. 1-3 (1961): 114-30.

———. "Over de betekenis van de wetsidee in de wijsbegeerte," *Philosophia reformata*, vol. 30 (1965).

ROSSMANN, KURT. "Die Philosophen und die Technik," *Studium generale* 4, no. 1 (1951): 59-64.

SCHILLING, KURT. *Philosophie der Technik; Die geistige Entwicklung der Menschheit von den Anfängen bis Gegenwart* [Philosophy of technology; spiritual development of mankind from the beginnings to the present]. Herford: Max-

imilian-Verlag, 1968. Pp. 244. "A philosophic consideration of technology . . . only has meaning when we view human technology fundamentally against the background of the technology of life" (p. 14). Divides the history of technology into three periods: the era of the destructive tool technology of primitive man and the wild predators, the constructive artisan technology of the agrarian world, and modern technology of Western civilization. These are three stages in the progressive replacement of the natural organs of human activity by artificial ones; technology is man's organic projection of natural life processes into human culture. Summary definition: technology is "the plan—put into operation through particular inventions, conceptually definable and restricted at any given time—of an artificial environment as a whole for a group of people or an age" (p. 205).

SCHRÖTER, MANFRED. *Philosophie der Technik.* Munich: Oldenbourg, 1934. Pp. 86. An examination of technology which, while it views National Socialism as the achievement of a "cultural-ethical balance" in technology, is not tainted by fanaticism. Technology is defined as practical constructive work on nature; machine technology is simply its latest and most perfect form. Argues against Dessauer for an immanentist metaphysics of technology.—R. J. R. and W. C. See also the author's *Die Kulturmöglichkeit der Technik als Formproblem der produktiven Arbeit* (Berlin-Leipzig: DeGruyter, 1920).

———. *Deutscher Geist in der Technik.* Cologne: Schaffstein Verlag, 1935. Pp. 64. For a few other articles pre-1956, see Dessauer's bibliography.

SCHUHL, PIERRE-MAXIME. *Machinisme et philosophie.* Paris: Alcan, 1938. Pp. 108. Begins with a historical study arguing that the oppositions of nature versus art and contemplation versus action interfered with the mechanistic approach to nature from Greek to modern times, and that these oppositions were resolved in the early modern period by Bacon and Descartes.

Then after an examination of the optimistic and pessimistic judgments on industrialization in the 19th century, the author concludes that, despite the problems created, it is neither possible nor desirable to give up machines. First, because history cannot be reversed. "Further, one could not prevent man from applying his genius to bettering his conditions of existence— even if some ill-omened and unexpected consequences must result from an effort which has always been in his nature." Instead, "the scientific and technological effort must ally itself with another effort . . . of justice and of generosity— in the Cartesian sense of the word—which would make the new and powerful means at our disposal serve the well-being of the generality of men" (p. 104). See also the author's "L'Homme et le progres scientifique et technique," *Revue philosophique de la France et de l'etranger* 84 (1959): 113-14.

SCHWIPPERT, HANS, ed. *Mensch und Technik: Erzeugnis, Form, Gebrauch* [Man and technology; products, forms, uses]. Darmstadt: Neue Darmstädter Verlagsanstalt, 1952. Pp. 225. Proceedings of the third Darmstadt conference. Widely commented upon in Germany. See W. deBoer, "Mensch und Technik; 3. Darmstädter Gespräch," *Merkur* 7, no. 1 (1953): 83-86; and A. Ostertag, "Zum Darmstädter Gespräch über Mensch und Technik," *Schweizerische Bauzeitung* 71, no. 14 (1953): 197-200. Dessauer also has a discussion of this and other German conferences on the philosophy of technology in *Streit um die Technik* (Frankfurt: Knecht, 1956).

SKOLIMOWSKI, HENRYK. "Praxiology." In *Polish Analytical Philosophy.* New York: Humanities, 1967. Originally published as "Praxiology—the Science of Accomplished Acting," *Personalist,* vol. 46, no. 3 (Summer 1965).

———. "The Structure of Thinking in Technology," *Technology and Culture* 7, no. 3 (Summer 1966): 371-83. Technology is an independent form of knowledge, not reducible to science. Science only studies reality; technology creates it. The key to the

idea of technology is in the understanding of technological progress. The criteria of technological progress are in essence the criteria of efficient action. Thus, praxiology, or the study of efficient action, is necessary for understanding technological progress and, consequently, the idea of technology itself. For a critique of this essay, see I. C. Jarvie's "The Social Character of Technological Problems: Comments on Skolimowski's Paper," *Technology and Culture* 7, no. 3 (Summer 1966): 384-90. Both are included in C. Mitcham and R. Mackey, eds., *Philosophy and Technology* (New York: Free Press, 1972).

———. "The 'Monster' of Technology," *Center Diary* (Center for the Study of Democratic Institutions) 18 (May-June 1967): 56-59. Technology "will destroy the human being in the process of making him happy. And this is because the means to an end grew out of all proportion, became the end in itself, and suppressed other ends."

———. "Technology and Philosophy." Pp. 426-37 in Raymond Klibansky, ed., *Contemporary Philosophy; a Survey,* Vol. 2: *Philosophy of Science.* Florence: La Nuova Italia Editrice, 1968. A short survey of the concept of technology in the theories of J. Ellul, M. McLuhan, D. Bell, M. Bunge, H. Skolimowski, J. Agassi, and A. Zvorykin.

———. "On the Concept of Truth in Science and in Technology." Pp. 553-59 in *Proceedings of the XIVth International Congress of Philosophy; Vienna, Sept. 2-9, 1968.* Vol. 2. Vienna: Herder, 1968. "Science aims at enlarging our knowledge by devising better and better theories. Technology aims at creating new artifacts through devising means of increasing effectiveness. Thus the aims and means of each are different."

———. "Towards a Humanistic Technology," *Research Management* 14, no. 5 (September 1971): 10-23. Discusses three basic approaches to technology: the pragmatic, the intellectual, and the dialectical. Argues that the kind of technological knowledge needed by contemporary society will have to come from technologists who choose to practice a new "humanistic technology." See also the author's review of P. Soleri's *Arcology* (Cambridge, Mass.: M.I.T. Press, 1970) in *Main Currents* 27, no. 3 (January–February 1971): 97–99.

_____. "Science and the Modern Predicament," *New Scientist* 53 (February 24, 1972): 435–37. Examires attitudes toward science today and questions whether science controls people, or people control science; defines the concepts of science—as an eternal truth, as an ideology, as pure knowledge, as a social institution, and as technology; concludes that science controls people only within the social institution and technology concepts; suggests that the answer to today's problems is not to curtail existing science but to create a new science.

SOUCY, CLAUDE. "Technique et philosophie," *Recherches et débats* 31 (June 1960): 109–23. "That the striking success of technique is accompanied by a malaise no less evideht, such is the paradox we are here attempting to clear up" (p. 109). "Technique increases disproportionately the power of human means. In order that this power find its application in the orientation toward its end, a corresponding effort is required of reason, which on the one hand illuminates this end and on the other ascertains the exact nature of the relation technique has with it" (p. 123). – W. C.

SPIELHOFF, ALFONS. "Die Technik als Problem der Kulturwissenschaft" [Technology as a problem for the humanities]. n. p.: 1949. Pp. 164. An unpublished dissertation which surveys the major German literature on the subject to 1949. Analyzes the prephilosophical views of F. Redtenbacher and F. Reuleaux; the materialistic and biological ideas about technology of E. Kapp, L. Erhard, and P. Krennhals; O. Spengler's idea of technology as man's striving for power; the idea of technology as the will to freedom from nature in W. Sombart and E. Zschimmer; the theories of invention of M. Eyth, F. Dessauer, R. Müller-Libenau, A. Wenzl, and J. Bernhart; the theories of C. Mat-

schoss, R. Weyrauch, M. Schneider, E. Spranger, H. Lilje, E. Diesel, P. R. Lehnert, F. von Gottl-Ottlilienfeld, and W. Moock on the relation between technology and economics; the futurology of E. Jünger; the ideas on the work ethic of technology of M. Schröter; D. Brinkmann's theory of technology as self-salvation; M. Bense and F. G. Jünger's theories about the perfection of technology through human intelligence; and E. Dvorak's theory of the autonomy of technology.

SPITALER, ARMIN, and ALFRED SCHIEB, eds. *Wissen und Gewissen in der Technik* [Knowledge and conscience within technology]. Graz: Verlag Styria, 1964. Pp. 311. Includes P. Wilpert's "Das Phänomen Technik" (pp. 11–30), H. Winkmann's "Der Mensch im technischen Zeitalter von heute und morgen" (pp. 31–67), H. Dolch's "Moderne Technik als Erfüllung des Schöpfungsauftrages an den Menschen" (pp. 71–93), P. Koessler's "Technik, Wirtschaft und Politik" (pp. 195–209), M. Braunfels's "Einbruch der Technik in die Kunst (pp. 228–48), and H. Hirschmann's "Die sittlichen Grenzen des technischen Fortschritts" (pp. 296–307).

STOVER, CARL F., ed. *The Technological Order*. Detroit: Wayne State Univ. Press, 1963. Pp. 280. Proceedings of the Encyclopaedia Britannica Conference on the Technological Order held March 1962 in Santa Barbara, California. First published in *Technology and Culture*, vol. 3, no. 4 (Fall 1962). Contents include: R. Watson-Watt's "Technology in the Modern World," J. Ellul's "The Technological Order," W. N. Clarke's "Technology and Man: a Christian Vision," A. Zvorikine's "The Laws of Technological Development," L. White's "The Act of Invention: Causes, Contexts, Communities, and Consequences," A. R. Hall's "The Changing Technical Act," S. Buchanan's "Technology as a System of Exploitation," W. Hartner's "The Place of Humanism in a Technological World," R. Calder's "Technology in Focus," A. Goldschmidt's "Technology in Emerging Countries," and R. Theobald's "Long-Term Prospects and Problems." Also comments and discussion by: W. Ong, G. Piel, W. E. Preece, R. Calder, M. Kranzberg, R. L. Meier, C. F. Stover, Vu Van Thai, A. Huxley, R. M. Hutchins, and R. W. Tyler. A major anthology. "Technik," *Studium generale*, vol. 4, no. 4 (1951). Contains G. Herberer's "Der phylogenetische Ort des Menschen," W. Fuchs's "Die Technik und die physische Zukunft des Menschen," G. Steiner's "Biologie und Technik," H. Böhm's "Vom lebendigen Rhythmus," J. P. Steffe's "Gottes Schöpfung in technischer Forschung und Arbeit," K. F. Steinmetz's "Wesen und Sendung des Ingenieurs," D. Brinkmann's "Kollektivismus und Technik," and K. Rossmann's "Die Philosophen und die Technik."

"La technique et l'homme," *Recherches et débats du centre Catholique des Intellectuels Français*, cahier 31 (1960). Pp. 248. Divided into two parts. The first, on the "Self-Examination of Technologists and Men of Science," contains a response by three scientists to a questionaire from the Catholic Union, L. Chevallier's "Nature et diversité des fonctions techniques," I. A. Caruso's "Crise du monde technique et psychologie," A. Leroi-Gourhan's "L illusion technologique," J.-L. Kahn's "Valeur culturelle de la technique," and P. Germain's "Idéal scientifique et idéal technique. The second, on "Philosophical and Theological Elements," contains: C. Soucy's "Technique et philosophie," C. A. Dondeyne's "Technique et religion," L. Chevallier's "Athéisme du monde technique?" G. Rotureau's "Conscience religieuse et mentalité technique," and M.-D. Chenu's "Vers une théologie de la technique."

TEICHMANN, D. "Technik und Philosophie," *Maschinenbau Technik* 8 (1959): 625-27. Other works by this author listed under Ethical and Political Critiques, Section C.

TELLEGEN, FRANCISCUS PHILIPPUS ANTONIUS. *Aard en zin van de technische bedrijvigheid* [Nature and meaning of technological industry]. Delft: Waltmann, 1953. Pp. 21.

_____. *Samen-leven in een technische tijd* [Living together in technical times]. Utrecht: Het Spectrum, 1957. Pp. 144. Revised ed., Baarn: Het Wereldvenster, 1969.

_____. "Filosofie van de techniek," *Wijsgerig Perspectief op Maatschappij en Wetenschap, Philosophical Perspective on Society and Science* (Amsterdam) 6, no. 2 (1965): 58-80.

TONDL, L. "On the Philosophy of Technology" (in Czech), *Filosofický časopis*, vol. 12, no. 3 (1964). See also the author's paper on "The Concept of Science as Cognitive Action," *Filosofický časopis*, vol. 16, no. 3 (1968).

"Toward a Philosophy of Technology," *Technology and Culture*, vol. 7, no. 3 (Summer 1966). An important special issue containing: L. Mumford's "Technics and the Nature of Man," J. K. Feibleman's "Technology as Skills," M. Bunge's "Technology as Applied Science," J. Agassi's "The Confusion between Science and Technology in the Standard Philosophies of Science," J. O. Wisdom's "The Need for Corroboration: Comments on Agassi's Paper," H. Skolimowski's "The Structure of Thinking in Technology," and I. C. Jarvie's "The Social Character of Technological Problems: Comments on Skolimowski's Paper."

"Toward a Philosophy of Technology," *Philosophy Today* 15, no. 2 (Summer 1971): 75-156. Special issue containing: H. Jonas's "The Scientific and Technological Revolutions," C. Mitcham and R. Mackey's "Jacques Ellul and the Technological Society," D. Brinkmann's "Technology as Philosophic Problem," and S. Moser's "Toward a Metaphysics of Technology."

TUCHEL, KLAUS. "Christentum und Technik; Über Grenzen und Sinn der technischen Arbeit" [Christianity and technology; on the limits and meaning of technological work], *VDI-Nachrichten* 15, no. 24 (1961): 5.

_____. "Kybernetik – die neue Phase der Technik," *Frankfurter Hefte* 16, no. 7 (January 1961): 452-60.

_____. "Gibt es ein Ethos der Technik?

Verschiedene Ursachen für ihre Vernachlassigung im philosophischen Denken" [Is there an ethos of technology? Diverse reasons for its neglect in philosophic thinking], *VDI-Nachrichten* 15, no. 29 (1961): 5.

————. "Technik als Gabe und Aufgabe; Die Geisteswissenschaftler müssen das Gespräch mit den Ingenieurwisschenschaftlern suchen" [Technology as gift and task; humanists must seek dialogue with engineering scientists], *VDI-Nachrichten* 16, no. 8 (1962): 9.

————. "Mensch und Technik in aller Welt" [Man and technology all over the world], *VDI-Nachrichten* 16, no. 40 (1962): 9.

————. "Vom Geist und Ethos der Technik," *Die berufsbildende Schule* (Wolfenbüttel) 14 (1962): 249-52.

————, ed. *Mensch und Technik in aller Welt.* Berlin: Ullstein, 1963. Pp. 150. Essays which first appeared in the *VDI-Nachrichten,* vol. 16, no. 1 (January 1962).

————. "Der Auftrag unserer Zeit an die Technik" [The mission of our time to technology], *ETZ: Elektrotechnische Zeitschrift,* Ausgabe B, 15, no. 19 (1963): 541-42.

————. *Die Philosophie der Technik bei Friedrich Dessauer; ihre Entwicklung Motive und Grenzen* [The philosophy of technology of Friedrich Dessauer; its development, motives, and limits]. Frankfurt: Knecht, 1964. Pp. 140. See also "Friedrich Dessauer, ein Philosoph der Technik," *VDI-Nachrichten* 17, no. 8 (1963): 9 ff.

————. "Technik als Aufgabe des Menschen" [Technology as task of man], *Die berufsbildende Schule* (Wolfenbüttel) 16 (1964): 680-83.

————. "Technik als Bildungsaufgabe; Gedanken zum philosophischen und pädagogischen Verständnis der Technik" [Technology as a task of education; thoughts on the philosophic and pedagogical understanding of technology], *VDI-Zeitschrift* 106, no. 22 (1964): 1113-18. Reprinted under a slightly different title in Heinrich Roth, ed., *Technik als Bildungsaufgabe der Schulen* (Hannover: Schroedel, 1965), pp. 69-84.

————. "Neue Formen der Humanität; die Technisierung beeinflusst unser Verständnis der Welt und des Menschen" [New forms of humanity; technicization influences our understanding of the world and man], *VDI-Nachrichten* 19, no. 38 (1965): 9-10.

————. "Zum Verhältnis von Technik und Wirtschaft" [On the relation between technology and economics], *Die berufsbildende Schule* (Wolfenbüttel) 17, no. 10 (1965): 680-85.

————. *Sinn und Deutung der Technik* [Meaning and interpretation of technology]. Stuttgart: E. Klett, 1966.

————. "Die Technik als Problem der Gegenwartsphilosophie" [Technology as a problem for contemporary philosophy], *Neue Zeitschrift für systematische Theologie und Religionsphilosophie* (Berlin) 8, no. 3 (1966): 266-88.

————. *Herausforderung der Technik; gesellschaftliche Voraussetzungen und Wirkungen der technischen Entwicklung* [The challenge of technology; social prerequisites and effects of technological development]. Bremen: Carl Schunemann Verlag, 1967. Pp. 317. An eighty-page essay on "Technical Development and Social Change" is followed by an anthology of "Documents on the Classification and Interpretation of Technology" of similar length. The remaining half of the book is a dictionary of "Key Terms—Concepts and Institutions" and a "Time Table of the History of Modern Technology." An extremely valuable introduction to German thought and research on the subject of technology.

————. "Der technische Fortschritt und die Zukunft des Menschen," *Die deutsche Berufs- und Fachschule* 64, no. 5 (1968): 321-27.

————. "Zum Verhältnis von Kybernetic, Wissenschaft und Technik" [The relation of cybernetics, science and technology]. Pp. 578-85 in *Proceedings of the XIVth International Congress of Philosophy; Vienna, Sept. 2-9, 1968.* Vol. 2. Vienna: Herder, 1968.

————. "Wissenschaftliche Erkenntnisse und technische Fortschritte in metaökonomischen Wertordnungen" [Scientific knowledge and technological progress in a meta-economic ordering of values], *VDI-Zeitschrift* 113, no. 1 (January 1971): 1-6. Science and technology alter human values, so that all efforts to control science-technology on the basis of traditional values fail. These efforts need to be replaced by the construction of value systems recognizing the plurality of present value consciousness. This would entail the cooperation of scientists, engineers, social scientists, and humanists.

VALLEE, R., and A. VERGEZ. "Technique et philosophie," *Structure et évolution des techniques*, vol. 8, nos. 45-46 (March-July 1956).

WENZL, ALOYS. *Die Technik als philosophisches Problem* [Technology as a philosophical problem]. Munich: Pflaum-Verlag, 1946. Pp. 32. Situates technology as one of four great problems of the modern age. "A philosophy of technology would be an account of the essence and meaning of technology in the realm of total reality." After noting the two prephilosophical attitudes toward technology, that is, optimism and pessimism, Wenzl seeks a definition. Compares technology of ancient and moderns, artists and scientists, and considers the relation between organic and mechanistic forms of organization. "Although Driesch proved that nature was not to be understood by physcio-chemical methods, the very fact that such a mechanistic interpretation could arise shows that organisms are *machine-like*." The entelechy of nature also possesses a technology; without technology life itself would be impossible. Technology is thus grounded in the essence of reality. Concludes with some ethical reflections. "Technology raises the level of responsibility as contrasted to non-technical epochs."—J. V.

———. "Technik und Ethik." Pp. 61-67 in P. Hastenfel, ed., *Markierungen.* Munich: Kösel-Verlag, 1964. See also the author's *Wissenschaft und Weltanschauung* (Leipzig: Meiner, 1936). 2d ed. 1949.

WIENER, NORBERT. *Cybernetics; or Control and Communication in the Animal and the Machine.* New York: Wiley, 1948. Pp. 194. 2d ed., Cambridge, Mass.: M.I.T. Press, 1961. Pp. 212. Supplementary chapter "On Learning and Self-reproducing Machines."

———. *The Human Use of Human Beings: Cybernetics and Society.* New York: Houghton Mifflin, 1950. Reprinted, Garden City, N.Y.: Doubleday Anchor, 1954. Pp. 199.

———. "My Connections with Cybernetics: Its Origins and Its Future," *Cybernetics* 1 (1958): 1-14.

———. "The Brain and the Machine (Summary)." In S. Hook, ed., *Dimensions of Mind.* New York: New York Univ., 1960. Reprinted, New York: Collier Books, 1961.

———. "Cybernetics." In *Encyclopedia Americana.* New York: Americana Corp., 1960. An article of the same title was first published in *Scientific American* 197, no. 5 (November 1948): 14-19, and is reprinted in Garrett Hardin, ed., *Science, Conflict and Society.* San Francisco: W. H. Freeman, 1969. On the popular level see also the interview with Wiener translated from *Voprosy filosofii* as "Cybernetics and Man" (Washington, D.C.: Joint Publications Research Service, Government Printing Office, November 16, 1960).

———. "Some Moral and Technical Consequences of Automation," *Science* 131 (May 6, 1960): 1355-58. Cf. also Arthur Samuel, "Some Moral and Technical Consequences of Automation: a Refutation," *Science* 132 (September 16, 1960): 741-42.

———. *God and Golem, Inc.; a Comment on Certain Points Where Cybernetics Impinges on Religion.* Cambridge, Mass.: M.I.T. Press, 1964. Pp. 99.

[NOTE: Wiener's work constitutes the single most important examination of cybernetics in English, and, because of the scope of his studies and the way he takes cybernetics as the essence of technology, his three main books constitute a philosophy of technology. Wiener's more technical mathematical works on the subject have not been cited. Two important articles written with others are: A. Rosenblueth and N. Wiener, "Purposeful and Non-purposeful Behavior," *Philosophy of Science* 17 (1950): 318-26; and A. Rosenblueth, N. Wiener, and J. Bigelow, "Behavior, Purpose and Teleology," *Philosophy of Science* 10 (January 1943): 18-24.]

WILLIAMS, GRIFFITH. "Towards a Creative Society: the Philosophy of Technology," *Journal of the Royal Society of Arts* (London) 114 (1966): 380-94.

YAMADA, KEIICHI. *Gendai Gizitsu-ron* [A history and philosophy of modern technology]. Tokyo: Asakura Shobō, 1964. Pp. 272. This textbook for Japanese science students is divided into three parts: the history of technology; technology and modern society; and questions on the basic nature of technology. Favorably reviewed by K. Yamamura in *Technology and Culture* 9, no. 1 (January 1968): 105-6.

ZSCHIMMER, EBERHARD. *Philosophie der Technik; Einführung in die technische Ideenwelt* [Philosophy of technology; introduction to the world of technological ideas]. Stuttgart: Enke, 1933. Pp. 76. First published in 1913 as *Philosophie der Technik; vom Sinn der Technik und Kritik des Unsinns über die Technik.* 2d ed., 1919. This third revised edition contains mild Nazi propaganda. Although Kant, Fichte, and Hegel are mentioned, the treatment remains superficial. Chapter titles: "The Idea of Technology," "Battles against the Technological Spirit," and "On Technological Creativity."—R. J. R. See also the author's *Technik und Idealismus* (Jena: Volksbuchhandlung, 1920); *Deutsche Philosophen der Technik* (Stuttgart: Enke, 1937); and "Vom Wesen des technischen Schaffens," *Zeitschrift für deutsche Philologie* (Stuttgart) 6 (1940): 231-38.

II. ETHICAL AND POLITICAL CRITIQUES

A. PRIMARY SOURCES

ALBRECHT, ULRICH. "Die Werturteilsfrage in der Technik" [The question of value judgment in technology],

Zeitschrift für Allgemeine Wissenschaftstheorie [Journal for general philosophy of science] 1, no. 2 (1970): 161-72. A critical review of the debate about whether engineering decision methodology is value neutral—with special reference to the East German argument that engineering decisions are in harmony with Marxist materialism. Followed by an analysis of the character of value judgments, and proposals for further discussion.

ALLEN, FRANCIS R., HORNELL HART, DELBERT C. MILLER, WILLIAM F. OGBURN, and MEYER F. NIMKOFF. *Technology and Social Change.* New York: Appleton-Century-Crofts, 1957. Pp. 529. A group of sociologists dedicate this introductory text to "the scientific study of social change." Part I on "Processes and Theories of Social Change" contains W. F. Ogburn's "The Meaning of Technology" and "How Technology Causes Social Change," H. Hart's "Acceleration in Social Change," M. F. Nimkoff's "Obstacles to Innovation," and D. C. Miller's "Theories of Social Change." Parts II and III contain essays on the social effects of selected technologies and on the influence of technology on social institutions. Part IV contains H. Hart's "The Hypothesis of Cultural Lag: a Present-Day View," and F. R. Allen's "Major Problems Arising from Social Change." Part V contains H. Hart's "Predicting Future Trends," "Planning in the Atomic Crisis," and "Human Adjustment and the Atom." The central concept of these essays is that progress in economic, political, and social planning and control has lagged behind progress in science and technology. The problems of technological change require advanced social science. Contains a good bibliography of the literature on the concept of cultural lag.

ALLEN, FRANCIS R. "Technology and Social Change: Current Status and Outlook," *Technology and Culture* 1, no. 1 (Winter 1959): 48-60.

ANDERS, GÜNTHER. *Die Antiquiertheit des Menschen; über die Seele im Zeitalter der zweiten industriellen Revolution* [The antiquitization of man; on the soul in the age of the second industrial revolution]. Munich: Beck, 1956. Pp. 353. See also the author's *Der Mann auf der Brücke; Tagebuch aus Hiroshima und Nagasaki* (Munich: Beck, 1963); *Philosophische Stenogramme* (Munich: Beck, 1965); and *Die Toten; Rede über die drei Weltkriege* (Cologne: Phal-Rugenstein, 1966).

———. "Commandments in the Atomic Age." Pp. 11-20 in *Burning Conscience*. Reinbeck bei Hamburg: Rowohlt Verlag, 1961. The technical age—and the atomic bomb—have reversed the classical relation between action and imagination. "We are unable to conceive what we can construct; to mentally reproduce what we can produce; to realize the reality which we can bring into being." Man's task is to expand imagination and feeling to cope with the increasing possibility for action. Included in C. Mitcham and R. Mackey, eds., *Philosophy and Technology* (New York: Free Press, 1972). For another version of this argument, see "Reflections on the H Bomb," *Dissent*, vol. 85, no. 2 (Spring 1956). See also the author's "The World as Phantom and as Matrix," *Dissent*, vol. 85, no. 1 (Winter 1956).

ARON, RAYMOND, ed. *World Technology and Human Destiny.* Translated by R. Seaver. Ann Arbor: Univ. Michigan Press, 1963. Pp. 246. From *Colloques de Rheinfelden* (Paris: Calmann-Lévy, 1960). The proceedings of the Basel-Rheinfelden Conference of 1959. Four papers with discussions. The papers: R. Aron's "Industrial Society and the Political Dialogues of the West," J. Hersch's "Brief Remarks about Raymond Aron's Text," E. Voegelin's "Industrial Society in Search of Reason," C. Morazé's "The Relationship between Thought and Action in the Three Worlds." Voegelin's paper and one of the discussions on messianism are probably the most important pieces.

BAIER, KURT, and NICHOLAS RESCHER, eds. *Values and the Future: the Impact of Technological Change on American Values.* New York: Free Press, 1969.

Pp. 527. 17 essays on the impact of technological development on American values which grew out of a joint investigation at the University of Pittsburgh during 1965-66 under grants from IBM and the Carnegie Corporation. The essays of philosophical importance are primarily concerned with the conceptual and methodological problems of change and value-relevant trends of technological development, such as advances in computers, transportation, communication, and biomedicine. The majority of the essays, however, are of sociological interest. They are concerned with "detection of trends of social change, such as those in sexual relations, the family, race relations, and the sphere of work and leisure," "the identification of economic trends, such as changes in productivity, organization of production, composition of product, and distribution of wealth." Another group of essays is concerned with "forecasting changes in the values of important social groups, such as scientists, educators, corporation managers, and professional women." In addition to the conceptual analyses of value and value change by Baier and Rescher, other representative articles are: J. B. Schneewind's "Technology, Ways of Living, and Values in 19th Century England," T. J. Gordon's "The Feedback between Technology and Values," B. de Jouvenel's "Technology as a Means," K. E. Boulding's "The Emerging Superculture," J. K. Galbraith's "Technology, Planning, and Organization," and B. Gold's "The Framework of Decision for Major Technological Innovation."

BALKE, SIEGFRIED. "Le développement technique et l'homme," *Société belge d'études et d'expansion* 53 (1954): 857-66.

———. "Technik und Gesellschaftsordung" [Technology and social order], *Das Gas- und Wasserfach* 99, no. 34 (1958): 27-33.

———. *Die imperfekte Perfektion der Technik.* Bericht aus dem Deutschen Museum. Munich: Oldenbourg, 1961 Pp. 32.

_____. "Der Ingenieur und die neue Technik," *Giesserei; Zeitschrift für das gesamte Giessereiwesen* (Düsseldorf) 49, no. 20 (1962): 668-74.

_____. *Vernunft in dieser Zeit; der Einfluss von Wirtschaft, Wissenschaft und Technik auf unser Leben* [Reason in our time; the influence of economy, science and technology in our life]. Düsseldorf: Econ-Verlag, 1962. Pp. 296. By a well-known German engineer and expert on atomic energy. Mainly popular sociology, but one chapter, on "The Technologists and the Spiritual Sciences" (pp. 211-26), gives a popular summary of German philosophical attitudes toward technology. – R. J. R.

_____. "Vom Idol zur Dienerin der Idee" [From idol to handmaid of the idea], *Europa* (Bad Reichenhall) 14, no. 3 (1963): 42-47.

_____. *Schattenseiten des technischen Fortschritts* [Dark sides of technological progress]. Bad Godesberg: Vereingung Deutscher Gewässerschutz, 1967. Pp. 15. Also in *Das Leben* (Hamburg) 4, no. 12 (1967): 290 ff.

_____. "Herausforderung durch die Technik" [Challenge by technology], *Jahrbuch der Schiffbautechnischen Gesellschaft* 62 (1968): 129.

_____. "Die technischen Entwicklungen unserer Zeit" [The technological developments of our time], *Universitas; Zeitschrift für Wissenschaft, Kunst und Literatur* 23, no. 4 (1968): 355-66. For a number of other articles pre-1956, see Dessauer's bibliography.

BOOKCHIN, MURRAY. *Post-Scarcity Anarchism.* Berkeley, Calif.: Ramparts, 1972. Unites modern technology with the ecological and utopian ideas of the countercultural revolt. See especially the chapter entitled "Toward a Liberatory Technology," which was first published as a pamphlet under the pseudonym Lewis Herber. New York: Anarchos, n.d.

BRICKMAN, WILLIAM W., and STANLEY LEHRER, eds. *Automation, Education, and Human Values.* New York: School & Society Books, 1966. Pp. 419. Contents includes: M. H. Goldberg's "Introduction: Automation, Education, and the Humanity of Man." Part I,

"Outlook," contains S. Lehrer's "Man, Automation, and Dignity"; W. G. Mather's "When Men and Machines Work Together"; A. B. Barach's "Changing Technology and Changing Culture"; G. E. Arnstein's "The Mixed Blessings of Automation"; M. Mead's "The Challenge of Automation to Education for Human Values"; and C. R. Bowen's "Automation: the Paradoxes of Abundance." Part II, "Humanistic Education," contains W. G. Rice's "The Humanities in the Sixties"; G. B. Dearing's "Education for Humane Living in an Age of Automation"; P. E. Siegle's "Education, Automation, and Humanistic Competence"; C. M. D. Peters's "The Education of the Executive as New Humanist"; E. F. Bacon's "Some Questions concerning Humanistic Education and Automation"; L. H. Harshbarger's "Technological Change, Humanistic Imperatives, and the Tragic Sense"; R. W. Burhoe's "The Impact of Technology and the Sciences on Human Values"; W. W. Brickman's "The Scholar-Educator and Automation"; and A. Heckscher's "Humanistic Invention in the Post-industrial Age." Part III, "Education and Technology: Varied Aspects," contains R. C. Buck's "Education, Technological Change, and the New Society"; L. K. Williams's "The Impact of Technology and the Need for a Dialogue"; J. A. Donovan, Jr.'s "Automation and Education: Reflections on the Involvement of Government"; W. L. Davis's "A Labor View of the Social and Educational Implications of Technological Change"; L. H. Evans's "Automation and Some Neglected Aspects of Society and Education"; H. H. Humphrey's "Education: the Central Need of 20th-Century America"; A. J. Celebrezze's "The Technological Revolution and Education"; F. Keppel's "Automation: Boon or Bane for Education?"; N. D. Kurland's "Stay-at-Home Classrooms for Space-Age Adults"; V. M. Rogers's "Education for the World of Work"; D. B. Harris's "The Role of Work in the Socialization of the Adolescent"; J. W. Ball, D. J. Lloyd, and A. J.

Croft's "Training 1974: the Revolution in Learning"; D. N. Michael's "Free Time: the New Imperative in Our Society"; and W. W. Brickman's "Work, Leisure, and Education." Part IV, "Man, Mind, and Soul in a Technological Age," contains J. MacIver's "Technological Change and Health"; S. F. Reed's "Automation, Education, and Creativity"; M. B. Bloy, Jr.'s "Technological Culture, the University, and the Function of Religious Faith"; and M. H. Goldberg's "Conclusion: and See It Whole." Contains bibliography.

BROZEN, YALE. "Value of Technological Change," *Ethics* 62 (July 1952): 249-65. A philosophical survey of various theories.

BRZEZINSKI, ZBIGNIEW. *Between Two Ages: America's Role in the Technetronic Era.* New York: Viking, 1970. Pp. 334. America is the first technetronic (technological and electronic) society, and is thus pioneering the solutions to the problems of such a society. See also "The American Transition," *New Republic* 157 (December 23, 1967): 18-21.

BUCHANAN, SCOTT. "Technology as a System of Exploitation." Pp. 151-59 in C. F. Stover, ed., *The Technological Order.* Detroit: Wayne State Univ. Press, 1963. A highly oracular discussion of art or technics and the forms of its organization. The suggestion seems to be that modern technology becomes an organized system of exploitation by abstracting technics from the arts, the rules for the actual making of things.

BURNS, TOM. "The Social Character of Technology," *Impact of Science on Society* 7, no. 3 (September 1956): 147-65. Two major changes have occurred in the social conditions affecting the introduction of innovations: the emergence of giant industrial firms and the emergence of a new professional group of technicians — the technologists — which has greatly modified the institutional structure of industrial firms. The technologist is at the focal point of technical innovation; but he belongs to a separate system which is in-dependent of both science and industry.

———. "Models, Images, and Myths." Pp. 11-23 in W. H. Gruber and D. G. Marquis, eds., *Factors in the Transfer of Technology.* Cambridge, Mass.: M.I.T. Press, 1969. A discussion of D. Price's "The Structure of Publications in Science and Technology" and S. Toulmin's "Innovation and the Problem of Utilization" in this same volume, underlining "the revolutionary significance of the move away from the model of technological transfer with which government and business operate." Stresses that "the mechanism of technological transfer is one of agents, not agencies." See also T. Burns and G. Stalker, *Management of Innovation* (Chicago: Quadrangle, 1961).

CALDER, NIGEL. *Technopolis: Social Control and the Uses of Science.* New York: Simon & Schuster, 1970. Pp. 376. Argues for popular democratic control of modern technology. Includes bibliography. Review essay by C. Mitcham scheduled for publication in *International Philosophical Quarterly,* vol. 13, no. 1 (March 1973). See also Calder's "Tomorrow's Politics: the Control and Use of Technology," *Nation* (January 4, 1965), pp. 3-5, reprinted in J. G. Burke, ed., *The New Technology and Human Values* (Belmont, Calif.: Wadsworth, 1966); and "Averting Technological Tyranny," *New Statesman* 79 (February 6, 1970): 177-78.

CAREY, JAMES, and JOHN QUICK. "The Mythos of the Electronic Revolution," *American Scholar* 39, nos. 2-3 (1970): 219-41 and 395-424. A critique of H. Innis, L. Mumford, M. McLuhan, and their sympathizers.

CASSERLEY, J. V. LANGMEAD. *In the Service of Man: Technology and the Future of Human Values.* Hinsdale, Ill.: Regnery, 1967. Pp. 204. Best part is chapter 3, "Axiology, or the Philosophy of Value"; but even this survey of different theories of value remains rather superficial.

DEWEY, JOHN. "Science and Free Culture." Pp. 131-54 in *Freedom and Culture.* New York: Putnam, 1939. The

Enlightenment faith that the advancement of science will of itself produce free culture is naïve. Still, "a culture which permits science to destroy traditional values but distrusts its power to create new ones is a culture which is destroying itself." Our problems cannot be blamed on science; men are called upon to accept their responsibilities.

──────. "Science and Society." Pp. 318-30 in *Philosophy and Civilization*. New York: Capricorn, 1963. First published, 1931. A pragmatist's argument for social control of the means of production. See also Samuel M. Levin, "John Dewey's Evaluation of Technology," *American Journal of Economics and Sociology* 15, no. 2 (January 1956): 123-36.

DIESEL, EUGEN. *Das Phänomen der Technik; Zeugnisse, Deutung und Wirklichkeit*. Leipzig: Reclam; Berlin: VDI-Verlag, 1940. Pp. 258. Subtitled "Documents, Interpretation and Reality." Divided into three main parts: "The Span and Path of Technology" (90 pp.), "Man as Creator of Technological Works" (120 pp.), and "The Pendulum of Judgment on Technology" (15 pp.). Consists largely of Diesel's summations and evaluations of others' thoughts regarding technology and technological man. An extraordinarily objective book for 1939 Germany, although it silently passes over the problem of technological weaponery. – R. J. R.

──────. "Die Tyrannei der Apparatur: der Weltaufruhr der Technik" [The tyranny of apparatus: the world-wide uprising of technology], *Gralswelt* 14 (1960): 57-61.

──────. *Menschheit im Katarakt* [Mankind in a cataract]. Griesbach: A. Winkler, 1963. Pp. 354. For a large number of other works pre-1956, see Dessauer's bibliography.

DOUGLAS, JACK D., ed. *Freedom and Tyranny; Social Problems in a Technological Society*. New York: Random House, 1970. Pp. 289. Good selection of material from the works of Marcuse, Ellul, Galbraith, Feuer, Boorstin, Boguslaw, Theobald, Mannheim, etc., all of it previously published.

──────, ed. *The Technological Threat*. Englewood Cliffs, N.J.: Prentice-Hall, 1971. Pp. 185. Essays by D. Bell, I. Taviss, R. Nisbet, K. Mannheim, D. Riesman, N. Wiener, C. R. Rogers and B. F. Skinner, C. W. Mills, L. K. Frank, and E. T. Chase.

DRUCKER, PETER F. *The Future of Industrial Man*. New York: John Day, 1942. Reprinted, New York: New American Library, 1965. Pp. 208. Industrial society – the owners, the workers, the managers – is unsupported by any legitimatizing theory; it is burdened by both earlier agrarian sociopolitical conventions and by an anachronistic mercantilism. A functioning postwar industrial order requires a free society and "a Conservative approach." Contains a brief theoretical discussion of the totalitarian implications of rationalist liberalism.

──────. *The Landmarks of Tomorrow*. New York: Harper & Row, 1957. Pp. 270. A rather broad discussion of "the post-modern world." Concerned with the philosophical shift from the Cartesian universe of mechanical cause to that of pattern, purpose, and process, the social impact of an educated society, the emergence of economic development, the decline of government by the nation-state, the collapse of non-Western culture, and the new spiritual reality of human existence. For a more sociological approach to the problems of business and management within the new industrial order, see the author's *The New Society: the Anatomy of the Industrial Order* (New York: Harper & Row, 1949).

──────. *Knowledge and Technology*. Dimensions for Exploration series. Oswego, N.Y.: Division of Industrial Arts and Technology, State Univ. College, 1964. Pp. 11. Technological education needs to be part of a general education because we are in danger of "taking that fairly small part of the human being that is his verbal-intellectual faculty and considering it the whole man." Verbal subjects lack the ingredient of doing. A technological education "is needed because the people with knowledge must learn how to do something with it. They had better know how one becomes a technologist."

———. *The Age of Discontinuity.* New York: Harper & Row, 1968. Pp. 394. Four major discontinuities will shape the closing decades of the 20th century: the growth of new knowledge-based technologies, major changes in the world's economy, conflicts between the individualistic society of the 18th-century liberal theory and the power concentration of modern socioeconomic organizations, and the rise of knowledge as the crucial resource of the economy.

———. *Technology, Management, and Society.* New York: Harper & Row, 1970. Pp. 207. A collection of essays. The Preface defines the theme of this book, and of all the author's work, as the argument that "technology is not about tools, it deals with how Man works." Includes "Work and Tools," "The First Technological Revolution and Its Lessons," and "The Technological Revolution: Notes on the Relationship of Technology, Science, and Culture"—all of which first appeared in *Technology and Culture.* Also contains the author's contributions to M. Kranzberg and C. W. Pursell, eds., *Technology in Western Civlization* (New York: Oxford Univ. Press, 1967) on "Technological Trends in the Twentieth Century" and "Technology and Society in the Twentieth Century." One other essay of special note: "Can Management Ever Be a Science?" Another relevant essay not collected in this volume: "Modern Technology and Ancient Jobs," *Technology and Culture* 4, no. 3 (Summer 1964): 277-81. See also M. Kranzberg, "Drucker as Historian of Technological Change," in Tony H. Bonaparte and John E. Flaherty, eds., *Peter Drucker: Contributions to Business Enterprise* (New York: New York Univ. Press, 1970), pp. 337-61.

———. "Saving the Crusade," *Harper's* (January 1972), pp. 66-71. Argues that the ecologist's crusade to do away with pollution cannot mean doing away with technology.

DUCHET, RENÉ. *Bilan de la civilisation tech-nicienne; anéantissement ou promotion de l'homme* [Balance sheet on technical civilization; annihilation or promotion of man]. Toulouse: Privat-Didier, 1955. Pp. 293. A survey of doubts about the technological world in general and the problem of the atomic age in particular, followed by chapters of philosophical analysis on the "Natural Milieu and Technical Milieu," "In the Service of the Machine," and "The Human Condition." Concludes with analyses of the romantic and existentialist critiques of technology and the argument that all men must become engaged if the present situation is to be altered.

DUNLOP, JOHN T., ed. *Automation and Technological Change.* Englewood Cliffs, N.J.: Prentice-Hall, 1962. Pp. 184. Contains R. L. Heilbroner's "The Impact of Technology: the Historic Debate," L. A. DuBridge's "Educational and Social Consequences," F. C. Mann's "Psychological and Organizational Impacts," M. Anshen's "Managerial Decisions," G. W. Taylor's "Collective Bargaining," W. A. Wallis's "Some Economic Considerations," E. Clague and L. Greenberg's "Employment," R. N. Cooper's "International Aspects," F. Bello's "The Technology behind Productivity," H. M. Wriston's "Perspective."

DURBIN, PAUL T. "Technology and Values; a Philosopher's Perspective," *Technology and Culture* 13, no. 4 (October 1972): 556-76. Surveys a spectrum of evaluations of technology using H. Marcuse, J. McDermott, E. G. Mesthene, J. K. Galbraith, K. Boulding, F. Lundberg, A. Hacker, and J. Ellul as examples. Then, after arguing that the evaluation of technology is properly part of political philosophy, Durbin develops a theory of value change based on the pragmatism of G. H. Mead and uses it to critique the authors in the initial spectrum.

FIEBLEMAN, JAMES K. "The Impact of Science on Society." Pp. 39-75 in *Studies in Social Philosophy.* Tulane Studies in Philosophy, 11. New Orleans: Tulane Univ., 1962. This article ranges over a number of topics:

the relation between pure and applied science in light of the different metaphysical assumptions of positivism, realism, and dialectical materialism; the characteristics of a society must have to advance science; the effects of science on art and the humanities. But by trying to cover so much, the discussion remains rather superficial. See also, the author's *Mankind Behaving: Human Needs and Material Culture* (Springfield, Ill.: Thomas, 1963).

_____. "The Technological Society." In *The Reach of Politics*. New York: Horizon, 1970.

FERKISS, VICTOR C. *Technological Man: the Myth and the Reality*. New York: Braziller, 1969. Pp. 336. Considers the following topics in an attempt to point out what man must do in order to meet the challenges of technological change: technology and industrial man, the prophets of the new, the existential revolution, technological change and cultural lag. Conclusion: technological man requires a new naturalistic, holistic, and immanentist philosophy.

FROMM, ERICH. *The Revolution of Hope: Toward a Humanized Technology*. New York: Harper & Row, 1968. Pp. 178. As an inspirational tract by a well-known humanistic psychologist, this work sometimes tends to oversimplify both the problems and the solutions. But see also "Toward a Humanized Technology," in S. E. Deutsch and J. Howard, eds., *Where It's At; Radical Perspectives in Sociology* (New York: Harper & Row, 1970).

FULLER, R. BUCKMINSTER. *Education Automation*. Carbondale: Southern Illinois Univ. Press, 1962. Pp. 88.

_____. *Untitled Epic Poem on the History of Industrialization*. Charlotte, N.C.: Heritage, 1962. Pp. 227. A poetic testament of hope in man, science, and technology. Fuller calls his poetry "ventilated prose." See also *No More Secondhand God and Other Writings* (Carbondale: Southern Illinois Univ. Press, 1963) and *Intuition* (Garden City, N.Y.: Doubleday, 1972), for more of this.

_____. *Ideas and Integrities; a Spontaneous Autobiographical Disclosure*. Edited by Robert W. Marks. Englewood Cliffs, N.J.: Prentice-Hall, 1963. Pp. 318. Perhaps Fuller's most comprehensive book.

_____. *Nine Chains to the Moon*. Carbondale: Southern Illinois Univ. Press, 1963. Pp. 375. One of his earliest works which was only published late.

_____. *Operating Manual for Spaceship Earth*. Carbondale: Southern Illinois Univ. Press, 1969. Pp. 143. Calls for a total anticipatory design science.

_____. *Utopia or Oblivion? The Prospects for Humanity*. New York: Bantam, 1969. Pp. 366. A collection of articles and addresses. Includes: "A Citizen of the 21st Century Looks Back" (from "Man with a Chronofile," *Saturday Review* (April 1, 1967), and "Summary Address at Vision 65" (from *American Scholar*, vol. 35, no. 2). Some other articles of interest which are not included in this collection: "Report on the 'Geosocial Revolution,'" *Saturday Review* (September 16, 1967), pp. 31-33 ff.; "The Age of Astro-Architecture," *Saturday Review* (July 13, 1968), pp. 17-19 ff.; "Letter to Doxiadis," *Main Currents* 25 (March-April 1969): 87-97. See also the interview in *Playboy* (February 1972).

_____. *The Buckminster Fuller Reader*. Edited with introduction by James Meller. London: Cape, 1970. Pp. 383.

_____. "Education for Comprehensivity." Pp. 3-77 in *Approaching the Benign Environment*. The Franklin Lectures in the Sciences and Humanities at Auburn University, 1st series. University: Univ. Alabama Press, 1970. Fuller's essay "offers a rambling, but imaginative and sometimes eloquent argument in behalf of man's capacity to command and integrate specialities, thus regenerating a society bent on extinction through overspecialization. Continually overwhelming his own pessimisms with a sense of man's immanent possibilities, he speaks for a 'design science revolution that will guide science and technology to humane ends and put mankind's common affairs in order." — C. F. Stover, *Technology and Culture* 12, no. 1 (January 1971): 140.

———. *I Seem to Be a Verb.* With Jerome Agel and Quentin Fiore. New York: Bantam, 1970. Pp. 192.

———. "Technology and the Human Environment." Pp. 174–80 in R. Disch, ed., *The Ecological Conscience.* Englewood Cliffs, N.J.: Prentice-Hall, 1970. [NOTE: Actually Fuller does not have a self-conscious philosophy of technology. Yet, imbedded in his complex private jargon is a collection of practical inventions and utopian futurology which contains perhaps the most optimistic vision of the technological society of anyone writing about the nature and meaning of modern technology. Technology is generally conceived as the means for making the most advantageous use of environmental energy and resources; it is distinguished from craft tools and industrial tools. Its meaning is that it creates a totally new relationship between man and the world, one oriented around energy and motion rather than substance and permanence.]

GEHLEN, ARNOLD. "Die Technik in der Sichtweite der philosophischen Anthropologie" [Technology in the perspective of philosophical anthropology], *VDI-Zeitschrift* 96, no. 5 (1954): 149–53. See also the author's "Anthropologische Ansicht der Technik," in Hans Freyer et al., eds., *Technik im technischen Zeitalter* (Düsseldorf: Schilling, 1965), pp. 101–18.

———. *Die Seele im technischen Zeitalter; sozialpsychologische Probleme in der industriellen Gesellschaft* [The soul in the technological age; problems of social psychology in the industrial society]. Hamburg: Rowohlt Verlag, 1957. Pp. 131. This is an extension of Gehlen's basic philosophical position which was first outlined in *Der Mensch; Seine Natur und seine Stellung in der Welt* (Berlin: Junker & Dunnkampt, 1940), where he projected an anthropology in which man is conceived as an active being who is not adjusted to a fixed environment. As a result, man must adjust his biological constitution to the means of existence until he can conduct his life prudently, take an independent view of his world, and actively convert it to his own use.

———. "Der Begriff Technik in entwicklungsgeschichtlicher Sicht" [The concept of technology from the perspective of evolutionary history], *VDI-Zeitschrift* 104 (1962): 674–77. For other articles pre-1956, see Dessauer's bibliography.

GILFILLAN, S. COLUM. *The Sociology of Invention.* Chicago: Follett, 1935. Pp. 191. A pioneering study of the social causes, social effects, and psychology of invention. Chap. 2, "Examining the Nature of Invention," approaches a metaphysical analysis. Reprinted, Cambridge, Mass.: M.I.T. Press, 1963. See also the author's *Supplement to the Sociology of Invention* (San Francisco: San Francisco Press, 1972), which updates the main theses and bibliographies. Gilfillan has also prepared a major bibliography on "Social Implications of Technical Advance" which is mentioned in the introduction.

GILLOUIN, RENÉ. *Man's Hangman Is Man.* Translated by D. D. Lachman. Mundelein, Ill.: Island Press, 1957. Pp. 113. From *L'homme moderne bourreau de lui-même* (Paris: Portulan, 1951). A French conservative's critique of modernity and the anthropology which takes man to be good by nature. The chapter on "Technique and Ethics, or Technique as Temptation" argues that as techniques increase in quantity they pose an inhuman temptation. There is an appendix on "Technique and Nationalism" arguing against the optimistic thesis of J. Benda et al. that scientific progress makes man more cosmopolitan.

GLASS, BENTLEY. *Science and Ethical Values.* Chapel Hill: Univ. North Carolina Press, 1965. Pp. 101. After giving what he calls "a natural history of value" and discussing "heredity and the ethics of tomorrow," the author argues in the third chapter on "the ethical basis of science" that "the problem of the future is the ethical problem of the control of man over his own biological evolution." It is the duty of scientists to encourage informed discussion in this area.

———. "Science: Endless Horizons or Golden Age?" *Science* 171 (January 8, 1971): 23–29.

GONSETH, FERDINAND. "The Humanization of Technics," *Philosophy Today* 1, no. 3 (Fall 1957): 196–201. From "De l'humanisation de la technique," *Dialectica* 10, no. 2 (June 15, 1956): 99–112. Argues that, "if technicians mean to take their responsibilities seriously in order that the age of technic remain fully human, they cannot dispense with a study, a 'complete' study of two complementary things: man and men, societies and civilizations."

GOODMAN, PAUL. "The Human Uses of Science," *Commentary* 30, no. 6 (December 1960): 461–72. Distinguishes between science and technology, then argues for the unfettered pursuit of science because science is an intellectual virtue. Technology, however, is not in itself a good and should be governed by humanistic criteria, such as utility, economy, function.

———. "Morality of Scientific Technology." Pp. 297–316 in *Like a Conquered Province*. New York: Random House, 1968. Control of technology "is lodged in managers who finally are not interested in efficiency, not to speak of prudence." Hence, "there ceases to be a morality of technique at all." A remedy "to restore morale to scientific technology" would be to "judge technology directly in terms of the moral criteria appropriate to it as a branch of practical philosophy."

———. "Can Technology Be Humane?" *New York Review of Books* 13, no. 9 (November 20, 1969): 27–34. "Technology is a branch of moral philosophy, not of science." Argues for "a kind of Jeffersonian democracy or guild socialism" in which "scientists and inventors and other workmen are responsible for the uses of the work they do, and . . . competent to judge these uses and have a say in deciding them." For a more popular treatment of the same theme, see "The Trouble with Today's Technology: a Social

Critic's View," *Innovation* 1, no. 2 (June 1969): 38–46. Also see "Neutral Science De-humanizes Man," *National Catholic Reporter* (April 16, 1969), p. 6. [NOTE: For a restatement of some of the main themes in these essays, see *The New Reformation* (New York: Random House, 1970) and *Utopian Essays and Practical Proposals* (New York: Random House, 1951), especially the "Introduction" and the essay on " 'Applied Science' and Superstition."]

GRANT, GEORGE P. *Technology and Empire.* Toronto: House of Anansi Press, 1969. Pp. 143. Important work by a philosopher influenced by Leo Strauss and Jacques Ellul. The first essay, "In Defense of North America," argues that the technological attitude toward the world is opposed to classical contemplation. This essay is included in C. Mitcham and R. Mackey, eds., *Philosophy and Technology* (New York: Free Press, 1972). See also *Philosophy in the Mass Age*, 2d ed. (Toronto: Copp Clark, 1959), a short account of some Western moral traditions and their relations to the religion of progress through technology; *Lament for a Nation; the Defeat of Canadian Nationalism* (Princeton, N.J.: Van Nostrand, 1965), a meditation on the impossibility of conservatism in the technological era; and "Critique of the New Left," *Our Generation* 3, no. 4, and 4, no. 1 (May 1966): 46–51.

GRUBER, WILLIAM H., and DONALD G. MARQUIS, eds. *Factors in the Transfer of Technology.* Cambridge, Mass.: M.I.T. Press, 1969. Pp. 289. Section I, "Innovation: the Development and Utilization of Technology," contains the following articles: T. Burns's "'Models, Images, and Myths"; S. Toulmin's "Innovation and the Problem of Utilization"; W. H. Gruber's "The Development and Utilization of Technology in Industry''; D. C. McClelland's "The Role of Achievement Orientation in the Transfer of Technology." Three other articles of note: D. Price's "The Structures of Publication in Science and Technology"; H. Reiss's "Human Factors at the Science-Technology Interface"; and W. H. Gruber and D. G.

Marquis's "Research on the Human Factor in the Transfer of Technology." Individual articles contain good bibliographic references to material in this field.

———. "Research on the Human Factor in the Transfer of Technology." Pp. 255–82 in W. H. Gruber and D. G. Marquis, eds., *Factors in the Transfer of Technology* (Cambridge, Mass.: M.I.T. Press, 1969). "If technical elements are brought together in a new way and a new technology results, this would be called an 'invention' until it is used to satisfy a demand, at which point an 'innovation' occurs. Research on the transfer of technology thus focuses on 'innovation' and 'diffusion' because the word 'technology' connotes a method of achieving a practical purpose or 'use.' " Follows this distinction with a four-stage diagram of technical advance which is defined as "an increase in the level of technical knowledge and/or an increase in the economic uses of technical knowledge." See also the authors' Introduction to this volume and their "Communications Patterns in Applied Technology," *American Psychologist*, vol. 21 (November 1966).

GUCHET, YVES. *Technique et liberté* [Technology and freedom]. Paris: Nouvelles Editions Latines, 1967. Pp. 339. Basically a survey of opinions on the problem of man in the technological society. Part I on the ideological approach to this problem contains, under the heading "L'optimisme technicien," analyses of the socialists (Saint-Simon, Pecqueur, Marx, Sorel, Berth), humanists (Mumford, Fourastie, Armand), and religious (Mounier, de Chardin); under the heading "Le volontarisme technicien," analyses of the socialists (Proudhon), humanists (Friedmann, Romains, Huxley, de Rougemont, Maritain), and religious (Siegfried); under the heading "'Le pessimisme technicien," analyses of the humanists (Duhamel), and religious (Bernanos, Weil, Marcel, Berdyaev, Ellul). Part II on the sociological approach contains more original analyses of the relations between technological society and the citizen, technological society and economic man, and technological society and the person.

HABERMAS, JÜRGEN. *Toward a Rational Society; Student Protest, Science and Politics.* Translated by J. J. Shapiro. Boston: Beacon, 1970. Pp. 132. The first three essays in this collection concern the student protest movement, especially in Germany, and are from *Protestbewegung und Hochschulreform* (Frankfurt: Suhrkamp, 1969). The last three essays are from *Technik und Wissenschaft als 'Ideologie'* (Frankfurt: Suhrkamp, 1968). Of these, the last essay, "'Technology and Science as 'Ideology,' " is the most relevant to the philosophy of technology. It presents a critique of the theories of Talcott Parsons and Herbert Marcuse and sketches out a new general theory of social evolution. See also the author's "Technischer Fortschritt und Soziale Lebenswelt," *Praxis* 2, nos. 1–2 (1966): 217–28. Two other books by Habermas have recently been published in Boston by Beacon Press; these are *Knowledge and Human Interests* and *Theory and Practice.* A number of articles on Habermas can be found listed under Ethical and Political Critiques, Section B.

HARDIN, GARRETT. "The Tragedy of the Commons," *Science* 162 (December 13, 1968): 1243–48. The problems created by technology are not amenable to technical solutions. Critique of Adam Smith's theory that the pursuit of self-interest naturally produces the common good. He argues instead that it leads necessarily to the destruction of the common good. Cf. B. L. Crowe, "Tragedy of the Commons Revisited," *Science* 166 (November 28, 1969): 1103–7. Hardin identified a set of problems with no technical solutions. Crowe says these are the same problems that political scientists have been waiting for technologists to solve.

Harvard University Program on Technology and Society 1964–72; a Final Review. Cambridge, Mass.: Harvard Univ., 1972. Pp. 285. A good summary of the projects sponsored by the Program which existed in 1964–72 with a grant from IBM "to undertake an inquiry in depth into the effects of technological change on the economy, on public policies, and on the character of society, as well as into the reciprocal effects of social change on the nature, dimension, and directions of scientific and technological developments." See also E. G. Mesthene, program director, "On Understanding Change: the Harvard University Program on Technology and Society," *Technology and Culture* 6, no. 2 (Spring 1965): 222-35; "An Experiment in Understanding: the Harvard Program Two Years After," *Technology and Culture* 7, no. 4 (Fall 1966): 475-92; and the symposium on "The Role of Technology in Society," *Technology and Culture* 10, no. 4 (October 1969): 489-536, which contains Mesthene's "Some General Implications of the Research of the Harvard University Program on Technology and Society," S. Ramo's "Comment: the Anticipation of Change," P. F. Drucker's "Comment: Is Technology Predictable?" A. Hunter Dupree's "Comment: the Role of Technology in Society and the Need for Historical Perspective," and Mesthene's "A Comment on the Comments." See also Mesthene's *Technological Change; Its Impact on Man and Society* (Cambridge, Mass.: Harvard Univ. Press, 1970), for a summary of some of the Program's conclusions about technology.

HEINEMANN, F. H. "Beyond Technology?" Pp. 14-29 in *Existentialism and the Modern Predicament*. New York: Harper & Row, 1953. Reprinted, Harper Torchbooks, 1958. First published in *Hibbert Journal* 51, no. 1 (October 1952): 37-46. Argues, first, that it is not the rise of mass society which is definitive of our time, but technology. Second, develops the notion of "technological alienation" with illustrations of how means usurp ends in modern art and philosophy. Third, calls for an existential transcendence of this alienation.

HEYDE, JOHANNES ERICH. "Technischer Fortschritt—menschliche Verantwortung" [Technical progress—human

responsibility], *Verhandlungen der deutschen Gesellschaft für Arbeitsschutz* 7 (1961-62): 1-29. See also two other articles by this same author: "Wissenschaft, Bildung, Technik," in *Wege zur Klarheit; Gesammelte Aufsätze* (Berlin: De Gruyter, 1960), pp. 384-401; and "Technik und Bildung; die Bedeutung der Bildung für unsere Hochschulen," in Richard Schwarz, ed. *Universität und moderne Welt; ein international Symposium* (Berlin: De Gruyter, 1962), pp. 242-76.

HINNERS, RICHARD. "Vietnam: Technology v. Morality," *Continuum* 5 (Summer 1967): 221-34. Argues from the case of Vietnam that technology is not morally neutral, but the prose is so turgid as to obscure the particulars of the discussion. Only the conclusions are clear: "What we must become conscious of is that technological development involves a profound change not so much in what we value as in the very structure of how we make our valuations" (pp. 230-31).

———. "The Ideological Turn and Its Problem for the History of Philosophy." Pp. 130-38 in *Proceedings of the American Catholic Philosophical Association*. Washington, D.C.: Catholic Univ. America Press, 1969. When it becomes ideological, philosophy ceases to be based in nature, reason, or God, and takes on a new base in human historicity. Marxism is not a "total ideology" (Mannheim) because it is explicitly grounded on the idea and realization of technicity (Heidegger). But non-Marxist thought, originating from within this project of technicity while insisting on its own trans-historical validity, is the most vulnerable to an unmasking as such a total ideology.

HOFFER, ERIC. "Automation Is Here to Liberate Us," *New York Times Magazine* (October 24, 1965), pp. 48 ff. A workingman's view of the value of automation and a critique of those intellectuals who criticize it. See also his *The Ordeal of Change* (New York: Harper & Row, 1963), and the general review essay by Edgar Z. Friedenberg,

New York Review of Books 12, no. 9 (May 8, 1969): 9-12.

HOMMES, JAKOB. *Der technische Eros; das Wesen der materialistischen Geschichtsauffassung* [The technological eros; the essence of the materialist interpretation of history]. Freiburg: Herder, 1955. Pp. 519.

———. *Dialektik und Politik; Vorträge und Aufsätze zur Philosophie in Geschichte und Gegenwart.* Cologne: J. P. Bachem, 1968. Pp. 376. See the third essay in this collection, "Technology — Crisis of Freedom," an important attempt to answer the question, What quality of technology is it that threatens man as man? — R. J. R.

"Human Dignity and Technology," *Philosophy Forum,* vol. 9, nos. 3-4 (June 1971). A special issue which includes the following articles: E. Lazlo's "Human Dignity and the Promise of Technology"; R. A. Peterson's "Technology: Master, Servant, or Model for Human Dignity"; R. A. Watson's "Human Dignity and Technology"; J. B. Wilbur's "Human Diginity and Technology: a study and Commentary"; G. K. Plochmann's "The God from the Machine May Soon Be Dead"; and L. Foss's "After Profits What? Human Dignity and Technology." Also a symposium of reviews on Bertalanffy's *Robots, Men and Minds* and Woolridge's *Mechanical Man* with responses by the authors, and a final review article of Koestler's *The Ghost in the Machine* by W. P. Alston.

JONAS, HANS. "Philosophical Reflections on Experimenting with Human Subjects," *Daedalus* 98, no. 2 (Spring 1969): 219-47. Uses a number of different ethical frameworks to try to approach the subject, and in so doing points up the gravity of the issues and, incidentally, the weaknesses of some of the frameworks. Thoughtful.

JORDAN, VIRGIL. *Manifesto for the Atomic Age.* New Brunswick, N.J.: Rutgers Univ. Press, 1946. Pp. 70. Sketches out the changes that have taken place since World War I and the ways they will shape the future. Basic changes divided into three classes: technological, biological, ideological. Technology frees man from any de-

pendence on earth. Biologically, population will level off and decline; and men are becoming genetically more uniform with less individuality. The central ideological fact is the replacement of the ideal of freedom with that of totalitarianism.

JOUVENEL, BERTRAND DE. "The Political Consequences of the Rise of Science," *Bulletin of the Atomic Scientists* 19, no. 10 (December 1963): 2-8. Sketches the transition from the dominance of the clergy in the Middle Ages to that of lawyers in early modern times and the scientist in the present. Then examines some of the conflicts between the scientific and the political mentalities and functions. See also "On the Evolution of Forms of Government," *Futuribles* (Geneva: Droz, 1963), 1:65-119. "Letter from France: the Technocratic Age," *Bulletin of the Atomic Scientists* 20 (October 1964): 27-29; and "Political Science and Prevision," *American Political Science Review* 59 (1965): 29-38.

JÜNGER, FRIEDRICH GEORG. *The Failure of Technology.* Translated by Fred D. Weick. Chicago: Regnery, 1949. Reissued, 1956. Pp. 186. Argues that technology fails to really relieve the miseries of mankind; its perfection only transforms their character. Selections from this volume first appeared as *The Price of Progress* (Chicago: Regnery, 1948). In both cases, however, what is involved is a translation in a poor scholarly format of the first complete edition of *Die Perfektion der Technik* (Frankfurt: Klostermann, 1946) (157 pp.). (This volume was partially through the press in 1940, and as a result of the war was published in incomplete form at that time.) 2d ed., 1949 (232 pp.). The 4th ed., 1953 (370 pp.), incorporates Jünger's *Maschine und Eigentum* (Frankfurt: Klostermann, 1949) (191 pp.).

KASS, LEON R. "The New Biology: What Price Relieving Man's Estate?" *Science* 174 (November 19, 1971): 779-88. A survey of new biomedical technologies, with a consideration of some basic ethical and social problems raised by the use of these technologies, and a discussion of some of' the fundamental philosophical questions toward which these problems point. Concludes with general reflections on what is to be done: "First, we sorely need to recover some humility in the face of our awesome power. . . . Because we lack wisdom, caustion is our urgent need."

KLAGES, HELMUT. "Marxismus und Technik." Pp. 137-50 in Hans Freyer et al. eds., *Technik im technischen Zeitalter*. Dusseldorf: Schilling, 1965. A scholarly attempt to elucidate Marx's attitude toward technology and technological progress. "Human history is essentially the history of labor, and 'progress' in this history cannot be gauged so much by *what* is produced as by *how* it is produced, while at the same time the level of production-technology achieved at any given time provides the decisive index" (p. 138).

KRAEMER, OTTO. "Seelische ,Wirkungen des 'technischen' Zeitalters" [Spiritual effects of the 'technological' age], *VDI-Zeitschrift* 97, no. 26 (1955): 907-16.

―――. "Der Weg des Menschen zur Technik" [The way of man to technology], *Ausbau; Studienhilfe zur Fortbildung für den technischen Nachwuchs* (Konstanz) 9 (1956): 3-4, 67-68, 195-96, 259-60, 323-24, 387-88, 451-52.

―――. "Wandelt die Technik den Menschen?" [Is technology changing man?), *Betriebswirtschaftliche Forschung und Praxis* 12 (1960): 303-5. Also in *Die berufsbildende Schule* (Wolfenbüttel) 15 (1963): 2-9.

―――. "Ursprung, Weg und Grenze der Technik" [Origin, course and limits of technology]. In Werner Boeck, ed., *Perspektiven für das letzte Drittel des 20. Jahrhunderts*. Stuttgart: W. Kohlmmer, 1968. By a German engineer. For a number of other articles pre- 1956, see Desauer's bibliography.

KRANZBERG, MELVIN. "Technology and Human Values," *Virginia Quarterly Review* 40, no. 4 (Autumn 1964): 578-92. Technology has increased the worth of the individual by improving the material conditions of life, increasing leisure, and advancing social democracy. Advancing technology has also "made possible man's cultivation of the arts" and possible fulfillment of the Christian social ethic through eliminating poverty and destitution with social welfare measures and international technical assistance programs.

LANE, ROBERT E. "The Decline of Politics and Ideology in a Knowledgeable Society," *American Sociological Review* 31, no. 5 (October 1966): 649-62. "There is increased application of scientific criteria for policy determination at the expense of the usual short-term political criteria and ideological thinking as well." Cf. Daniel Bell, *The End of Ideology* (New York: Free Press, 1960).

LASSWELL, HAROLD D. "Must Science Serve Political Power?" *American Psychologist* 25, no. 2 (February 1970): 117-23. The "historic subordination" of science to "the institutions of war and oligarchy" is discussed and ways of transforming the role of science to better serve the "commonwealth" are suggested.

LEISS, WILLIAM. "The Social Function of Knowledge," *Social Theory and Practice* (Fall 1970), pp. 1-12. A critique of the technocratic and end of ideology theories of Galbraith, Drucker, Brzezinski, Bell, and others.

―――. *The Domination of Nature*. New York: Braziller, 1972. Pp. 242. A historico-philosophical study of the idea of the domination of nature with an analysis of the aims and limitations of this concept. Contains a good survey of neo-Marxist European literature. Originally a doctoral dissertation of the same title, University of California at San Diego, 1969. The substance of the following articles has also been incorporated in this book: "Husserl and the Mastery of Nature," *Telos* 5 (1970): 82-97; a review of "Husserl's Crisis of European Sciences,*" Telos*, vol. 8 (1971); "Max Scheler's Concept of Herrschaftswissen," *Philosophical Forum* 2, no. 3 (Spring 1971): 316-31; "The Social Consequences of Technological Progress: Critical Comments on Recent Theories," *Canadian*

Public Administration 13 (1970): 246-62; "Utopia and Technology: Reflections on the Conquest of Nature," *International Social Science Journal* 22 (1970): 576-88.

LESTER, JULIUS. "The Revolution: Revisited." In James Y. Holloway, ed., *Introducing Jacques Ellul*. Grand Rapids, Mich.: Eerdmans, 1970. Argument by a black thinker that black political radicals have failed to adequately appreciate the problems posed by technology. Black and white cultural radicals offer greater hope of overcoming technological tyranny.

LEWIS, C. S. *The Abolition of Man*. New York: Macmillan, 1947. Paperback reprint, 1965. Pp. 121. This is a more formal presentation of some of the ideas that animate the author's science fiction trilogy, *Out of the Silent Planet* (1938), *Perelandra* (1944), and *That Hideous Strength* (1946). Starting with an examination of an English textbook, Lewis develops a general critique of nihilism in both its political and metaphysical dimensions, after which he deals specifically with what he takes to be one consequence of this nihilism — namely, the technological conquest of nature. According to Lewis, man's power over nature is really just some men exercising power over other men using nature as an instrument. In the end, the human conquest of Nature (in a value-laden sense) results in the conquest of man by nature (in the modern scientific sense).

LEY, HERMANN. "Die dialektische Einheit von Technik, Persönlichkeit mit Humanismus" [The dialectic unity of technology, personality, with humanism], *Neue Hütle; Technisch-wissenschaftliche Zeitschrift für die gesamte Berg- und Hüttenwesen* (Berlin) 4 (1959): 5-9.

———. "Mensch und Technik auf dem Prüffeld des Lebens" [Man and technology as the testing ground of life], *Sonntag; Wochenzeitung für Kultur, Politik und Unterhaltung* (Berlin) 15, no. 17 (1960): 10-11.

———. *Dämon Technik?* Berlin: Deutscher Verlag Wissenschaften, 1961. Pp. 428. A strenuous Marxist rejec-

tion of contemporary questioning of technology. The scope of the book is best indicated by an abbreviated table of contents: "Introduction, the Working Class Creates the Foundations for the Highest Spiritual Culture by Means of Modern Technology"; Part I, "Regularized Technology and Automation Beget Nervousness and Fear in the Imperialist Countries"; Part II, "The 'Second Industrial Revolution' as an Opiate — West German Social Democracy between Hope and Reality"; Part III, "New Witch's Trial against Technology"; Part IV, "Between the Fear of Demons, Fear of Communism, and Atom-Agression."

———. "Materialität und Modell," *Wissenschaftliche Zeitschrift der Humboldt-Universität; Gasellschafts- und sprachwissenschaftliche Reihe* 13, no. 7 (1964): 801-7. This is the initial article to a symposium on scientific and technological problems. See also the author's "Schlusswort," ibid., p. 826.

———. "Struktur und Prozess," *Wissenschaftliche Zeitschrift der Humboldt-Universität; Mathematisch-naturwissenschaftliche Reihe* 16, no. 6 (1967): 855-64.

———. "Kategorien der Operations forschung und ihre Anwendung in der sozialistischen Volkswirtschaft" [Categories of operations research and their application in socialist economy], *Deutsche Zeitschrift für Philosophie* 16, no. 2 (1968): 170-90.

———. "Natur und Technik im Verständnis von Karl Marx," *Wissenschaftliche Zeitschrift der Humboldt-Universität; Gesellschafts- und sprachwissenschaftliche Reihe* 17, no. 4 (1968): 455-74. English summary, p. 474.

———. "Zu einigen Fragen, die mit dem Wesen materieller und ideeler Faktoren der wissenschaftlich-technischen Revolution in Beziehung stehen" [On some questions concerning the nature of material and ideal factors of the scientific-technological revolution], *Wissenschaftliche Zeitschrift der Universität Rostock* 14, nos. 5-6 (1965): 577-84.

———. "Die technische Revolution in der westdeutschen Ideologie," *Deutsche*

Zeitschrift für Philosophie 13, no. 9 (1965): 1057-72.

———. *Technik und Weltanschauung.* Berlin: Urania-Verlag, 1970. Pp. 132.

———. "Zum Klassencharakter der Funktion von Wissenschaft" [On the class character of the function of science], *Deutsche Zeitschrift für Philosophie* 18, no. 10 (1970): 1250-69.

———. "Geist und Technik." Pp. 52-63 in *Proceedings of the XIV International Congress of Philosophy: Vienna, Sept. 2-9, 1968.* Vienna: Herder, 1971. Vol. 6. A number of other articles can be found in Dessauer's and Herlitzius's bibliographies.

LITT, THEODOR. *Technisches Denken und menschliche Bildung* [Technological thought and human education]. Heidelberg: Quelle & Meyer, 1957. Pp. 95. By a neo-Hegelian philosopher. For an essay in English, see "Science and Objectivity," *Science and Freedom* (Boston: Beacon, 1955). For other works pre-1956, see Dessauer's bibliography.

LOMBROSO, GINA. *The Tragedies of Progress.* Translated by C. Taylor. New York: Dutton, 1931. Pp. 329. From *Le tragedie del progresso* (Turin: Bocca, 1930). A work which has not been given as much attention in the English-speaking world as it perhaps deserves. The French translation, *La rançon du machinisme* (Paris: Payot, 1931), is often cited in European literature. Also, this is just one of a number of works from the 1930s and 1940s on the relation between man and machines. Some others of note: Stuart Chase, *Men and Machines* (1929); Silas Bert, *Machine Made Man* (1930); Edward A. Filene, *Successful Living in the Machine Age* (1931); Ralph E. Flanders, *Taming Our Machines* (1931); Walter N. Polakov, *The Power Age* (1933); S. McKee and Laura Rosen, *Technology and Society* (1941), and Carl Becker, *Progress and Power* (1949). All of these are listed separately under Ethical and Political Critiques, Section B.

MCDERMOTT, JOHN. "Technology: the Opiate of the Intellectuals," *New York Review of Books* 13, no. 2 (July 31, 1969): 25-35. Extended review of the Fourth Annual Report of the Harvard Program on Technology and Society. The contention is that just as "religion was formerly the opiate of the masses" so today "technology is the opiate of the educated public" since "no other single subject is so universally invested with high hopes for the improvement of mankind." In fact, however, there is developing a new class conflict between the technologically oppressed and the technologically liberated.

MCLUHAN, HERBERT MARSHALL. *Understanding Media; the Extensions of Man.* New York: McGraw-Hill, 1964. Pp. 359. Extremely influential and widely reviewed, this book helped spawn a whole field of media studies. Technology is understood as an organ projection. Mechanical technology exteriorizes the human body; electronic technology exteriorizes the nervous system. Electronic technology also produces a more humane environment. An earlier work, *Guttenberg Galaxy* (Toronto: Univ. Toronto Press, 1962), is also important. Both works variously reprinted. See also "Cybernetics and Culture," in C. R. Dechert, ed., *The Social Impact of Cybernation* (New York: Simon & Schuster, 1967), pp. 95-108. There are many articles on McLuhan, but see, especially, three collections: Gerald Emanuel Stearn, ed., *McLuhan: Hot and Cool* (New York: Dial, 1967); Harry H. Crosby and George R. Bond, eds. *The McLuhan Explosion* (New York: American Book, 1968); Raymond Rosenthal, ed., *McLuhan: Pro and Con* (New York: Funk & Wagnalls, 1968).

MACPHERSON, C. B. "Technical Change and Political Decision; Introduction," *International Social Science Journal* 12 (1960): 357-68. "The net effect of the technical changes we have considered . . . appears to be to diminish the democratic quality of political decisions." But this need not be so, especially since technical changes in the productive process offer "new possibilities for human freedom."

———. "Democratic Theory: Ontology and Technology." Pp. 203-20 in D.

Spitz, ed., *Political Theory and Social Change*. New York: Atherton, 1967. Argues that the theory of man (for which he uses the word "ontology") upon which Western technological society presently rests must be changed from man as a consumer of utilities to man as an exerter of his uniquely human capabilities, in order to keep democracy viable. Included in C. Mitcham and R. Mackey, eds., *Philosophy and Technology* (New York: Free Press, 1972).

MANICAS, PETER T. "Men, Machines, Materialism, and Morality," *Philosophy and Phenomenological Research*, vol. 27, no. 2 (December 1966). "I believe that the thesis of materialism comes down to the question: Would we or rather should we admit a 'sufficiently' complex 'robot' to the moral community?"

MARCUSE, HERBERT. *One-dimensional Man: Studies in the Ideology of Advanced Industrial Society*. Boston:Beacon, 1964. Pp. 260. A Marxist-revisionist analysis of the technological society examining the real possibilities for historical change given the death of the working class as a revolutionary force. Since Marcuse attributes this fact primarily to the ideology of post-industrial society rather than material conditions, the work is primarily a critique of prevalent modes of thought.

———. *An Essay on Liberation*. Boston: Beacon, 1969. Pp. 91. Against the traditional Marxist rejection of utopianism, Marcuse (by applying the Freudian analysis of fantasy as ideal gratification of the id to utopian thought) argues that, given the fact that technology can now realize almost all possibilities, what is needed is thought which is utopian in its aims. Reviews: Peter Clecak, "Marcuse: Ferment of Hope," *Nation* (June 16, 1969), pp. 765-68; and John Sparrow, "Marcuse: Gospel of Hate," *National Review* (October 21, 1969), pp. 1068-69. For another similar argument on the present need for utopian thought, cf. Paul Goodman in the introduction to *Utopian Essays and Practical Proposals* (New York: Random House, 1962).

———. *Five Lectures*. Translated by J. J. Shapiro and S. W. Weber. Boston: Beacon, 1970. See also the author's "Radical Perspectives 1969," *Our Generation* 6, no. 3 January 1969): 9-14.

MARCUSE, HERBERT, ROBERT PAUL WOLFF, and BARRINGTON MOORE, JR. *Critique of Pure Tolerance*. 2d ed. Boston: Beacon, 1969). Three essays loosely dealing with the same topic, the political nature of tolerance in advanced industrial society: Wolff's "Beyond Tolerance," Moore's "Tolerance and the Scientific Outlook," and Marcuse's "Repressive Tolerance." For some discussion of the issues raised in this work, see David Spitz, "Pure Tolerance: a Critique of Criticism," *Dissent* 13, no. 5 (September-October 1966): 510-25, and the exchange of views in succeeding issues.

[NOTE: For important background, see Marcuse's *Eros and Civilization* (Boston: Beacon, 1955; reprint with new preface, New York: Vintage, n.d.). Paul A. Robinson, *The Freudian Left* (New York: Harper & Row, 1969) contains a discussion of Marcuse. See also the review of this book by Richard Poirier, *New York Times Book Review* (October 26, 1969), pp. 66-69. Some other important secondary material on Marcuse: Daniel Callahan, "Resistance and Technology," *Commonweal* 88, no. 12 (December 22, 1967): 377-81; Maurice Cranston, "Herbert Marcuse," *Encounter* 32 (March 1969): 38-50; Donald Duclow, "Marcuse and 'Happy Consciousness,'" *Liberation* (October 1969), pp. 7-15; P. Eidelberg, "The Temptation of Herbert Marcuse," *Review of Politics* 31 (October 1969): 442-58; Kurt Glaser, "Marcuse and the German New Left," *National Review* 20 (July 2, 1968): 649-54; Herbert Gold, "California Left: Mao, Marx, et Marcuse!" *Saturday Evening Post* (October 19, 1968), pp. 56 ff.; L. Goldmann, "Understanding Marcuse," *Partisan Review* 38, no. 3 (1971): 247-62 (followed with a reply by Marcuse); Andre Gorz, "Call for Intellectual Subversion," *Nation* 198 (May 25, 1964): 534-37; Allen Graubard,

"One- dimensional Pessimism: a critique of Herbert Marcuse," *Dissent* 15 (May-June 1968): 216-28; Richard Greenman, ''A Critical Re-Examination of Herbert Marcuse's Works," *New Politics* 6, no. 4 (Fall 1967): 12-23; Irving Howe, "Herbert Marcuse or Milovan Djilas? The Inescapable Choice of the Next Decade," *Harper's* (July 1969), pp. 84-92; George Kateb, "The Political Thought of Herbert Marcuse," *Commentary* 49, no. 1 (January 1970): 48-63; Carl D. Schneider, "Utopia and History: the Logic of Revolution," *Union Seminary Quarterly Review* 24, no. 2 (Winter 1969): 155-69, and *Philosophy Today* 12, no. 4 (Winter 1968): 236-45; and Sol Stern, "The Metaphysics of Rebellion," *Ramparts* 6, no. 12 (June 29, 1968): 55-60. See also Paul Breines, ed., *Marcuse: Critical Interruptions* (New York: Herder & Herder, 1970); and Alasdair MacIntyre, *Herbert Marcuse: an Exposition and a Polemic* (New York: Viking, 1969). A bibliography of Marcuse's writings to 1967 can be found in Kurt H. Wolff and Barrington Moore, Jr., eds., *The Critical Spirit: Essays in Honor of Herbert Marcuse* (Boston: Beacon, 1967). Some of Marcuse's remarks at a symposium are printed in the *New York Times Magazine* (May 26, 1968), and portions of an interview are given in the *New York Times Magaxine* (October 27, 1968). Some foreign critiques of Marcuse are listed separately in the Ethical and Political Critiques, Section B of this bibliography.]

MENDELSOHN, EVERETT J., JUDITH P. SWAZEY, and IRENE TAVISS, eds., *Human Aspects of Biomedical Innovation.* Cambridge, Mass.: Harvard Univ. Press, 1971. Pp. 234.

MESTHENE, EMMANUEL G. "Can Only Scientists Make Government Science Policy?" *Science* 145 (July 17, 1964): 237-40. Reprinted in William R. Nelsen, ed., *The Politics of Science* (New York: Oxford Univ. Press, 1968), pp. 457-65.

_____. "Social Heracliteanism: the Problem of Educating People without Teaching Them Anything in Particular," *Harvard Graduate School of Educa-*tion *Association Bulletin* 10, no. 1 (Spring 1965): 12-17.

_____. "Technological Change and Social Development," *Harvard Business School Bulletin* 42, no. 3 (May—June 1966): 7-11.

_____. *Technology and Social Change.* Indianapolis: Bobbs-Merrill, 1967. Pp. 64. A small collection of essays centering around the question, "What are the implications of today's rapid advances in science and technology?" Articles by: H. A. Simon, J. Lederberg, H. Brooks, A. M. Weinberg, J. Ellul, E. G. Mesthene, and J. R. Pratt. Mesthene's article, "Technology and Wisdom"—which is based on two previous articles in the *Saturday Review,* "Learning to Live with Science" (July 17, 1965) and "What Modern Science Offers the Church" (November 19, 1966)—is the most important piece in the book. It argues that modern technology or invention calls for a "recovery of nerve" and an increase in wisdom. Also included in C. Mitcham and R. Mackey, eds., *Philosophy and Technology* (New York: Free Press, 1972).

_____. "The Impacts of Science on Public Policy," *Public Administration Review* 27, no. 2 (June 1967): 97-104. Discusses the impact of science-technology on policy options which have resulted from its ability to change the physical world.

_____. "How Technology Will Shape the Future," *Science* 161 (July 12, 1968): 135-43. Reprinted, Cambridge, Mass.: Harvard Univ. Program on Technology and Society, n.d. (Reprint no. 5) and in William R. Ewald, Jr., ed., *Environment and Change: the Next Fifty Years* (Bloomington: Indiana Univ. Press, 1968). Technology is the possibility for bringing about or inhibiting change in physical nature. In turn, it affects social and individual values, politics, and education by altering the conditions of choice. Included in C. Mitcham and R. Mackey, eds., *Philosophy and Technology* (New York: Free Press, 1972).

_____. "Technology and Humanistic Values," *Computer and the Humanities* 4, no. 1 (September 1969): 1-10.

_____. "Technology and Human Val-

ues," *Science Journal* (England) 5A, no. 4 (October 1969): 45-50. "Human values are not absolute and unchanging but rather originate in a society's choices. Technology's main effect on them is to open up fresh options." Reprinted in M. Kaplan and P. Bosserman, eds., *Technology, Human Values, and Leisure* (Nashville, Tenn.: Abingdon, 1971). See also Mesthene's "Technology and Values," in K. Vaux, ed., *Who Shall Live?* (Philadelphia: Fortress, 1970).
————. *Technological Change: Its Impact on Man and Society.* Cambridge, Mass.: Harvard Univ. Press, 1970; New York: New American Library, 1970. Pp. 127. Summarizes the author's and the Harvard Program's conclusions on the subject of technological change. Includes brief, but extensively annotated bibliography. [NOTE: A more complete list of this author's publications can be found in *Harvard University Program on Technology and Society 1964-1972: a Final Review* (Cambridge, Mass.: Harvard Univ., 1972).]

MEYNAUD, JEAN. "Les techniciens et le pouvoir," *Revue française de science politique* (Fondation nationale des science politiques; Association française de science politique, Paris) 7, no. 1 (1957): 5-37
————. "Qu'est-ce que la technocratie?" *Revue économique* 11 (1960): 497-526.
————. "A propos de la technocratie," *Revue française de science politique* (Fondation nationale des sciences politiques; Association française de science politique, Paris) 11, no. 3 (1961): 671-83.
————. *Technocracy.* Translated by P. Barnes. New York: Free Press, 1969. Pp. 315. From *Technocratie, mythe ou réalité?* (Paris: Payot, 1964). Technocracy is that demand that "politics be reduced to technics" and results from the "pressure of technology on the political system." An important analysis of how this happens, and the consequences. Argues for the need to rescue politics from technocrats and "electronic fascism" and return it to politicians who are responsible to the people. Includes brief bibliography.

MICHAEL, DONALD N. *Cybernation: the Silent Conquest.* Santa Barbara, Calif.: Center for the Study of Democratic Institutions, 1962. Pp. 49. Discusses what social technology or engineering is, its future growth, the coming pressures that will apply on it, the problems of forecasting studies, and the place of social technology in the conduct of such studies. See also the author's "Technology and the Human Environment," *Public Administration Review* 28, no. 1 (January-February 1968): 57-60.

MORGENTHAU, HANS J. "Modern Science and Political Power," *Columbia Law Review* 64, no. 8 (December 1964): 1386-1409. Discusses the effects on advanced nations of governmental accumulation of technological power: rising number of coup d'états and decline of popular revolutions, the rise of scientific elites to leadership positions, the advance of military control, and the decline of democratic participation in policy decisions. See also the author's *Science: Servant or Master?* (New York: New American Library, 1972).

MOUNIER, EMMANUEL. "The Case against the Machine." In *Be Not Afraid* (London: Rockliff, 1951). A valuable survey of some ethical and political (as well as religious) critiques of technology, with a positive reply to the critics.

MUELLER, GUSTAV E. *Sinister Savior: Two Essays on Man and the Machine.* Norman, Okla.: Cooperative Books, 1941. Pp. 28. The first essay, "The Machine-Age and Its Illusions," argues agianst that fascile liberalism which believes machines can be used simply to provide material necessities while leaving traditional culture and values intact. "*We cannot have the benefits of machines without also taking the evils.*" The second essay, "This Tragic Hour," is a Hegelian attempt to understand why "modern or Renaissance-Bourgeois culture" naturally led to the present historical crisis. Conclusion: "Machine-civilization is in danger of losing human life and the ends of life in the process of acquiring the means and tools of power. Its means become ends in themselves. To restore means to their servant-place is a matter of philosophical insight and religious love, not of political or economic organization alone."

MULLER, HERBERT J. *The Children of Frankenstein: a Primer on Modern Technology and Human Values.* Bloomington: Indiana Univ. Press, 1970. Pp. 448. "The title is misleading because the reference to the monster which destroyed his creator made me expect another easy condemnation of technology, a doomsday story about how science and technology inevitably destroy man and his humanity. But Muller is too knowledgeable and too wise [for this] and has written a thoughtful analysis of the role of technology in our society. He first retraces the major changes brought about by industrialization and urbanization and then considers, one by one, the impact of technology on our contemporary institutions, on culture, and on people."—J. Schmandt, *Technology and Culture* 12, no. 1 (January 1971): 137–38. Marked by an absence of rhetorical overstatement; critical of both Mumford and Ellul. The weaknesses are that the balanced historical analysis lacks a clear theoretical framework and sometimes tends to "a certain blandness." Also reviewed by C. M. Cipolla, "Monsters: Technology and Man," *Virginia Quarterly Review* 47, no. 1 (Winter 1971): 156–60. Includes bibliography.

NIEL, MATHILDE. *Le phénomène technique; liberation ou nouvel esclavage?* Paris: Editions "Courrier du Livre," 1965. Pp. 63. According to the author, who is a socialist, technology, despite its many liberating effects, has "a natural tendency to enslave man." Thus "humanist socialism cannot be limited to changing the property system," but must educate the people to cope with the problems of technology. An abridged version of this essay has been translated and included in Erich Fromm, ed., *Socialist Humanism* (Garden City, N.Y.: Doubleday, 1965).

NISBET, ROBERT A. "The Impact of Technology on Ethical Decision-Making." Chap. 10 in *Tradition and Revolt: Historical and Sociological Essays.*

New York: Vintage, 1970. First printed in R. Lee and E. Marty, eds., *Religion and Social Conflict* (New York: Oxford Univ. Press, 1964), pp. 9–23. Technology influences morals by becoming part of the normative environment, which it influences in four ways: abstraction, generalization, individuation, and rationalization. As a result, "technology becomes one of the forces in society whereby the individuality and concreteness of ethical norms—norms such as honor, guilt, loyalty—become tenuous and indistinct."

OETTINGER, ANTHONY G., with SEMA MARKS. *Run, Computer, Run: the Mythology of Educational Innovation.* Cambridge, Mass.: Harvard Univ. Press, 1969. Pp. 322. Argues that the problems of American education are political and philosophical rather than simply technological. See also the authors' "Educational Technology: New Myths and Old Realities," with a Discussion and Reply, *Harvard Educational Review*, vol. 38, no. 4 (Fall 1968), and Oettinger's "A Vision of Technology and Education," *Communications of the ACM*, vol. 9, no. 7 (July 1966).

OGBURN, WILLIAM F. "How Technology Changes Society," *Sociology and Social Research* 36 (November 1951): 75–83. On the role of technology in social change by the first sociologist to use the concept of the "cultural lag." See also Ogburn's *Social Change with Respect to Culture and Original Nature* (New York: Viking, 1922; new ed. 1950); and *Living With Machines* (Chicago: American Library Association, 1933). Pp. 16. See also Francis R. Allen et al., *Technology and Social Change* (New York: Appleton-Century-Crofts, 1957).

————. "Technology as Environment," *Sociology and Social Research* 41 (1956): 3–9.

————. *On Culture and Social Change; Selected Papers.* Edited by Otis Dudley Duncan. Chicago: Univ. Chicago Press, 1964. Pp. 360. Chapters include "Social Evolution," "Social Trends," "Short-Run Changes," and "Methods." Editor has emphasized understanding the nature of social change. Further references to Ogburn's works can be found in the Gilfillan bibliography mentioned in the introduction.

OSTERTAG, ADOLF. "Die Wandlung des Menschen durch die Technik" [The transformation of man through technology], *Schweizerische Bauzeitung* 71, no. 21 (1953): 301-3.

––––––. "Betrachtungen über die Technik" [Reflections on technology], *Schweizerische Bauzeitung* 72, no. 16 (1954): 223-28; 72, no. 23 (1954): 334-38; 72, no. 24 (1954): 350-54.

––––––. "Die Technik im Dienst der Weltordnung" [Technology in the service of world order), *Schweizerische Bauzeitung* 75, no. 30 (1957): 469-73.

––––––. *Über Förderung und Pflege des technischen Nachwuchses* [Concerning the aid and care of technical programs]. Zurich: Polygrapjischer Verlag, 1960. Pp. 26.

––––––. "Freiheit und Verantwortung des Ingenieurs" [Freedom and responsibility of engineers], *Schweizerische Bauzeitung* 79 (1961): 909-12 and 925-28.

––––––. "Sinn der Wirtschaft und technischer Fortschritt" [Meaning of economics and technological progress], *Schweizerische Bauzeitung* 80, no. 51 (1962): 851-55; and 80, no. 52 (1962): 874-79.

––––––. "Technik als menschliches Problem" [Technology as a human problem], *Österreichische Ingenieur-Zeitschrift; Zeitschrift des Österreichischen Ingenieur- und Architekten-Vereines* (Vienna) 6 (1963): 381-86.

––––––. "Zweifel am technischen Fortschritt" [Doubts about technological progress], *Internationales Gewerbearchiv* (St. Gallen) 11, no. 4 (1963): 145-67.

––––––. "Technik und christlicher Glaube" [Technology and Christian faith], *Reformatio* 13, nos. 11-12 (1964): 714-31. For a large number of other articles pre-1956, see Dessauer's bibliography.

OZEBEKHAN, HASAN. *The Triumph of Technology: "Can" Implies "Ought."* Research paper SP-2830. Santa Barbara, Calif.: System Development Corp.

1967. Pp. 17. Also in S. Anderson, ed., *Planning for Diversity and Choice; Possible Futures and Their Relations to the Man-controlled Environment* (Cambridge, Mass.: M.I.T. Press, 1968). A critique of that type of future planning which takes ends as extrinsically given and considers only an efficiency of means. The result is that "feasibility, which is a strategic concept, becomes elevated into a normative concept, with the result that whatever technological reality indicates we *can* do is taken as implying that we must do it."

POLANYI, MICHAEL. *Pure and Applied Science and Their Appropriate Forms of Organization*. Occasional Pamphlet no. 14. Oxford: Society for Freedom in Science, December 1953. A critique of Marxian notions about science and technology which first formulates some conceptual distinctions between science and technology, and then shows how different social organizations are appropriate to the pursuit of each. Reprinted in *Science and Freedom* (Boston: Beacon, 1955), with comments by S. Hook, E. Shils, A. Weissberg-Cybulski, C. Courty, H. Mehlberg, G. F. Nicolai, R. Apéry, F. A. Hayek, F. G. Houtermans, A. Tarski, and a reply by the author. See also "The Republic of Science: Its Political and Economic Theory," *Minerva* 1, no. 1 (Fall 1962): 54-73.

POSSONY, STEFAN T. "Technology and the Human Condition," *American Behavioral Scientist* 11, no. 6 (July-August 1968): 43-48. Concerned with the relationship among three processes: scientific-technological-economic, cultural, and social-political. Argues that "the predominant characteristic of our times . . . is that in the face of the growing capabilities of organized war, government has been deteriorating, and not that technology has been developing at so rapid a rate. Technology is only a tool and, by itself, neither helps nor hinders human progress." The real problem is that "modern man has forgotten not only the pragmatic lessons of experience but also spiritual insights that he once possessed." See also Possony and J. E. Pournelle's *The Strategy of Technology; Winning the Decisive War* (New York: Dunellen, 1970).

Progrès technique et progrès moral. Recontres internationales de Geneve. Neuchatel: La Baconnière, 1947. Pp. 485. A Collection of nine papers and seven discussions from a conference at Geneva in 1947. The authors and their papers: A. Siegfried's "Historique de la notion de progrès," M. Prenant's "Le progrès humain vu par un biologiste," E. d'Ores's "Du paternel et du fraternel," N. Berdiaeff's "L'homme dans la civilisation technique," J. B. S. Haldane's "Influence de progrès technique sur le progrès moral," G. de Ruggiero's "La fin et les moyens," T. Spoerri's "Eléments d'une moral créatrice," S. Siddheswarananda's "La conscience humaine et l'angoisse de la civilisation," and E. Mounier's "Le christianisme et l'idée de progrès." As might be expected, the most important papers are those of Berdiaeff (Berdyaev), Haldane, Ruggiero, and Mounier. Although Berdyaev is reformulating ideas presented earlier in "Man and Machine," *The Bourgeois Mind* (London: Sheed & Ward, 1934), the treatment here is not so balanced because there is much post-World War II pessimism. Haldane argues (against Berdyaev) that technology does not so much create evils as moral challenges which should not be shirked. Ruggiero (in what must have been one of the philosopher's last works) argues that, whereas ends dominate means in art, in technology means alone dominate. Mounier's essay is available in English as "The Case against the Machine," in *Be Not Afraid* (London: Rockliff, 1951).

RABINOWITCH, EUGENE. *The Dawn of a New Age; Reflections on Science and Human Affairs.* Chicago: Univ. Chicago Press, 1964. Pp. 400. A Collection of the author's articles from *Bulletin of the Atomic Scientist.* The central theme: Despite the great political upheavals of our age, technology rather than ideology will shape our future.

————. "Science Popularization in an Atomic Age," *Impact of Science on Society* 17, no. 2 (1967): 107-13. "Traditionally concerned with intellectual initiation or with the dissemination of new findings for practical application, science popularization now is faced with a third and urgent responsibility – that of helping nations to appreciate the dangers which science has created for their future as a result of both the destructive powers and the capacity for production it has placed in man's hands."

READ, HERBERT. "The Redemption of the Robot." Pp. 144-72 in *The Redemption of the Robot.* New York: Trident, 1966. The argument, representative of much of English socialism, is that technology is humanized by education and art. "Man must become an artist and fill his new-found leisure with creative activity"; and this is made possible by means of popular education. Also somewhat representative of the Bauhaus attitude toward technology.

REISS, HOWARD. "Human Factors at the Science-Technology Interface." Pp. 105-16 in W. H. Gruber and D. G. Marquis, eds., *Factors in the Transfer of Technology.* Cambridge, Mass.: M.I.T. Press, 1969. Rejects the common conceptions of science and technology in favor of a distinction between two kinds of technical activity: (1) science, and (2) engineering development and applied science. These two are distinguished, first, on the basis of the kinds of activities performed in each and then on the basis of their respective subject matters. After being illustrated with examples from industry, they are used to propose solutions to some practical problems of technology transfer.

RICKOVER, HYMAN G. "A Humanistic Technology," *Nature* (London) 208 (November 20, 1965): 721-26. Viewed as autonomous, technology becomes a Frankenstein monster. "But when it is viewed humanistically, in other words as a means to human ends, it can be made to produce maximum benefit and do minimum harm . . . [and] enable man to become more truly human than it has ever been possible for him to be. Of technology it can be truly said that it is not 'either good or bad, but thinking makes it so.' " A critique of the tendency to look upon technology in any other way. Slightly different versions reprinted in *American Behavioral Scientist* 14 (January 1965): 3–8; Charles R. Dechert, ed., *The Social Impact of Cybernetics* (Notre Dame, Ind.: Univ. Notre Dame Press, 1966); *American Forests* 75 (August 1969): 12, 14, 42–44; and *Humanist* 29 (September-October 1969): 22–24. See also the author's "Can Technology Be Humanized – in Time?" *National Parks Magazine* (July 1969), pp. 4–7.

ROSZAK, THEODORE. *The Making of a Counter Culture; Reflections on Technocratic Society and Its Youthful Opposition.* Garden City, N.Y.: Doubleday 1969. Pp. 303. The best of the youth culture books. Argues that the "counter culture" offers a viable alternative to the scientific and technological attitude toward the world. Chap. 7, "The Myth of Objective Consciousness," is the most ambitious part of the book. See also Roszak's review-survey of some recent literature on this subject in "Technocracy: Despotism of Beneficient Expertise," *Nation* (September 1, 1969), pp. 181–88.

―――. *Where the Wasteland Ends; Politics and Transcendence in Postindustrial Society* (New York: Doubleday, 1972).

ROTENSTREICH, NATHAN. "Technology and Politics," *International Philosophical Quarterly,* vol. 7, no. 2 (June 1967). Politics, as "the set of means by which man puts to use the forces inherent in social organization" is both indirectly and directly influenced by technology as "the set of means by which man puts the forces and laws of nature to use." Included in C. Mitcham and R. Mackey, eds., *Philosophy and Technology* (New York: Free Press, 1972). See also the author's *Spirit and Man: an Essay on Being and Value* (The Hague: Martinus Nijhoff, 1963), especially the section on "Tools," pp. 91 ff., and the discussion of the problems of cybernetics and computing machines, pp. 149 ff.

―――. "Le sens de l'humanisme dans l'ère technologique," *Revue philosophique de la France et de l'étranger* 157 (July-September 1967): 337–48.

RUESCH, JURGEN. "Technology and Social Communication." Pp. 452–70 in L. Thayer, ed., *Communication: Theory and Research.* Springfield, Ill.: Thomas, 1967.

―――. "Technological Civilization and Human Affairs," *Journal of Nervous and Mental Disease* 145 (September 1967): 193–205. A brief survey of the history of technology, followed by the detailed argument that "in the atomic age the person orientation that has dominated the western world for the last few thousand years is slowly giving way to the system orientation." Concludes by suggesting the need to reassert the validity of the personal and nontechnical.

RUSSELL, BERTRAND. *The Impact of Science on Society.* London: Allen & Unwin, 1952. Pp. 102. Science is power and "offers the possibility of far greater well-being for the human race than has ever been known before. It offers this on certain conditions: abolition of war, even distribution of ultimate power, and limitation of the growth of population." "If human life *is* to continue in spite of science, mankind will have to learn a discipline of the passions which, in the past has not been necessary." There are individual essays on the effect of scientific techniques on values, war, tradition, democracies, and oligarchies. But see the author's more pessimistic *Icarus, or the Future of Science* (London: Kegan Paul, 1925) – a response to J. B. S. Haldane's optimistic *Daedalus, or Science and the Future* (London, 1923) – where he says, "Much as I should like to agree with his forecast I am compelled to fear that science will be used to promote the power of dominant groups rather than to make men happy."

RYAN, JOHN. "The Hope for Humanization," *Cross Currents* 21, no. 1 (Winter 1971): 1–24. The present dehumanizing situation in technological society is contrasted with a humanized ideal, followed by considerations on how to move from one toward the other. Contains a thoughtful analysis of the distinctions between science, moral, and artistic or technical methods, along with a discussion of the nature of truly humanized technical operations. This article is drawn from Ryan's book, *The Humanization of Man* (New York: Newman Press, 1972).

SCHELLER, HERMANN. "Philosophische Probleme der Entwicklung der Technik" [Philosophical problems of the development of technology]. Pp. 13-31 in *Forschen und Wirken; Festschrift aur 150-Jahr-Feier der Humboldt-Universitat Zu Berlin*. Vol. 3. Berlin, 1960.

SCHON, DONALD A. "The Fear of Innovation," *International Science and Technology* 59 (November 1966): 70-78. Discusses corporate resistance to technological change which is destructive of corporate stability.

————. *Technology and Change; the New Heraclitus.* New York: Delacorte, 1967. Pp. 256. A critical analysis of innovation and technological change in American industry and its impact on American society. See especially chap. 3, "An Ethic of Change."

————. "Forecasting and Technological Forecasting," *Daedalus* (Summer 1967), pp. 759-70. Defines technological forecasting and outlines attempts to do it, many of which are not so much forecasts of technology itself as technology-related items (such as industrial growth or the direction and diffusion rates of invention). Considers, as well, some obstacles to effective forecasting.

————. *Beyond the Stable State.* New York: Random House, 1971. Pp. 254. "If we rank . . . authors . . . as ranging between optimists and pessimists, undoubtedly the chief of the former is Donald Schon. . . . His unstated premise is that it is possible to analyse any social phenomenon . . . according to techniques which are a grandiose version of operations research, systems analysis, and cybernetics." – "Quarrelling with Technology," *Times Literary Supplement* (London) 70 (August 13, 1971): 957.

SCHUMACHER, E. F. "A Saner Technology," *Liberation* (August 1967), pp. 15-19. Argues for the pursuit of a "decentralized," "intermediate technology." Also appears in *Good Word* 30 (Fall 1967): 106-12, under the title "Intermediate Technology." See also A. Latham-Koenig, "The Church and Intermediate Technology," *Clergy Review* 54 (July 1969): 519-26.

SCHUMANN, FRIEDRICH KARL. *Mythos und Technik.* Cologne: Westdeutschen Verlag, 1958. Pp. 60. Summaries in French and English.

SCHWARTZ, EUGENE S. *Overskill: the Decline of Technology in Modern Civilization.* Chicago: Quadrangle, 1971. Pp. 338. An indictment of technology which cites the problems it has created and argues that technological solutions will only increase those problems. Emphasizes the structure of the technologial process and its effect on environmental issues. Includes bibliography.

SELIGMAN, BEN B. *Most Notorious Victory: Man in the Age of Automation.* New York: Free Press, 1966. Pp. 441. A provocative study of the socioeconomic consequences of automation. See also Seligman's "Automation and the State," *Commentary* 37 (April 1964): 49-54.

SHAPIRO, J. J. "One-Dimensionality: the Universal Semiotic of Technological Experience." Pp. 136-86 in P. Breines, ed., *Critical Interruptions; New Left Perspectives on Herbert Marcuse.* New York: Herder & Herder, 1970.

SHRIVER, DONALD W., JR. "Man and His Machines: Four Angles of Vision," *Technology and Culture* 13, no. 4 (October 1972): 531-55. Describes first four basic perspectives: technology as a neutral means for the realization of human ends (H. Cox and engineers in general); technology as the determiner of human values (M. McLuhan and J. Ellul); technology as the instrument of class interests (B. Seligman and J. McDermott); and technology as the evolutionary product of a plurality of historical causes (V. Ferkiss). Then analyzes some basic philosophical questions underlying the differences between these perspectives.

SIBLEY, MUMFORD Q. "Socialism and Technology," *New Politics* 1, no. 1 (Fall 1961): 203-13. Criticizes traditional technological optimism in both its socialist and nonsocialist versions, and argues for social controls over the utilization of technology.

SIMON, HERBERT A. *Administrative Behavior; a Study of Decision-making Processes in Administrative Organization*, 2d ed. New York: Free Press, 1957. Pp. 259. First published 1945. A basic description of rationalized production and the kind of business behavior closely associated with technology. See also James G. March and Herbert A. Simon, *Organizations* (New York: Wiley, 1967), for further discussions of Taylor's theory of scientific management and planning and innovation in organizations.

———. "Effects of Technological Change in a Linear Model." Pp. 260-76 in Tjalling C. Koopmans, ed., *Activity Analysis of Production and Allocation; Proceedings of a Conference*. Cowles Commission for Research in Economics, Monograph no. 13. New York: Wiley, 1951. See also the informative comments on this essay by Ansley Coale and Yale Brozen, pp. 277-81.

———. *The Shape of Automation for Men and Management*. New York: Harper & Row, 1965. Pp. 111. Contains sections on the long-range economic effects of automation, the degree to which corporations will be managed by machines, and the new science of management decision. See also Herbert A. Simon and Allen Newell's "Heuristic Problem Solving: the Next Advance in Operations Research," *Operations Research* 6 (January-February 1958): 1-10.

SIMON, YVES, R. "Democracy and Technology." Pp. 260-322 in *Philosophy of Democratic Government*. Chicago: Univ. Chicago Press, 1951. A careful analysis by a Thomist, seeking to answer the question whether farming or industrialization is more congenial to democratic life. Portion included in C. Mitcham and R. Mackey, eds., *Philosophy and Technology* (New York: Free Press, 1972). See also the review by Leo Strauss which is reprinted in

What Is Political Philosophy? And Other Studies (New York: Free Press, 1959). See also Simon's posthumous *Work, Society and Culture* (New York: Fordham Univ. Press, 1971).

SKINNER, B. F. *Beyond Freedom and Dignity*. New York: Knopf, 1971. See especially chap. 1, "A Technology of Behavior." Cf., too, the review by Noam Chomsky, "The Case against B. F. Skinner," *New York Review of Books* 17, no. 11 (December 30, 1971): 18-24; and the symposium on "The Skinnerian Challenge," in *Center Magazine* (Center for the Study of Democratic Institutions) (March-April 1972), pp. 33-65, which contains J. Platt's "A Revolutionary Manifesto," M. Black's "A Disservice to All," A. Toynbee's "An Uneasy Feeling of Unreality," and an interview with Skinner.

SNOW, C. P. *The Two Cultures and the Scientific Revolution*. New York: Cambridge Univ. Press, 1959. See also *The Two Cultures: a Second Look* (New York: Cambridge Univ. Press, 1963). Then compare F. R. Leavis, *Two Cultures? The Significance of C. P. Snow; with "An Essay" by Michael Yudkin* (New York: Pantheon, 1963). Leavis, a literary critic, and Yudkin, a biochemist, both agree that the humanities, not science, are primary. See also Lionel Trilling, "Science, Literature, and Culture: a Comment on the Leavis-Snow Controversy," *Commentary* 33, no. 6 (June 1962): 461-77; Loren Eiseley, "The Illusion of the Two Cultures," *American Scholar* 33, no. 3 (Summer 1964): 387-99; and Paul Goodman, "The Human Use of Science," *Commentary* 30, no. 6 (December 1960): 461-72. Cf. also the interview with Snow in *General Electric FORUM* 9, no. 3 (July-September 1966): 5-6, entitled "Science and Technology: Friend or Foe?" and the author's "Scientists and Decision Making," in Martin Greenberger, ed., *Computers and the World of the Future* (Cambridge, Mass.: M.I.T. Press, 1962). This last article contains an important qualification of the author's original position.

STANLEY, MANFRED. "Technicism, Liberalism, and Development: a Study in Irony as Social Theory." In M. Stanley, ed., *Social Development: Critical Perspectives.* New York: Basic, 1972. Explores "the place of social theory in moral rhetoric." Examines the theories of M. Heidegger, J. Ellul, H. Arendt, H. Marcuse, L. Mumford, F. von Hayek, L. von Mises, W. Sypher, and Wolin.

STARR, CHAUNCEY. "Social Benefit versus Technological Risk," *Science* 165 (September 19, 1969): 1232-38. Sketches the framework for a risk-benefit assessment of technological innovations.

TEICH, ALBERT H., ed. *Technology and Man's Future.* New York: St. Martin's, 1972. Pp. 274. An anthology in four parts: I, "Scientific Views of Advancing Technology"; II, "Philosophers of the Technological Age"; III, "An Attempt at Synthesis"; IV, "The Movement toward Control: Technology Assessment."

"Valeur humaine de la technique," *Nova et vetera* (Fribourg), new series 1 (1950): 1-23. An anonymous article by a Carthusian monk often referred to in European literature.

VEREIN DEUTSCHER INGENIEURE. *Der Mensch im Kraftfeld der Technik* [Man in the force-field of technology]. Düsseldorf: VDI-Verlag, 1955. Pp. 155. From a conference at Munster, 1955. One of a series of German Engineer Society yearbooks, filled with a good deal of mutual admiration. Essays by F. Dessauer, W. Vogel, F. Kesselring, P. Koessler, and O. Kraemer. Dessauer's essay is a good introduction to his thought on a popular level. The whole collection tends to be more serious in tone that others in this series.—R.J.R. [NOTE: Three earlier conferences which are part of this series: "Über die Verantwortung des Ingenieurs" (Kassel, 1950); "Mensch und Arbeit im technischen Zeitalter" (Marburg, 1951); and "Die Wandlungen des Menschen durch die Technik" (Tubingen, 1953).]

_____. *Die Technik prägt unsere Zeit* [Technology: the characteristic of our time]. Düsseldorf: VDI-Verlag, 1956. Pp. 155. Essays by M. Pfender, Vieweg, Finkelnburg, Kuhnke, and Strug-

ger largely in praise of technology. Little of philosophical importance. —R. J. R.

_____. *Die Technik im Dienst der Weltordnung* [Technology in the service of world order]. Düsseldorf: VDI-Verlag, 1958. Pp. 188. From a conference at Freiburg, 1957. Essays by H. Goeschel, H. E. Holthusen, F. Baade, S. Balke, P. Wigny, F. Etzel, W. Pohl, and G. Wirsing. The best essay is by Holthusen, who tries to clarify the kind of "order" involved. Argues that technological perfection and order are not necessarily the same thing; technological unity is not world agreement and harmony of thought and interests. The other essays mainly discuss such topics as the population explosion, problems of underdeveloped nations, the European Common Market, etc., but with much duplication of material. Only slightly more sophisticated than last year's volume.—R. J. R.

_____. *Der Ingenieur und seine Aufgaben in neuen Wirtschaftsräumen* [The engineer and his tasks in the new spheres of business]. Düsseldorf: VDI-Verlag, 1959. Pp. 202. From a conference at Aachen, 1959. Continues with the broad theme of the 1958 volume, i.e., primarily economic issues. [NOTE: See also the three volumes entitled *Mensch und Technik; Veröffentlichungen* 1961/62 (Düsseldorf: Verein Deutscher Ingeniere, 1963); ibid., 1963/64 (Düsseldorf: Verein Deutscher Ingeniere, 1965); and ibid., 1965/66 (Düsseldorf: Verein Deutscher Ingeniere, 1967).]

WEAVER, RICHARD M. "Humanism in an Age of Science," *Intercollegiate Review* 7, nos. 1-2 (Fall 1970): 11-18. A strong conservative critique of the scientific and technological milieu. Included in C. Mitcham and R. Mackey, eds., *Philosophy and Technology* (New York: Free Press, 1972).

WEIHE, CARL. *Kultur und Technik—ein Beitrag zur Philosophie der Technik* [Culture and technology—a contribution toward a philosophy of technology]. Frankfurt: Selbstverlag des Verfassers, 1935. Pp. 137. See also the author's discussion entitled "Die Philosophie der Technik—II," *Technik und Kultur* (Verband deutscher diplom-ingenieure, Berlin) 18 (1927): 97. For a large number of other works by this author from the 1920s and 1930s, see Dessauer's bibliography.

WEINBERG, ALVIN M. "Criteria for Scientific Choice," *Minerva* 1, no. 2 (Winter 1963): 159–71. Argues that "The most valid criteria for assessing scientific fields come from without rather than from within the scientific discipline that is being rated" (p. 170). Furthermore, because of the present needs of the world, "science can hardly be considered its major business" (p. 171). Reprinted in *Physics Today* 17 (March 1964): 42–48. See also "Criteria for Scientific Choice, II: The Two Cultures," *Minerva* 3 (Autumn 1964): 3–14; and "Science Choice, and Human Values," *Bulletin of the Atomic Scientists* 22, no. 4 (April 1966): 8–12, which argues for the need to limit the scientific establishment in favor of the humanities. All three articles are included, along with other discussions of political issues raised by the existence of large mission-oriented government laboratories in the author's *Reflections on Big Science* (Cambridge, Mass.: M.I.T. Press, 1967), a book which is perhaps the best introduction to contemporary questions and problems centering around the politics of science. On this literature specifically, see the author's review of Daniel S. Greenberg's *The Politics of Pure Science* (New York: New American Library, 1967), in "Scientific Choice and the Scientific Muckrakers," *Minerva* 7, nos. 1–2 (Autumn–Winter 1968–69): 52–63. Finally, see also the author's "The Axiology of Science," *American Scientist* 58 (1970): 612–17, where he argues that the urgent question of scientific priorities has helped to promote a growing concern with value in science.

———. "Can Technology Replace Social Engineering?" *Bulletin of the Atomic Scientists* 22, no. 10 (December 1966): 4–8. Explores the concept of the "technological fix"—which accepts man's weaknesses and, instead of trying to reform human nature, circumvents or capitalizes upon it for socially useful ends (e.g., the technology of contraception, which replaces the practical need for chastity). Although "technological solutions to social problems tend to be incomplete and metastable," they do buy time and open up new options for social action. Included in E. G. Mesthene, ed., *Technology and Social Change* (Indianapolis: Bobbs-Merrill, 1967). See also "Can Technology Stabilize the World Order?" *Public Administration Review* 27, no. 5 (December 1967): 460–64. "First military technology can make war a totally irrational undertaking; second modern civil technology can help stabilize the mutual deterrent by eliminating inequities between rich and poor." On this same topic see, again, the author's "Social Problems and National Socio-Technical Institutes," in *Applied Science and Technological Progress*, report prepared by the National Academy of Sciences for the committee on Science and Astronautics, U.S. Congress, House (Washington, D.C.: Government Printing Office, 1967). Cf., too, Amitai Etzioni, " 'Shortcuts' to Social Change?" *Public Interest* 12 (Summer 1968): 40–51; and Amitai Etzioni and Richard Remp, *Technological "Short-Cuts" to Social Change* (New York–Washington, D.C.: Center for Policy Research, April 1971), where a similar idea is given a different name. This last essay can also be found in *Science* 175 (January 7, 1972): 31–38.

———. "In Defense of Science," *Science* 167 (January 9, 1970): 141–45. Answering some of the current attacks on science and technology, the author argues that both are more necessary today than ever before. See also the author's "Prudentia et Technologia; die Antwort eines Technologen auf die Prophezeiung vom Untergang," *VDI-Zeitschrift* 113 (June 1971): 557–62, where he argues against ecological and countercultural critics of technology, and especially that atomic energy solves the energy crisis.

WEIZSÄCKER, CARL F. VON. *Ethical and Political Problems of the Atomic Age.*

London: SCM Press, 1958. Pp. 22. See also the author's *Gedanken über unsere Zukunft; 3 Reden* (Göttingen: Vandenhoeck & Ruprecht, 1966).

WHEELER, HARVEY. "Means, Ends and Human Institutions," *Nation* (January 2, 1967), pp. 9-16. A critique of J. Ellul's notion of an ends-means inversion and his extension of the idea of technology beyond the arts and crafts. Concludes by rejecting Ellul's theories but arrives at equally pessimistic conclusions: "Science, being cumulative, poses progressively more complex ethical and political problems, but ethics and politics, being noncumulative, face a rapidly growing disparity between the problems they can solve and those with which they are confronted" (p. 16).

WILKINSON, JOHN. *The Quantitative Society, or What Are We to Do with Noodle?* Santa Barbara, Calif.: Center for the Study of Democratic Institutions, 1964. Pp. 23. A polemical attack on "those cultural systems in which the erstwhile means [i.e., technologies] have become ends-in-themselves, to which man is *forced* to accomodate himself with no more freedom of choice than that available to a slave." By the translator of J. Ellul's *The Technological Society*.

――, ed. *Technology and Human Values*. Santa Barbara, Calif.: Center for the Study of Democratic Institutions, 1966. Pp. 42. Occasional papers by Wilkinson, G. Sykes, D. Gabor, M. B. Bloy, M. Grotjahn, T. Roszak, and B. de Jouvenel. Of more popular interest.

ZVORIKINE, ALEXSANDR A. (also spelled Sworykin, Zvorikin, Zvorykin, Zvorykine, etc.). "O nekotorych voprosach istorii tekhniki" [On several questions in the history of technology], *Voprosy filosofii* 7, no. 6 (1953): 32-45.

――. *Sovremennyi kapitalizm i tekhnika* [Contemporary capitalism and technology]. Vsesoiûznoe obshchestvo po raspotraneniiy politicheskikh nauchnikh znanii, series 2, no. 47. Moscow: Znanie, 1954. Pp. 47.

――. "Tekhnicheskii progress v period razvernutogo stroitel'stva kommunizma" [Technological progress in the period of developed building of communism], *Kommunist* (Moscow) 36, no. 7 (1959): 26-37.

――. "The Material and Technological Basis of Communism" (in Russian), *Voprosy filosofii*, vol. 14, no. 5 (1960).

――. "Concerning a Unified Concept of Technical Progress: Zvorikine on Drucker's 'Work and Tools,'" *Technology and Culture* 2, no. 3 (Summer 1961): 249-53. Followed by Drucker's reply.

――. "Science as a Direct Productive Force," *Impact of Science on Society* 13, no. 1 (1963): 49-60.

"Technical Progress and Society." Pp. 322-39 in G. S. Métraux and Francois Crouzet, eds., *The Evolution of Science*. New York: New American Library, 1963. An extended Marxist discussion of the relationship between technological development and social conditions. Argues that the contradictions of technological development results from capitalist relations of production, not from technology itself. Includes reference to W. Ogburn, A. L. White, F. Dessauer, S. Lilley, G. Harrison, and K. Jaspers. Reprinted from *Journal of World History* 6, no. 1 (1960-61): 183-98.

――. "The Laws of Technological Development." Pp. 59-74 in Carl F. Stover, ed., *The Technological Order*. Detroit: Wayne State Univ. Press, 1963. "Technology may be defined as the means of work, the means of human activity developing within a system of social production and social life. The means of work become technology only within a system of social production. . . ." In a broad sense, the means of work "include all the material conditions necessary to enable the production process to take place at all." Includes sections on the natural-scientific and social bases of technology, the role of technology in present society, dialectical materialism and technology, science as a direct productive force, and the change of man's place and role in production. A shorter version of this essay appeared as "Technology and the Laws of Its Development," in *Proceedings of the XIVth International Congress of Philoso-*

phy; Vienna, Sept. 2-9, 1968 (Vienna: Herder, 1968), 2: 609-14.

———. *Les conséquences sociales de la mécanisation et de l'automatisation en URSS.* Paris: UNESCO, 1963. Pp. 207.

———. "Über einige Besonderheiten der Entwicklung der Technik,"*Technik* 19, no. 11 (1964): 705-8.

———. *Nauka, proizvodstvo trud* [Science, production, labor]. Moscow, 1965. Pp. 259.

———. "The Culture of the Twentieth Century," *Journal of World History* 12, nos. 1-2 (1970): 37-62. Since "culture is that which is created by man as contrasted to that which is created by nature," this discussion has implications for technology.

ZVORIKINE, ALEXSANDR A., and EUGENE I. RABINOVICH. "Technology and Society," *Journal of World History* 12, nos. 1-2 (1970): 103-26. On the role of technology in industrialized social systems with some discussion of the socioeconomic implications of automation in the theories of W. Ogburn, J. Burnham, J. Ellul, L. Mumford, and G. Friedmann. "Technology has two types of opposites: those of technology and social conditions, and the opposites inherent in technology itself. At certain states of social development, technology, as an element of productive forces, enters into contradiction with the social conditions of its development, which leads to changes in the relations of production and to changes within technology itself."

———. "Science, société et personalité," *Journal of World History* 12, no. 3 (1970): 353-412. A general examination of the influence of the scientific and technological revolution on society, artistic culture, and human life.

B. SECONDARY SOURCES

[GENERAL NOTE: In the 1950s and 1960s there developed a large debate about what has become known as the "politics of science" (also called "science and public affairs" and "science policy"). Because adequate biblio-graphies of this material already exist (see, for instance, those on science and public affairs mentioned in the Introduction), works in this area have been included only on a very selective basis. But see James L. Penick, Jr., Caroll W. Pursell, Jr., Morgan B. Sherwood, and Donald C. Swain, eds, *The Politics of American Science: 1939 to the Present* (Chicago: Rand McNally, 1965); and William R. Nelson, ed., *The Politics of Science: Readings in Science, Technology, and Government* (New York: Oxford Univ. Press, 1968).]

ABAD CARRETERO, LUIS. *Instantes, inventos y humanismo* [Instants, inventions and humanism]. Mexico: Herrero, 1966. Pp. 227.

ABT, CLARK. *The Impact of Technological Change on World Politics.* Washington, D.C.: World Future Society, 1967. Pp. 24. Concerned with the political impact on world politics for the next ten years of weapons, transportation, communications, information processing, education, medicine, and agriculture technologies. Discusses what technology will and will not do, criticizing utopian and antiutopian theories. Concludes that, since there is no halting technology, it is necessary to respond in a productive way and set up countermeasures.

ACKOFF, RUSSELL LINCOLN. *Scientific Method: Optimizing Applied Research Decisions.* New York: Wiley, 1962. Pp. 464. See also the author and Maurice W. Sasieni, *Fundamentals of Operations Research* (New York: Wiley, 1968).

ADORNO, THEODOR W., ed., *Sprätkapitalismus oder Industriegesellschaft* [Late capitalism or industrial society]. Proceedings of the 16th Meeting of German Sociologists, April 8-11, 1968, Frankfurt. Stuttgart: Ferdinand Enke, 1969. Pp. 300. Papers principally concerned with the problem of developing an adequate post-Marxist sociological theory of contemporary society.

ALEJANDRO, JOSÉ M. "Técnica y humanismo," *Razón y fé* 159 (1959): 253-70.

ANDERSON, EUGENE N. "In Defense of Industrialism," *Diogenes* 11 (1955): 1-17.

ANGEL, FRANCISCO LUIS. "Valorización de la técnica," *Franciscanum* (Bogota) 9 (1967): 52–59.

ANSHEN, RUTH N., ed. *Science and Man; Twenty-four Original Essays.* New York: Harcourt Brace, 1942. Two essays of note: A. H. Compton's "Purpose of Science" (pp. 121–34) and K. T. Compton's "Man and Technology" (pp. 309–24). Other essays by F. H. Knight, R. B. Perry, C. L. Becker, E. G. Conklin, L. Mumford, H. D. Lasswell, J. Huxley, A. Hardlicka, J. Maritain, K. Koffka, B. Blanshard, C. Jung, R. Niebuhr, J. Piaget, B. Malinowski, W. Kaempffert, H. Kelsen, P. C. Jessup, W. B. Cannon, J. T.Shotwell, A. E. Cohn, H. C. Urey, and R. N. Anshen.

Applied Science and Technological Progress. Prepared by National Academy of Sciences for Science and Astronautics Committee, House, U.S. Congress. Washington, D.C.: Government Printing Office, June 1967. Pp. 434. Deals with the effective applications of the resources of science to advances in technology. Contents include: H. Brooks's "Applied Research: Definitions, Concepts, Themes"; H. W. Bode's "The Systems Approach"; R. A. Bauer's "Application of the Behavioral Sciences"; R. W. Gerard's "Shaping the Mind: Computers in Education"; C. N. Kimball's "Technology Transfer"; and A. Weinberg's "Social Problems and National Socio-Technical Institutes."

Are Engineering and Science Relevant to Moral Issues in a Technological Society? New York: Engineers Joint Council, October 1969. Pp. 30. Report from the Engineering Manpower Commission of the Engineers Joint Council. Four papers: R. B. Helfgott's "Technology, Social Science, and Moral Choices"; G. M. Newcombe's "Engineering, a Modern Profession in a Moral Society"; T. F. Malone's "Moral Issues in a Modern Technological Society"; and S. Wallen's "Social Problems and the Engineering Profession."

ARENDT, HANNAH. *The Human Condition.* Chicago: Univ. Chicago Press, 1958.

Reprinted, Garden City, N.Y.: Doubleday Anchor, n.d. Pp. 385. An analysis (by a student of Jaspers and Heidegger) of the differences between ancient and modern views of the nature of labor, work, and political action, stressing the technological implications of the modern understanding. Cf. also the author's *On Violence* (New York: Harcourt, Brace & World, 1970), where she argues that student protest is a reaction against the popular worship of scientific and technological progress.

———. "The Conquest of Space and the Stature of Man." Pp. 265–80 in *Between Past and Present.* New York: Viking, 1961.

ARMAND, LOUIS. "Machines, Technology and the Life of the Mind," *Impact of Science on Society* 3, no. 3 (Autumn 1952): 155–70. "Modern technology and the machine, far from having killed the poetry and charm of living, have on the contrary freed man's mind from physical drudgery and have thus contributed to the enrichment of his intellectual life. Technology, too, has its poetry."

ARMER, PAUL. "The Paul Principle: When Technology Outgrows the Man," *Geriatrics* 24, no. 6 (December 1970): 20–30, 34.

ARON, RAYMOND. "Political Action in the Shadow of Atomic Apocalypse." Pp. 445–57 in *The Ethic of Power: the Interplay of Religion, Philosophy, and Politics.* New York: Harper & Row, 1962. See also the author's *War and Industrial Society* (London: Oxford Univ. Press, 1958).

———. *The Epoch of Universal Technology.* Encounter pamphlet no. 11. London: Encounter, 1964. Pp. 23. Not particularly insightful. Also published as chap. 1 of *The Industrial Society* (New York: Praeger, 1967).

ASHBY, ERIC. "Technological Humanism," *Impact of Science on Society* 9, no. 1 (1958): 45–58. Also published in *Journal of the Institute of Metals* 85 (1957): 461–67.

———. *Technology and the Academics; Essay on Universities and the Scientific Revolution.* New York: St. Martin's, 1958.

Pp. 117. Includes short bibliography. See also "Science and Antiscience," *Nature* (London) 230 (April 2, 1971): 283–86.

"Automation et avenir humain," *Recherches et débats* du centre Catholique des Intellectuels François, Cahier 20 (1957). Contains R. Perrin, G. Levard, M. Demonque, and L. Devauh's "Automation et avenir humain" (pp. 77–98), and G. Tournier's "L'automation, espoir au menace pour l'homme?" (pp. 99–108).

AXELOS, KOSTAS. *Marx penseur de la technique.* Paris: Les Editions de Minuit, 1961. Pp. 324. Reviewed by Danilo Pejović, *Praxis* 1, no. 1 (1965): 121–22.

AYRES, ROBERT U. *Technological Forecasting and Long-Range Planning.* New York: McGraw-Hill 1969. Pp. 237. Describes and appraises the major techniques of forecasting, and considers the relationship between forecasting and planning. Includes bibliography. See also Ayres's "The Forces That Change Technology," *Futurist* 3 (October 1969): 132–33.

BAGDIKIAN, BEN H. *The Information Machines; Their Impact on Men and the Media.* New York: Harper & Row, 1971.

BAGRIT, SIR LEON. *The Age of Automation.* New York: New American Library, 1965. Pp. 128. BBC lectures by a British industrialist. A popular discussion of the general industrial, economic, and political consequences.

BAHR, HANS-DIETER. *Kritik der "politischen Technologie"; eine Auseinandersetzung mit Herbert Marcuse und Jürgen Habermas* [Critique of "Political Technology"; an exposition of Herbert Marcuse and Jurgen Habermas]. Frankfurt: Europäische Verlagsonst; Vienna: Europa-Verlag, 1970. Pp. 107.

BAHRDT, HANS PAUL. "Die Stellung der 'Technischen Intelligenz' in der Gesellschaft" [The place of "Technological Intelligence" in society], *Frankfurter Hefte* 15 (1960): 241–48.

———. "Helmut Schelskys technischer Staat," *Atomzeitalter* 9 (1961): 195–200.

BAKER, ELIZABETH FAULKNER. *Technology and Woman's Work.* New York: Columbia Univ. Press, 1964. Pp. 460. For an interesting comparison, see Catherine Ester Beecher and Harriet Beecher Stowe, *The American Woman's Home: or, Principles of Domestic Science; Being a Guide to the Formation and Maintenance of Economical, Healthful, Beautiful, and Christian Homes* (New York: J. B. Ford, 1869). For a related discussion see the special issue of *Impact of Science on Society,* vol. 20, no. 1 (January–March 1970), on "Women in the Age of Science and Technology."

BARAM, MICHAEL S. "Social Control of Science and Technology," *Science* 172 (May 7, 1971): 535–39. Argues for training new scientists and technologists in professional responsibilities.

BARBUY, HERALDO. *Cultura e processo técnico* [Culture and technological progress]. São Paulo: Faculdade de Ciências Econômicas e Administrativas de Universidade de São Paulo, 1961. Pp. 145.

BARDET, GASTON. "Die Sackgasse der Technik; der Arbeiter als Funktion und als Mensch," *Wort und Wahrheit* 5 (1950): 673–84.

BARLOW, JOHN S. "Cybernetics, Technology, and the Humanities," *Technology Review* 71, no. 1 (October–November 1968): 40–45.

BARNES, HARRY ELMER, and GERALD WENDT. "Cybernetics and Technical Change," *Humanist* 25, no. 5 (Fall 1965): 231–32.

BARNETT, H. G. "Invention and Culture Change," *American Anthropologist* 44, no. 1 (January–March 1942): 14–30.

———. *Innovation, the Basis of Cultural Change.* New York: McGraw-Hill, 1953. Pp. 462.

BARUCHET, ANDRÉ. *La hantise de l'abondance, un mal inavouable* [The obsession of affluence, a shameful evil]. Paris: Editions du Scorpion, 1965. Pp. 192. Discusses economic policy and technology and civilization.

BASIUK, VICTOR. *Technology and World Power.* Headline Series no. 200. New York: Foreign Policy Association,

April 1970. Pp. 63. Discusses the trends and the impact of the technological revolution in the United States, U.S.S.R., and Western Europe in relation to world power.

BAUER, RAYMOND A. *Second-Order Consequences: a Methodological Essay on the Impact of Technology.* With Richard S. Rosenbloom, Laure Sharp, and the assistance of others. Cambridge, Mass.: M.I.T. Press, 1969. Pp. 240. Generalizes from an analysis of the space program.

BAUMGARTNER, LEONA. "Medicine and Society—New Issues," *Proceedings of the American Philosophical Society* 115, no. 5 (October 15, 1971): 335–40.

BEARD, W. "Technology and Political Boundaries," *American Political Science Review* 25 (1931): 557–72.

BECKER, CARL. *Progress and Power.* New York: Knopf, 1949. Pp. 116. Three lectures delivered at Stanford University in 1935 on "Tools and the Man," "The Sword and the Pen," and "Instruments of Precision." Progress equals the accumulation of technological power.

BELL, DANIEL. "The Bogey of Automation," *New York Review of Books* 5, no. 2 (August 26, 1965): 23–25.

———. "The Post-industrial Society: a Speculative View." Pp. 154–70 in Edward and Elizabeth Hutchings, eds., *Scientific Progress and Human Values.* New York: American Elsevier, 1967.

———. "Toward the Year 2000: Work in Progress," *Daedalus,* vol. 96, no. 3 (Summer 1967). Reprinted, Boston: Beacon, 1969. Futurology, with articles and discussions by Bell, H. Kahn and A. J. Wiener, F. C. Iklé, D. A. Schon, M. Shubik, L. J. Duhl, H. S. Perloff, D. P. Moynihan, L. K. Frank, S. R. Graubard, H. Orlons, E. Mayr, G. C. Quarton, K. Stendahl, E. H. Erikson, M. Mead, H. Kalven, Jr., G. A. Miller, D. Riesman, J. R. Pierce, E. V. Rostow, S. P. Huntington, and J. de S. Pool. [NOTE: See the philosophical critique of this work by Paul G. Kuntz, "Goals and Values in Transition," *International Philosophical Quarterly* 9, no. 2 (June 1969): 278–96. For other philosophical comments on futurology, see Paul T. Durbin, "Philosophy and the Futurists," *Proceedings of the American Catholic Philosophical Association* (Washington, D.C.: Catholic Univ. America Press, 1968), 42:62–73; Robert L. Heilbroner, "Futurology," *New York Review of Books* (September 26, 1968); George Lichtheim, "Ideas of the Future," *Partisan Review* 33, no. 3 (Summer 1966): 396–410; Robert A. Nisbet, "The Year 2000 and All That," *Commentary* (June 1968); and Irene Taviss, "Futurology and the Problem of Values," *International Social Science Journal* 11, no. 4 (1969): 574–84.]

[GENERAL NOTE: For more futurology books, see entries under R. Boguslaw, M. Brown, N. Calder, S. Chase, W. R. Ewald, Jr., O. K. Flectheim, D. Gabor, B. de Jouvenel, R. Jungk, H. Kahn and A. J. Wiener, J. McHale, D. N. Michael, and R. W. Prehuda. Three relevant works on technological forecasting: R. V. Ayres, R. A. Bauer, and M. J. Cetron. For bibliographies in this field, see Annette Harrison's *Bibliography on Automation and Technological Change and Studies of the Future.* (Santa Monica, Calif.: RAND Corp., March 1968), Billy Rojas's "Future Studies Bibliography" (Amherst: Univ. Massachusetts Program for the Study of the Future in Education, January 1970), and Bettina Huber's "Studies of the Future: a Selected and Annotated Bibliography" (New Haven, Conn.: Yale Univ.).]

———. "Notes on the Post-industrial Society: I and II," *Public Interest* (Winter 1967), pp. 24–35 and (Spring 1967), pp. 102–18.

———. "The Measurement of Knowledge and Technology." In E. B. Sheldon and W. E. Moore, eds., *Indicators of Social Change.* New York: Russell Sage, 1968.

———. "The Balance of Knowledge and Power," *Technology Review* 71, no. 8 (June 1969): 38–44. "It is in the interplay of the technical and the moral that rationality has to be sought."

BELL, GARETT DE, ed. *The Environmental Handbook.* New York: Ballantine,

1970. Pp. 367. Reprints two essays directly relevant to the philosophy of technology, both of which are cited elsewhere: Lynn White, jr., "The Historical Roots of Our Ecologic Crisis," and Garrett Hardin, "The Tragedy of the Commons." Includes brief bibliography. A good bibliography on ecology covering both books and films is in *Paperbound Books in Print* 15, no. 3 (March 1970): 4-15.

BENDIX, REINHARD. *Work and Authority in Industry.* New York: Wiley, 1956. Pp. 464. Historically analyzes the way in which the entrepreneurial class responded to the creation and management of an industrial work force in several widely differing types of industrial societies. As such it is a general study of the history of managerial attitudes and ideologies in both developed and underdeveloped countries.

BENJAMIN, CORNELIUS A. *Science, Technology, and Human Values.* Columbia: Univ. Missouri Press, 1965. Pp. 296. One of the very few textbooks in the philosophy of science to treat technology at all. But although technology is given some independent consideration in the latter chapters of the book, it is primarily just used as a means of getting into a discussion of the social consequences of science, e.g., atom bombs and automation. As with virtually all philosophies of science, technology is facilely identified with applied science. Despite the title, the index does not even make technology a separate reference point.

BENKO, FRANÇOIS. *La sociedad ante el imperativo technológico.* Caracas: Ediciones Universidad Central de Venezuela, 1961. Pp. 96.

BENN, ANTHONY WEDGWOOD. "We and Them—the People versus the Machines," *Proceedings of the Royal Institution of Great Britain* 43 (August 1970): 170-89.

BENTHEM, WALTER VAN. *Das Ethos der technischen Arbeit und der Technik; ein Beitrag zur personalen Deutung.* Essen: Ludgerus-Verlag, 1966. Pp. 183.

BERGMANN, JOACHIM. "Technologische Rationalität und spätkapitalistiche

Okonomie," in J. Habermas, ed., *Antworten auf Herbert Marcuse.* Frankfurt: Suhrkamp, 1968.

BERGSTRÄSSER, A. "Aufgaben der Technik in der Kultursituation der Gegenwart" [Tasks of technology in the cultural situation of the present], *VDI-Zeitschrift* 101 (1959): 1264-69.

BERLINGER, RUDOLPH. *Das Werk der Freiheit; zur Philosophie von Geschichte, Kunst und Technik* [The work of freedom; on the philosophy of history, art, and technology]. Frankfurt: Klostermann, 1959. Pp. 138.

BERNARD, STÉPHANE. *Les conséquences sociales du progrès technique: méthodologie.* Brussels: Les Editions du Parthenon S.P.R.L., 1956. Pp. 211. A critical analysis of sociological concepts connected with the social implications of technological progress.

BERT, SILAS. *Machine Made Man.* New York: Farrar & Rinehart, 1930. Pp. 341. Argues that, despite the evident justice of much criticism of the Machine Age, the "items in the debt column against the machine are being diminished or erased." The machine is bringing to birth a new civilization.

BEVILLE, GILBERT. *Technocratie moderne.* Paris: Librairie générale du droit et de jurisprudence, 1964. Pp. 200.

BILLY, JACQUES. *Les techniciens et le pouvoir.* Paris: Presses Universitaires de France, 1963. Pp. 126. A discussion of technocracy.

BIRNBAUM, NORMAN. *The Crisis of Industrial Society.* New York: Oxford Univ. Press, 1969. Pp. 185. A wide-ranging analysis which explores the effect of technology upon political attitudes and argues that people are no nearer political control than a century ago. By one of the founders of *New Left Review.* See also "Staggering Colossus," *Nation* (September 2, 1968), pp. 167-75.

———. "Is there a Post-Industrial Revolution?" *Social Policy* 1, no. 2 (July-August 1970): 3-13. Analyzes the notion of a post-Industrial Revolution to see what basis it has in historical fact and to inquire into its political implications. Deals especially with manpower and work change, new politics, youth, high culture, and the transition from pre- to post-industrial society.

BLANSHARD, BRAND, ed. *Education in the Age of Science.* New York: Basic, 1959.

———. *The Life of the Spirit in a Machine Age.* Northampton, Mass.: Smith College, 1967. Pp. 28. "To live aright in the machine age, one cannot be wholly of it." Yet "surely the way to deal with machines is neither to flee from them nor to surrender to them, but to use them . . . as means to ends appointed . . . by ourselves. . . . I submit that the so-called machine age provides . . . the best conditions in history for a widely-lived life of the spirit. . . . Machines are not hateful monsters; they make us possible."

BLATTNER, H. "Gespräche über Mensch und Technik am Internationalen Kongress für Philosophie der Wissenschaften, Zürich, August 1954," *Schweizerische Bauzeitung* 72, no. 49 (1954): 715-19.

BLENKE, H. "Wesen, Entwicklung und Auswirkungen der Technik" [Essence, development and the effects of technology], *Technische Rundschau* (Bern) 53, no. 30 (1961): 1-3.

BOAS, GEORGE. "In Defense of Machines," *Harper's* (June 1932), pp. 93 ff.

BOBER, HARRY, ed. *Mensch, Technik, Gesellschaft als Leitungsproblem: eine philosophische-soziologische Studie aus der Sicht von heute und morgen* [Man, technology, society as a problem of planning; a philosophical-sociological study from the viewpoint of today and tomorrow]. Berlin: Verlag Tribüne, 1968. Pp. 276.

BOEHLER, DIETRICH. "Zum Problem des 'Emanzipatorischen Interesses' und seiner gesellschaftlichen Wahrnehmung," *Man and World* 3 (May 1970): 26-53. A discussion of J. Habermas's theories about technology.

BÖHLER, EUGEN. "Ist die technische Entwicklung zwangsläufig?" [Is technological development necessary?], *Schweizerische Bauzeitung* 76, no. 18 (1958): 265-70. Also in *Industrielle Organisation; Schweizerische Zeitschrift für Betriebswissenschaft* 27, no. 3 (1958): 63-70.

BOGUSLAW, ROBERT. *The New Utopians; a Study of System Design and Social Change.* Englewood Cliffs, N.J.: Prentice-Hall, 1965. Pp. 213.

BOISDÉ, RAYMOND. *Technocratie et démocratie.* Paris: Plon, 1964. Pp. 251.

BOITEN, R. *Vijand, vriend of bondgenoot?* [Enemy, friend or ally?]. Delft: Waltman, 1957. Pp. 20.

BONIFACE, JEAN. *Les misères de l'abondance.* Paris: Les Editions Ouvières, 1968. Pp. 108. An informal commentary on the general aspects of contemporary consumer society, especially in France, drawing frequently from statistics and recent events.

BONO, EDWARD DE. *Technology Today.* London: Routledge & Kegan Paul, 1971. Pp. 144. Contents: E. de Bono's "Technology Today," F. Warner's "Production Technology," W. Gunston's "Technology for Man's Survival," C. Leichester's "Social Technology: Information for Decisions," R. Anderton's "Technological Change: the Impact of Large Technical Systems."

BORGSTROM, G. "La dilemme de la technique moderne," *Société belge d'études et d'expansion* 57 (1958): 381-93.

BORKENAU, FRANZ. *Der Übergang vom feudalen zum bürgerlichen Weltbild: Studien zur Geschichte der Philosophie der Manufakturperiode.* Paris: Félix Alcan, 1934. Pp. 559.

———. "Will Technology Destroy Civilization?" *Commentary* 11, no. 1 (January 1951): 20-26. A critique of post-World War II pessimistic theories. Main point: Insofar as technology brings about chaos in existing civilization, it does so only as a prelude to a new cultural order.

BORN, MAX. "Blessings and Evils of Space Travel," *Bulletin of the Atomic Scientists* 22 (October 1966): 12-14. Suggests that, although the technology of space travel is of value to the pure scientist (and to the military), it is not so valuable to the majority of men. Cf. also Freeman Dyson, "Human Consequences of the Exploration of Space," ibid., 25, no. 7 (September 1969): 8-13.

———. *La responsibilité du savant dans le*

monde moderne. Paris: Payot, 1967. Pp. 192. Investigation of problems involved in international social science research projects, etc. See also the author's *Physics and Politics* (New York: Basic, 1962).

BORSODI, RALPH. *This Ugly Civilization.* New York: Simon & Schuster, 1929. Pp. 468. An early and vigorous indictment of modern technological civilization by an economist. Has become something of an underground classic. See also the author's *Flight from the City* (New York: Harper, 1933).

BÖSS, WALTER. "Mensch und Technik," *Gralswelt* 16 (1962): 233–38.

BOTTOMORE, TOM B. "Machines without a Cause," *New York Review of Books* 17, no. 7 (November 4, 1971): 12–19. A review of five recent books, but primarily Mesthene's *Technological Change* and Richta's *La civilisation au carrefour*, arguing that, while it is now generally recognized that technology needs to be controlled, what has not been sufficiently discussed is how and by whom. The weakness of current discussions of the relation between technology and society rests on an inadequate theory of the nature of society itself and the lack of any clearly conceived social end.

BOULDING, KENNETH. *The Meaning of the Twentieth Century: the Great Transition.* New York: Harper & Row, 1969. Pp. 208. The transition in question is from civilization to post-civilization. Technology is intimately involved.

———. "The Scientific Revelation," *Bulletin of the Atomic Scientists* 26, no. 7 (September 1970): 13–18. "Those countries which have been most successful in accepting the scientific super-culture, and in generating the kind of economic development which is based on it, are also societies which have had a strong and vigorous folk culture. . . . Where the folk culture produces an ethic which is ill-adapted to the modern world, as it seems to be in the Arab states, the very impact of the super-culture disorganizes a society rather than moving it toward development."

BOWLES, EDMUND A., ed. *Computers in Humanistic Research; Readings and Perspectives.* Englewood Cliffs, N.J.: Prentice-Hall, 1967. Pp. 264. Papers and talks on six topics from a series of regional conferences held in 1964 and 1965. The topics: (1) "Introduction," (2) "Computers in Anthropology and Archeology," (3) "Computers in History and Political Science," (4) "Computers in Language and Literature," (5) "Computers in Musicological Research," and (6) "Man and the Machine."

BOXER, L. M. "Interaction of Technologies," *Nature* (London) 207 (September 11, 1965): 1121–25. Argues against specialization and for the educational stimulation of interaction both among technologies and between the technologies and the humanities.

BRADEN, WILLIAM. *The Age of Aquarius: Technology and the Cultural Revolution.* Chicago: Quadrangle, 1970. Pp. 306. Argues that the future will be an escalating struggle between humanism and technologists.

BRADY, ROBERT A. *Organization, Automation and Society: the Scientific Revolution in Industry.* Berkeley: Univ. California Press, 1961. Pp. 481. "The scientific revolution in industry, often called the 'Second Industrial Revolution,' is confronting peoples throughout the world with this question: Can our enormous industrial production be truly scientifically organized?" Discusses scientific revolutions in chemistry, standards and specifications, automation, and the systems of energy supply.

BRANCAFORTE, ANTONIO. *Umanità e disumanizzazione della tecnica.* Catania: Edingraf, 1967. Pp. 209.

BRANSCOMB, LEWIS M. "Taming Technology," *Science* 171 (March 12, 1971): 972–77. A plea for national regulation in a social context.

BRECHT, FRANZ-JOSEF. "Die moderne Technik als Mitgestalterin unserer Zeit" [Modern technology as the creator of our age]. Pp. 13–30 in *Der Mensch und die Kerntechnik.* Bonn: Deutsches Atomforum, 1962.

BRIGHT, JAMES R., ed. *Research, Development, and Technologic Innovation; an Introduction.* Homewood, Ill.: Irwin, 1964. Pp. 783. A textbook for business administration courses.

———, ed. *Technological Forecasting for Industry and Government; Methods and Applications.* Englewood Cliffs, N.J.: Prentice-Hall, 1968. Pp. 483. "This book is based upon the revised papers and audience contributions of the First Annual Technology and Management Conference: Technological Forecasting for Industry (Lake Placid, New York, May 1967). The conference was the first attempt in the United States to bring together the work of the best practitioners in the newly emerging art of technological forecasting. . . it is an excellent summary of the 1967 state of the art in a new field. . . . But, most important, the individual chapters demonstrate concretely how the best 1967 practitioners actually made technological forecasts."—J. B. Quinn, *Technology and Culture* 12, no. 2 (April 1971): 376–78.

———. "Can We Forecast Technology?" *Industrial Research* 10, no. 3 (March 1968): 52–56. A general overview of technological forecasting which tries to dispel some of the confusion as to its need, objectives interpretation, and application. Discusses the techniques of forcasting: forecasts of the past, analogy, and forecasts based on dynamic models. Closes with a listing of the areas in which technological forecasting may be useful.

BRINCKLOE, W. D. "Automation and Self-Hypnosis," *Public Administration Review* 26, no. 3 (September 1966): 149–55. Brief survey of the literature of the social implications of automation on civilization. Recommends that no thesis about the use of automation should be accepted without careful systematic analysis.

BROCHMANN, GEORG. *Mennesket og maskinen; Oiblikksbilleder av verden ved inngangen til en ny tidsalder.* 2 vols. Oslo: Aschehoug, 1937.

BROEZE, J. J. "Ethiek en technische wetenschappen." Pp. 272–98 in *Om de*

mens; Ethiek in wetenschap en beroep. Leiden: Sijthoff, 1968.

BRONOWSKI, JACOB. "Real Responsibilities of the Scientist," *Bulletin of the Atomic Scientists,* vol. 12 (January 1956).

———. *The Identity of Man.* Garden City, N.Y.: Doubleday, 1965. Pp. 107. Contains a good discussion of the differences between men and machines.

———. *Science and Human Values.* Revised ed., with a new dialogue, "The Abacus and the Rose." New York: Harper & Row, 1965. Pp. 119.

———. "The Disestablishment of Science," *Encounter* 37, no. 1 (July 1971): 8–16. "The time has come to consider how we might bring about a separation, as complete as possible, between Science and Government in all countries."

———. "Technology and Culture in Evolution," *American Scholar* 41, no. 2 (Spring 1972): 197–211.

BRONWELL, A. B., ed. *Science and Technology in the World of the Future.* London: Wiley, 1970.

BROOKS, HARVEY. "Can Science Be Planned?" Harvard Univ. Program on Technology and Society, Reprint no. 3. From *Problems of Science Policy: Seminar at Jouy-en-Josas on Science Policy.* Paris: OECD, 1967.

———. "Applied Science and Technological Progress," *Science* 156 (June 30, 1967): 1706–12. A condensed version of the author's "Applied Research: Definitions, Concepts, Themes," in *Applied Science and Technological Progress* (Washington, D.C.: National Academy of Sciences, 1967). Discusses the categorization of research into "basic" and "applied," arguing that for a "mission-oriented" organization this dichotomy focuses attention in the wrong direction. Recognizes a spectrum of activities ranging from pure research to technological development.

———. "Physics and the Polity," *Science* 160 (April 26, 1968): 396–400. "Today physics is experiencing the effects of the new disenchantment with science and technology earlier than most of the other basic sciences are," because it is most closely associated with those aspects of technology that have become suspect in the public mind.

―――. *Technology: Processes of Assessment and Choice.* Prepared by Panel on Technology Assessment, National Academy of Sciences. Washington, D.C.: Government Printing Office, 1969. Pp. 163.

BROOKS, HARVEY, and RAYMOND BOW-ERS. "The Assessment of Technology," *Scientific American* 222, no. 2 (February 1970): 13-21. A report on a panel convened by the National Academy of Sciences which recommended new federal mechanisms to consider the broad social consequences of advancing or retarding particular technological developments.

[NOTE: for other articles on technology assessment, see entries under J. R. Bright, C. Freeman, H. P. Green, R. G. Kasper, D. M. Kiefer, as well as those cited under the titles *Technology Assessment* and *Study of Technology Assessment.* See also some items of futurology referred to in the notes on Daniel Bell, ed., *Toward the Year 2000,* especially those on technological forecasting.]

―――. "Can Science Survive in the Modern Age?" *Science* 174 (October 1, 1971): 21-30. Asks whether modern society is not in fact generating a cultural climate which is no longer hospitable to the cultivation of a "true science," and whether the absence of such a science will destroy our ability to control the technology which science has helped to create, and which is essential to modern civilization. Originally delivered as the C. P. Snow lecture at Ithaca College, Ithaca, N.Y., January 19, 1971.

BROWN, MARTIN, ed. *The Social Responsibility of the Scientist.* New York: Free Press, 1971. Pp. 282. Sixteen essays by prominent scientists dealing with particular problems. Brief bibliography. For earlier discussions of this theme, see O. T. Benfey's "The Scientist's Conscience," *Bulletin of the Atomic*

Scientists 12, no. 5 (May 1956): 177-78; Dael Wolfle's "Social Responsibility of Science," *Science* 125 (January 25, 1957): 141, an editorial followed by the preliminary report of the American Association for the Advancement of Science Interim Committee on the Social Aspects of Science, 143-47; Bertrand Russell's "Social Responsibilities of Scientists," *Science* 131 (February 12, 1960): 391-92, and the letters in response, *Science* 131 (June 17, 1960): 1816-18; Warren E. Olson, "Responsibility: an Escape and an Approach," *Bulletin of the Atomic Scientists* 19, no. 3 (March 1963): 2-6; and Eugene Rabinowitch, "Responsibility of Scientists in Our Age," *Science* 25, no. 9 (November 1969): 2-3; John Ziman, "Social Responsibility: the Impact of Social Responsibility on Science," *Impact of Science on Society* 21, no. 2 (April-June 1971): 113-22; Milton Leitenberg, "Social Responsibility: the Classical Scientific Ethic and Strategic-Weapons Development," *Impact of Science on Society* 21, no. 2 (April-June 1971): 123-36; Steven Rose and Hilary Rose, "Social Responsibility: Myth of the Neutrality of Science," *Impact of Science on Society* 21, no. 2 (April-June 1971): 137-50.

BROWN, MURRAY. *On the Theory and Measurement of Technical Change.* Cambridge: Cambridge Univ. Press, 1966. Pp. 214. Reviewed by B. F. Massell, *American Economic Review* 57, no. 1 (March 1967): 231-32.

BROWNING, R. P. "Computer Programs as Theories of Political Processes," *Journal of Politics* 24 (1962): 562-82.

BRUMBAUGH, ROBERT S. "Automation and Intemperance." In *Plato for the Modern Age.* New York: Collier-Macmillan, 1962. Reprinted, Collier paperback, 1964.

BRYEN, STEPHEN DAVID. *The Application of Cybernetic Analysis to the Study of International Politics.* The Hague: Martinus Nijhoff, 1971. Pp. 135.

BRYSON, LYMAN, LOUIS FINKELSTEIN, and R. M. MACIVER, eds. *Conflicts of Power in Modern Culture; 7th Conference on Science, Philosophy and Religion.* New

York: Harper, 1947. Pp. 703. Contains three papers of marginal significance: J. K. Finch's "Science, Engineering and Western Civilization," E. D. Chapple and C. S. Coon's "Technological Change and Cultural Integration," and K. W. Deutsch's "The Crisis of Peace and Power in the Atom Age."

――――. "Technology and Freedom." Pp. 167–78 in Frederick Ernest Johnson, ed., *Wellsprings of the American Spirit.* New York: Harper, 1948.

BUCHANAN, ROBERT ANGUS. *Technology and Social Progress.* New York: Pergamon, 1965. Pp. 163. Studies the growth and impact of technology on society. Chapters on "Prelude to the Industrial Revolution," "The Growth of Modern Industry," "The Development of Transport," "The Development of Communications," "The Changing Pattern of Life," "The Political Implications of Technology," "The New Society," and a bibliography.

BUCHWALD, KONRAD. *Die Zukunft des Menschen in der industriellen Gesellschaft und Landschaft* [The future of man in industrial society and environment]. Braunschweig: Stolle, 1966. Pp. 43. A sociologist argues for the formative importance of a technological environment on man. Optimistic and of minimal importance.—R. J. R.

BURKE, JOHN G., ed. *The New Technology and Human Values.* Belmont, Calif.: Wadsworth, 1966. Pp. 408. A book of readings for adult education courses. Contains articles from a wide variety of printed sources, intended to explore the effects on human values of scientifically based technology. Good as a survey of popular opinion but no more.

――――. "Technology and Values." Pp. 190–235 in R. M. Hutchins and M. J. Alder, eds., *The Great Ideas Today, 1969.* Chicago: Encyclopaedia Britannica, 1969.

BUSCH, HEINZ. "Die Rangstellung einer Nation wird weitgehend von ihrer technologischen und wissenschaftlichen Leistung bestimmt" [The rank order of a nation is largely determined by its technological and scientific efficiency], *VDI-Zeitschrift* 112, no. 8 (April 1970): 477–81.

CALDER, NIGEL. *Eden Was No Garden: an Inquiry into the Environment of Man.* New York: Holt, Rinehart & Winston, 1967. Pp. 240. Deals with man's influence on the environment, food supply, and technology and civilizations. Published in England as *The Environment Game.* See also the author's *The World in 1984* (Baltimore: Penguin, 1965).

CALDER, RITCHIE. "Tyranny of the Expert," *Philosophical Journal; Transactions of the Royal Philosophical Society of Glasgow* 2, no. 1 (January 1965): 1–9.

――――. "Mortgaging the Old Homestead," *Foreign Affairs* 48 (January 1970): 207–20.

――――. "The Uses and Misuses of Technology," *Humanist* 30 (November–December 1970): 14–20.

CANETE, ANTONIO BERMUNDEZ. "Cultura y tecnica," *Revista de Occidente* 20 (April–May–June 1928): 378–97. A review article of some early literature.

CAPOCACCIA, ANTONIO. "Umanesimo della macchina," *Humanitas* 7 (1953): 165–75.

CAPPON, D. *Technology and Perception.* Springfield, Ill.: Thomas, 1971.

CARDONE, DOMENICO ANTONIO. *L'ozio, la contemplazione, il giuoco, la tecnica, l'anarchismo* [Leisure, contemplation, games, technology, anarchism]. Rome: Ricerche filosofiche, 1968. Pp. 144. A collection of essays.

CARLETON, WILLIAM G. *Technology and Humanism; Some Exploratory Essays for Our Times.* Nashville, Tenn.: Vanderbilt Univ. Press, 1970. Pp. 300. A mistitled collection of essays. The only essay of interest is "The Century of Technocracy."

CAROVILLANO, ROBERT L., and JAMES W. SKEHON, eds. *Science and the Future of Man.* Cambridge, Mass.: M.I.T. Press, 1970. Pp. 196. A diverse collection of speeches from a meeting of the American Association for the Advancement of Science. Reviewed by Irene Tavis, *Technology and Culture* 13, no. 2 (April 1972): 329–30.

CARREL, ALEXIS. *Man, the Unknown.* New

York: Harper, 1935. Pp. 346. Ambitious attempt by a scientist to base a nonbehaviorist, nontechnological theory of man on the discoveries of modern science. Undertaken because "humanity's attention must turn from the machine and the world of inanimate matter to the body and the soul of man" (p. xiv). Thus one of the book's main themes is the paradox of the way technology, which is supposed to liberate man, often winds up oppressing him. Calls for a comprehensive reform of civilization. "Today, the principles of industrial civilization should be fought with the same relentless vigor as was the *ancien regime* by the encyclopedists. But the struggle will be harder because the mode of existence brought to us by technology is as pleasant as the habit of taking alcohol, opium, or cocaine" (p. 294).

CARROLL, JAMES D. "Participatory Technology," *Science* 171 (February 19, 1971): 647–53. Analyzes the emerging "inclusion of people in the social and technical processes of developing, implementing, and regulating a technology" as an alternative to technological alienation.

CARTER, ANNE P. "The Economics of Technological Change," *Scientific American* 214, no. 4 (April 1966): 25–31.

———. "Technological Forecasting and Input-Output Analysis," *Technological Forecasting* 1, no. 4 (Spring 1970): 331–45.

CARTER, C. F. "Government and Technology," *Nature* (London) 206 (1965): 652–54.

CASAS, MANUEL GONZALO. "Ser y técnica." Pp. 429–34 in *Actas de las Segundas Jornadas Universitarias de Humanidades.* Mendoza: Instituto de Filosofía, 1964.

CASINI, PAOLO. "L'eclissi della Scienza," *Rivista de filosofia* 61 (July–September 1970): 239–62. A critique of Habermas and Marcuse.

CATTEPOEL, DIRK. "Auch die Technik ist Kultur; Sie ist eine schöpferische Leistung und erfordert eine Strenge sittliche Verantwortung" [Even technology is culture; it is a creative

achievement and calls for a strict moral responsibility], *Handelsblatt; westdeutsche Wirtschaftszeitung* 14 (1959): 13.

CAUSSIN, ROBERT. "The Transfer of Functions from Man to Machine," *Diogenes* 28 (Winter 1959): 107–25.

CÉRÉZUELLE, DANIEL. "Essai sur la nature du développement technique." Bordeaux: Faculté des Lettres et Sciences Humaines, Université de Bordeaux, 1970. Pp. 104. An unpublished dissertation.

CETRON, MARVIN J. *Technological Forecasting: a Practical Approach.* New York: Gordon & Breach, 1969. Pp. 345. "This book is an elaborate study of the recent and most approved methods for prediction, especially along technical lines, both the exploratory kind, which seeks to ascertain what *will* happen, and the normative type which asks what concrete plans *ought* to be formulated and carried through by governmental or large commercial planning and research departments. And it is to such planners that the book is largely directed. The discussion is often highly mathematical, dealing with the best methods of extrapolation of measured past trends."—S. C. Gilfillan, *Technology and Culture* 11, no. 4 (October 1970): 659–61.

The Challenge of Technology. New York: National Industrial Conference Board, 1967. Pp. 72. Report on a conference in New York on November 30, 1966. Contains these addresses: J. Wiesner's "The Nature of the Challenge," W. S. Melahn's "The New Information Technology," F. Keppel's "New Goals and New Tools in Education," E. Mesthene's "Response to the Challenge," E. Hilburn's "New Sights for the Private Sector," Hollomon's "New Patterns of Industry Government Partnership in the Coming Decade."

CHAPIRO, MARC. "Liberty and the Machine," *Diogenes* 40 (Winter 1962): 43–60.

CHAPLET, A. *Le règne de l'artificiel.* Paris: Nouvelles Editions Latines, 1951. Pp. 189. Positive defense of artifacts and the artificial. Man has always made artifacts; in fact, it is natural for him to do so. Moreover, artifacts are qualitatively superior to natural products and are also quantitatively more available.

CHASE, EDWARD T. "Politics and Technology," *Yale Review* 52 (1963): 331–39. Rapid technological change is "calling into question the viability of our political institutions." Tomorrow's political convulsion "derives from the cumulative impact of technology, an impact that is impersonal, nonideological, relentless, and possibly overwhelming." Examples are given of conflicts between technological changes and American institutions. A condensed version included in John G. Burke, ed., *The New Technology and Human Values* (Belmont, Calif.: Wadsworth, 1966), pp. 386–93.

CHASE, STUART. *Men and Machines*. New York: Macmillan, 1929. Pp. 354. A survey of opinion at the time, followed by an attempt at a philosophical analysis of the nature of machinery, and then a lengthy consideration of the influence of machinery on various aspects of modern life. Concludes with an optimistic call for men to squarely face the challenge. Of more than historical value.

———. *The Most Probable World*. New York: Harper & Row, 1968. Pp. 239. A popular sociological account of ten current technological trends—energy, nationalism, the arms race, automation, computers, etc.—and an attempt to project them into the next few decades. There are economic, social, biological, and physical limitations on technological advancement and "conditions governing growth will change, and force the rate of growth downward." See also the author's *Prosperity: Fact or Myth* (New York: Boni, 1929); *Waste and the Machine Age* (New York: League for Industrial Democracy, 1931); *The Promise of Power* (New York: John Day, 1933); *Technocracy: an Interpretation* (New York: John Day, 1933); *The Economy of Abundance* (New York: Macmillan, 1934).

CHASZAR, EDWARD. *Science and Technology in the Theories of Social and Political Alienation*. Program of Policy Studies in Science and Technology, staff discussion paper no. 410. Washington, D.C.: George Washington Univ. Press, 1969. Pp. 65.

CHOISY, E. G. "Technique et humanisme," *Schweizerische Bauzeitung*, vol. 71, no. 45 (1953).

Civilization and Science: In Conflict or Collaboration? A CIBA Foundation Symposium. Amsterdam, 1972. Pp. 220.

CLARK, HENRY. *The Irony of American Morality*. New Haven, Conn.: College and Univ. Press, 1971. Pp. 255. General thesis of the book: "Radical changes in society demand courageous and imaginative changes in theology and ethics" (p. 84). See especially chap. 3, "Human Values and Advancing Technology," and chap. 11, "Cybernation, Abundance and Human Fulfillment."

COLUMBIA UNIVERSITY SEMINAR ON TECHNOLOGY AND SOCIAL CHANGE, 1962–1965. This project has produced four main books: (1) Eli Ginzberg, ed., *Technology and Social Change* (New York: Columbia Univ. Press, 1964); (2) Aaron Warner et al., eds., *The Impact of Science on Technology* (New York: Columbia Univ. Press, 1965); (3) Dean Morse and Aaron Warner, eds., *Technological Innovation and Society* (New York: Columbia Univ. Press, 1966); (4) *Obsolescence and Updating of Engineer's and Scientist's Skills*, Final Report for the Office of Manpower Policy, Evaluation and Research, U.S. Department of Labor (New York: Columbia Univ. Press, 1966).

COMMONER, BARRY, *Science and Survival*. New York: Viking 1966. Pp. 150. A well-known biologist's argument that science and technology have gotten out of hand. See also the author's *The Closing Circle: Nature, Man and Technology* (New York: Knopf, 1971).

COMPTON, WALTER DALE, ed. *The Interaction of Science and Technology*. Urbana: Univ. Illinois Press, 1969. Pp. 137. Papers by E. R. Piore, J. E. Goldman, C. W. Sherwin, W. J. Price, M. Tanenbaum, D. Alpert, G. E. Pake, and W. K. Linvill arguing against D. J. de Sola Price's thesis that science and technology have developed as separate, largely independent intellectual traditions; on the occasion of the dedication of the University of Illinois Coordinated Space Laboratory.

CONANT, JAMES B. "Science and Politics in the Twentieth Century," *Foreign Affairs* 28 (January 1950): 189–202.

————. *Scientific Principles and Moral Conduct.* Cambridge: Cambridge Univ. Press, 1967. Pp. 48. See also the author's "Scientific Principles and Moral Conduct," *American Scientist* 60 (September 1967): 311–28. The author favors an unconventional approach to science and a conventional approach to ethical questions. See also the author's *Our Future in the Atomic Age* (New York: Foreign Policy Assoc., November–December 1951).

CORNFORTH, MAURICE. "Social Implications of Modern Technology." Pp. 345–52 in *The Open Philosophy and the Open Society.* New York: International Publishers, 1968. A Marxist analysis.

COTTA, SERGIO. *La sfida tecnologica* [The challenge of technology]. Bologna: Il Mulino, 1968. Pp. 204.

COTTIER, JEAN LOUIS. *La technocratie, nouveau pouvoir* [Technocracy, new power]. Paris: Editions du Cerf, 1959. Pp. 142.

COTTRELL, W. F. "Death by Dieselization: a Case Study in the Reaction to Technologial Change," *American Sociological Review* 16, no. 3 (June 1951): 358–65.

CROZIER, MICHEL. "La civilisation technique," *Temps modernes* 7 (1952): 1497–1516.

CURTH, WERNER, et al., eds. *Menschen, Maschinen, Energien; ein Sammelwerk zur polytechnischen Erziehung* [Men, machines, energies; an anthology for polytechnical training]. Berlin: Verlag Neues Leben, 1959. Pp. 462.

DADDARIO, EMILIO Q. "Technology Assessment," *Technology Review* 70, no. 2 (December 1967): 15–20.

————. "Technology and the Democratic Process," *Technology Review* 73, no. 9 (July–August 1971): 18–23.

DAEVES, KARL. "Die Rolle der Einzelpersönlichkeit für den technischen Fortschritt" [The role of individual personality for technological development], *VDI-Zeitschrift* 98, no. 3 (1956): 98–101.

DAINVILLE, FRANCOIS DE, and FRANCOIS RUSSO. "Culture, technique et spécialisation," *Etudes* 278 (1953): 158–71; 279 (1953): 45–54.

DANIELS, FARRINGTON, and THOMAS M. SMITH, eds. *The Challenge of Our Times; Contemporary Trends in Science and Human Affairs as Seen by Twenty Professors at the University of Wisconsin.* Minneapolis: Burgess, 1953. Pp. 364. A syllabus for an interdisciplinary course in contemporary trends inaugurated at the University of Wisconsin in 1941. Part I identifies science and technology and, using atomic energy as a case study, argues that "Science Is Everybody's Business." Most of the material is now either dated or popularly recognized. New edition under the lead title alone, Kennikat, 1971.

DANIELS, GEORGE H. "The Pure-Science Ideal and Democratic Culture," *Science* 156 (June 30, 1967): 1699–1705. Discusses the early justification of science, the rise of the professional and of organizations dedicated to the advancement of science, and the difficulties which the scientific community faces in relationship with the federal government.

DANTZIG, GEORGE B. "Management Science in the World of Today and Tomorrow," *Management Science* 13, no. 6 (February 1967): C107–11. "Operations Research or Management Science, two names for the same thing, refers to the science of decision and its application. In its broad sense, the word 'cybernetics,' the science of control, may be used in its place. This science is directed toward those tasks that humans have not yet delegated to machines. Tasks involving human energy and . . . simple human control have already been conceded to machines. . . . It is the automation of higher order human decision processes that is the last citadel. . . . We are witnessing a computer revolution in which nearly all tasks of man . . . are being reduced to mathematical terms and their solution delegated to computers."

DAVENPORT, WILLIAM H. "Engineering and the Applied Humanities: the Recent Past," *Technology and Society* 5, no. 1 (July 1969): 24-32. "It becomes clearer and clearer that the engineer's responsibilities do not stop with engineering. More and more he will find himself making awesome political decisions or becoming enmeshed in the nation's business."

——. *The One Culture.* New York: Pergamon, 1970. Pp. 182. Uses a critique of C. P. Snow's "two cultures" to develop a philosophy of "one culture" in the age of technology.

——. "Resource Letter TLA-1 on Technology, Literature, and Art since World War II," *American Journal of Physics* 38, no. 4 (April 1970): 407-14. An annotated bibliography of about a hundred books and articles.

DAVENPORT, WILLIAM H., and DANIEL ROSENTHAL, eds. *Engineering: Its Role and Function in Human Society.* New York: Pergamon, 1967. Pp. 284. Anthology for a course to establish a dialogue between the "two cultures." Part I, "The View of the Humanist"; II, "Attitudes of the Engineer"; III, "Man and Machine"; IV, "Technology and the Future." A good selection of historical and contemporary essays by J. H. Newman, S. Butler, E. Ashby, Aldous Huxley, H. Hoover, L. White, Vitruvius, N. Wiener, R. L. Heilbroner, V. Packard, R. Carson, J. Ellul, H. Margenan, J. von Neumann, H. Brown, etc.

DAVID, AUREL. *La cybernétique et l'humain.* Paris: Gallimard, 1965. Pp. 185. Study by a student of Pierre Ducassé.

DAVID, J. "Menschsein und Technik; Grösse und Gefahren im gesellschaftlichen Raum" [Human existence and technology; bigness and perils in the social realm], *Schweizerische Bauzeitung* 76, no. 18 (1958): 277-81.

DeCARLO, C. R. "Technology and Value Systems," *Educational Technology* 7, no.

3 (March 15, 1967): 1-8. From an address to the National Conference of State Legislators, December 4-6, 1966, Washington, D.C. A society "subject to continuous institutional and social change must teach its children a profound commitment to deep and enduring human values."

DECHERT, CHARLES R., ed. *The Social Impact of Cybernetics.* Notre Dame, Ind.: Univ. Notre Dame Press, 1966. Pp. 206. Paperback reprint, New York: Simon & Schuster, 1967. Contents: J. Diebold's "Goals to Match Our Means," C. R. Dechert's "The Development of Cybernetics," R. Theobald's "Cybernetics and the Problems of Social Reorganization," U. Neisser's "Computers as Tools and as Metaphors," M. McLuhan's "Cybernation and Culture," H. G. Rickover's "A Humanistic Technology," M. W. Mikulak's "Cybernetics and Marxism-Leninism," and J. J. Ford's "Soviet Cybernetics and International Development."

DeGELDER, G. M. *De mensch en de techniek.* Haarlem: Tjeenk Willink, 1936.

DeGROEN, P. A., and F. P. J. VAN GRUNSVEN. *Computers, mensen en systemen.* Amsterdam: Agon Elsevier, 1967. Pp. 159.

DEMCZYNSKI, S. *Automation and the Future of Man.* London: Allen & Unwin, 1964. Pp. 238. Reviews past social and philosophical consequences of automation, then speculates on those to come. Especially concerned with the relationship between men and machines. But tends to remain on a popular level.

DEMOLL, REINHARD, ed. *Im Schatten der Technik; Beiträge zur Situation des Menschen in der modernen Zeit* [In the shadows of technology; essays concerning the situation of man in the modern age]. Munich: Bechtle Verlag, 1960. Pp. 329. A comprehensive, although somewhat popularized, anthology. Contributions by a composer, doctors, a zoologist, architects, psychologists, a historian, a journalist, a sociologist. Stresses peace-time problems of technology, with articles on children, child development and technology, city planning, air pollution, music, eros, etc.—R. J. R.

"The Demonism of Technology," *Innovation* 26 (November 1971): 59-60. Discusses Dennis Wrong's sociological theories about the causes of rising opposition to technology, and calls for pollution and population controls.

DENBIGH, KENNETH G. *Science, Industry, and Social Policy.* Edinburgh: Oliver & Boyd, 1963. Pp. 103.

DERISI, OCTAVIO NICOLAS. "Técnica y espiritu." Pp. 435-38 in *Actas de las Segundas Jornadas Universitarias de Humanidades.* Mendoza: Instituto de Filosofía, 1964.

DESMOND, T. C. "Engineers in Politics," *Technology Review,* vol. 57 (February 1955).

DEUTSCH, K. W. "The Impact of Science and Technology on International Politics," *Daedalus* 99 (Fall 1959): 666-85. The impact will be to increase the need for high-quality political leadership: "The age of thermonuclear weapons may well have made obsolete the famous statement of General Douglas MacArthur to the effect that there is no substitute for military victory, and it may have replaced it by the proposition that there is no substitute for international perceptiveness, political mass support, and economic growth."

DIEBOLD, JOHN. *Man and the Computer; Technology as an Agent of Social Change.* New York: Praeger, 1969. Pp. 157. Collects the following lectures and essays: "The Profound Impact of Science and Technology," "Educational Technology and Business Responsibility," "International Disparities," "The Training of Managers," and "The Long-Term Questions." A short list of selected readings is found on pp. 155-57. By the author of *Automation: the Advent of the Automatic Factory* (New York: Van Nostrand, 1952), and *Beyond Automation: Managerial Problems of an Exploding Technology* (New York: McGraw-Hill, 1964).

DIPPEL, C. J. "Is er een wezenlijk conflict en wanbegrip tussen moderne exacte techniek en de huidige cultuur?" *Wetenschap en Samenleving* 22 (1968): 2-16.

DIXON, RUSSELL A., with E. K. EBERHART. *Economic Institutions and Cultural Change.* New York: McGraw-Hill, 1941. Pp. 543. A textbook. See especially the chapters on "Science as the Basis of Technology," "The Structure of Technology," and "The Social Effects of Technology."

DORE, P. "Cultura e technica." Pp. 55-80 in *Scienza e filosofia.* Milan: Bocca, 1942.

DOREMUS, ANDRE. "Note on the Coherence of the American Phenomenon," *Diogenes* 65 (Spring 1969): 49-73. See especially section 4 of this essay, "Technics and Ideology."

DRACAULIDES, N. N. "Les répercussions déshumanisantes du machinisme d'aujourd'hui," *Psyché* 5 (1951): 729-32.

DUBARLE, DOMINIQUE. "Technique et avenir," *Vie intellectuelle* 18 (1950): 142-63.

———. "Técnicas modernas y problemas de civilización," *Cuadernos hispanoamericanos* 7 (1951): 39-51.

DUBOS, RENÉ. *Mirage of Health: Utopias, Progress, and Biological Change.* New York: Harper, 1959. Pp. 236.

———. *The Dreams of Reason: Science and Utopias.* New York: Columbia Univ. Press, 1961. Pp. 167.

———. *Man Adapting.* New Haven, Conn.: Yale Univ. Press, 1965. Pp. 527.

———. "A Social Design for Science," *Science* 166 (November 14, 1966): 823. An editorial.

———. *So Human an Animal.* New York: Scribners, 1968. Pp. 267. Perhaps the best introduction to the author's views on technology. See especially chap. 6, "The Science of Humanity," which puts forth the vision of technology as part of Western man's "active wooing of the earth."

———. *Reason Awake: Science for Man.* New York: Columbia Univ. Press, 1970. Pp. 280.

DUBREUIL, HYACINTHE. *Robots or Men?* Translated by F. and M. Merrill. New York: Harper, 1930. Pp. 248. A French worker and union official's observations of the effect of advanced American industrial methods on personality and culture.

DUBRIDGE, LEE A. "Science Serves Society," *Science* 164 (May 30, 1969): 1137-40.

DURBIN, PAUL T. "Philosophy and the Futurists." Pp. 62-73 in *Proceedings of the American Catholic Philosophical Association.* Vol. 42. Washington, D.C.: Catholic Univ. America, 1968. There is a need for a radically future-oriented philosophy, an active team effort oriented to the social future of mankind. Brief discussion of D. Bell, B. de Jouvenel, and the American Institute of Planners.

"East-West Conference on Technology, Development and Values," *Center Report* (Center for the Study of Democratic Institutions) 5, no. 3 (June 1972): 8-9.

EBBINGHAUS, JULIUS. "Die Atombombe und die Zukunft des Menschen," *Studium generale* 10, no. 3 (1957): 144-53. A discussion of K. Jasper's theories.

EBENSTEIN, WILLIAM. "Technology and Totalitarianism." Pp. 27-30 in *Totalitarianism: New Perspectives.* New York: Holt, Rinehart & Winston, 1962. A brief argument that, although the one new element in 20th-century totalitarianism is technology, still "both totalitarian and liberal tendencies in a society can be intensified and accelerated by modern science and technology." Further discussion of the nature of tyranny and its relationship to technology can be found in Leo Strauss, *On Tyranny* (New York: Free Press, 1963), and George Grant, "Tyranny and Wisdom," *Technology and Empire* (Toronto: House of Anansi, 1970).

EBERHARD, JOHN P. "Man-centered Standards for Technology," *Technology Review* 71, no. 9 (July-August 1969): 50-55.

EDDISON, ROGERT T. "Social Applications of Operational Research," *Impact on Science on Society,* no. 2 (1953): pp. 61-82.

EICHELBERG, GUSTAV. "Fachliche Grundlegung technisch-schöpferischen Bauens mit einer unfas-

senden Ausweitung innermenschler Bildung," *Humanismus und Technik* 3, no. 2 (December 25, 1955): 63-71. Professional basis of technical-creative construction goes with a comprehensive inner growth of man.

ELGOZY, GEORGES. *Automation et humanisme.* Paris: Calmann-Levy, 1968. Pp. 382.

ENGLEHARDT, WOLF VON. *Der Mensch in der technischen Welt.* Cologne-Graz: Böhlau, 1957. Pp. 47.

Esigenze umane e progresso technologico nella societa contemporanea. [Human exigences and technological progress in contemporary society]. Rome: Stab. A. Staderini, 1959. Pp. 665. Proceedings of the Convegno nazionale per la civilita del lavoro, Rome, October 17-19, 1958.

"Ethical Aspects of Experimentation with Human Subjects," *Daedalus* (Spring 1969). Includes: H. Jonas's "Philosophical Reflections on Experimenting with Human Subjects," H. L. Blumgart's "The Medical Framework for Viewing the Problem of Human Experimentation," H. K. Beecher's "Scarce Resources and Medical Advancement," P. A. Freund's "Legal Frameworks for Human Experimentation," T. Parsons's "Research with Human Subjects and the 'Professional Complex,' " M. Mead's "Research with Human Beings: a Model Derived from Anthropological Field Practice," G. Calabresi's "Reflections on Medical Experimentation in Humans," L. L. Jaffe's "Law as a System of Control," and D. F. Cavers's "The Legal Control of the Clinical Investigation of Drugs: Some Political, Economic, and Social Questions."

"Ethique et technique," *Revue d'histoire et de philosophie religieuses,* vol. 44, no. 2 (1964). Contents: W. Vischer's "Foi et technique," P. Burgelin's "De l' 'homo faber' a l' 'homo laborans,' " J. Brun's "Technique et aliénation," and A. Dumas's "Ethique et société industrielle."

EWALD, WILLIAM R., JR., ed. *Environment for Man: the Next Fifty Years.* Bloomington: Indiana Univ. Press, 1967. Pp. 308.

————, ed. *Environment and Policy: the*

Next Fifty Years. Bloomington: Indiana Univ. Press, 1968. Pp. 459.
———, ed. *Environment and Change: the Next Fifty Years.* Bloomington: Indiana Univ. Press, 1968. Pp. 397.
[NOTE: These last three books are based on papers commissioned for the American Institute of Planners' two-year consultation on "The Next Fifty Years." The first contains 13 papers "concerned mainly with the contribution of science and the professions toward creating optimum human environments defined in physiological, psychological, and sociological terms." The second volume consists of 14 policy-program resource papers. The third is concerned with "the philosophy needed to cause a creative development of our environment." It is this volume—which includes P. Bertaux's "The Future of Man," B. de Jouvenel's "On Attending to the Future," E. G. Mesthene's "How Technology Will Shape the Future," C. Oglesby's "The Young Rebels," C. Brown's "The Effective Society," R. B. Fuller's "An Operating Manual for Spaceship Earth"—which is most philosophically relevant.]

FABUN, DON. *The Dynamics of Change.* Englewood Cliffs, N.J.: Prentice-Hall, 1967. "Hard-headed but creative answers to the questions raised by the 'dehumanizing' effects of modern technology. Far from signalling the 'death-of-man,' the new technology is seen as the herald of a fuller superior existence—provided man moves ahead unfettered by the trappings of outworn traditions."

FABUN, DON, et al. *Dimensions of Change.* New York: Collier-Macmillan, 1971.

FAHRBACH, GEORG, ed. *Der Mensch zwischen Natur und Technik* [Man between nature and technology]. Stuttgart: Fink, 1967. Pp. 100.

FAIN, GAËL. Progrès technique et régression humaine," *Revue politique et parlementaire* 54 (1952): 245–63.

FALLOT, JEAN. *Marx et la machinisme.* Paris: Editions Cujas, 1966. Pp. 226.

FANIZZA, FRANCO. "Polemica sulla scienza e coscienza técnica," *Aut Aut; Rivista di filosofia e di cultura* (Milan) 85 (January 1965): 60–74.

FAUNCE, WILLIAM A. *Problems of Industrial Society.* New York: McGraw-Hill, 1968. Pp. 189. Concerned not with problems *in* industrial society, which could include virtually all social ills, but problems *of* industrial society as industrial society.

FEINBURG, GERALD. *The Prometheus Project; Mankind's Search for Long-Range Goals.* New York: Doubleday, 1969. Pp. 215. Argues that possibilities created by science and technology require the conscious formulation of long-range goals by humanity as a whole. The "Prometheus Project" is the author's proposal for "a cooperative effort by humanity to choose its long-term goals" on a democratic basis. Suggests that a suitable goal would be "the creation of universal consciousness."

FERNAND-LAURENT, CAMILLE JEAN. *Moral et tyrannies.* Paris: Les Editions Ouvrieres, 1967. Pp. 127. A call, in the face of technological tyranny, for a rehabilitation of morals. "The object of this little book is not to make an ordered presentation of the latest works of specialists" but "in a personal fashion" to summon man "to look beyond the social phenomenon . . . in order to discover in himself those frequently belittled energies which, when liberated, permit of moral thinking and moral action."

FERRARIO, ARTEMIO. *Le macchine nella vita moderna* [The machine in modern life]. Milan: Vallardi, 1957. Pp. 330. A popular survey with pictures.

FERRAROTTI, FRANCO. *Macchina e uomo nella società industriale.* Turin: ERI-Edizioni RAI Radiotelevisione italiana, 1963. Pp. 174.

FERRY. W. H. "Must We Rewrite the Constitution to Control Technology?" *Saturday Review* (March 2, 1968), pp. 50–54.
———. "The Technophiliacs," *Center Magazine* 1, no. 5 (July 1968): 45–49. Questions the assumption that science and technology equal progress and human welfare. Drawing on J. Ellul's hypothesis that people can assert their freedom by "upsetting the course of this evolution," the author wants technology to be slowed down,

"placed under regulation," and to be reappraised according to "People first, Machines second."

———. "The Unanaswerable Questions," *Center Magazine* (July 1969), pp. 2-7. Can we cope with a scientific-technological way of life?

FILENE, EDWARD A. *Successful Living in This Machine Age.* New York: Simon & Schuster, 1931. Pp. 274. An optimistic account. Mass production "is liberating the masses . . . from the struggle for their existence and enabling them, for the first time in human history, to give their attention to more distinctly human problems" (p. 1).

FINK, EUGEN. "Technische Bildung als Selbsterkenntnis" [Technical education as self-knowledge], *VDI-Zeitschrift* 104 (1962): 381-93.

———. "Vom Sinn der Arbeit in unserer Zeit" [On the meaning of work in our time], *VDI-Zeitschrift* 110 (1968): 573-80. Work must be considered the basic phenomenon of human existence. Yet it has the dual character of creating shelter and alienating man. Ancient Greek philosophy found the meaning of existence outside work. Only in modern philosophy, beginning with Hegel and Marx, has the decisive significance of work and technology been recognized. The never-ending reshaping of the world creates a unique sovereignty of man, from which arises a meaningful interpretation of work.

FISCHERHOF, HANS. "Technologie und Jurisprudenz," *Neue juristische Wochenschrift* 22, no. 28 (1969): 1193-97.

FISHLOCK, DAVID. *Man Modified, an Exploration of the Man-Machine Relationship.* London: Cape, 1969. Pp. 215.

FLANDERS, RALPH E. *Taming Our Machines: the Attainment of Human Values in a Mechanized Society.* New York: Richard R. Smith, 1931. Pp. 244. Sample chapter titles: "The Machine or Personal Devil," "Present Stake and Contemporary Criticism," "Instability of Our Mechanized Society," "Beauty in the Age of Machines," "Machinery and Moral Problems." Argues for an optimistic approach to the problems.

FLECTHEIM, OSSIP K. *History and Futur-ology.* Meisenheim: Verlag Anton Hain, 1965. Pp. 126. This is a collection of essays written between 1914 and 1952, about equally divided between history and futurology. "A slim but scholarly work by a distinguished political scientist. The portion dealing with futurology is very readable and should interest most futurists."

FLOWERS, BRIAN. *Technology and Man.* The first Leverhulme Memorial Lecture. Liverpool: Liverpool Univ. Press, 1972. Pp. 29. ". . . technology, which merely means whatever is done by scientific methods to make practical processes more efficient, whether it be in industry, in the laboratory, or in the home. Technology is what we *do*, whether we are scientists or engineers or housewives. It is an activity."

FOECKE, HAROLD A. "Engineering in the Humanistic Tradition," *Impact of Science on Society* 20, no. 2 (April-June 1970): 125-35. "The accusations that have been levelled at the scientist and technologist for the anti-social consequences of their work are misdirected. The real culprit is the engineer. Generally speaking, the engineer has always been largely concerned only with the technical factors involved in design projects. But it is his social responsibility to take into account, as well, the social and human factors."

FOLSON, RICHARD G. "Technology and Humanism," *Mechanical Engineering* 88, no. 1 (1966): 20-23.

FORSTHOFF, ERNST. "Technischer Prozess und politische Ordnung" [Technological process and political order], *Studium generale* 22, no. 8 (1969): 849-56.

FORUM, ROBERTO SIMONSEN. *Desenvolvimento industrial e tarefas do pensamento* [Industrial development and the problem of thought]. São Paulo, 1959. Pp. 202.

FOSTER, DAVID M. "Computers and Social Change: Uses—and Misuses." *Computers and Automation* 19, no. 8 (August 1970): 31-33. In modern society people have a tendency to look toward technology to solve social problems without considering the structural changes that are necessary to improve society. Explores various structural changes that must be made before computers can have any great impact in solving our problems.

FOSTER, GEORGE MCCLELLAND. *Traditional Cultures, and the Impact of Technological Change.* New York: Harper & Row, 1962. Pp. 292.

FOURASTIÉ, JEAN. *La civilisation de 1960.* Paris: Presses Universitaires de France, 1947. Pp. 118.

————. *Le progrès technique et le niveau de vie en France et à l'étrange; Conference* [Technical progress and the level of life in France and foreign countries; a lecture]. Paris: Centre de perfectionnement technique et administratif, 1947. Pp. 42.

————. *Le grand espoir du XXe siècle; progrès technique, progrès économique, progrès social* [The great hope of the 20th century; technological progress, economic progress, social progress]. Paris: Presses Universitaires de France, 1949. Pp. 223.

————. *Pourquoi nous travaillons* [Why we work]. Paris: Presses Universitaires de France, 1959. Pp. 126.

————. *The Causes of Wealth.* Translated and edited by T. Caplow. New York: Free Press, 1960. Pp. 246. This is the American edition, revised, adapted, and supplemented from *Machinisme et bien-être; niveau de vie et genre de vie en France de 1700 a nos jours* (Paris: Editions de Minuit, 1951).

————. *La grande metamorphose du XXe siècle* [The great metamorphosis of the 20th century]. Paris: Presses Universitaires de France, 1961. Pp. 223.

————. *Les 40.000 heures* [The 40,000 hours]. Paris: Laffont, 1965. Pp. 246.

————. *Essais de morale prospective.* Paris: Gonthier, 1966. Pp. 199. "The development of scientific knowledge, the rising standard of living, the amelioration of the way of life present grave challenges to traditional morals." Moves from concrete observations to a more theoretical discussion of traditional and contemporary moral situations and finally to the basis for a system of morals relevant to a scientific society.

FOX, RENÉE C. "A Sociological Perspective on Organ Transplantation and Hemodialysis," *New Dimensions in Legal and Ethical Concepts for Human Research* (Annals New York Academy of Sciences) 169, no. 2 (January 1970): 406-28. Also in Harvard Univ. Program on Technology and Society, Reprint no. 7. Condensed version, "Menschliche Probleme bei Organverpflanzung und Kunstlicher Niere," published in *Umschau in Wissenschaft und Technik* 26 (1970): 851-53.

FOX, W. T. "Science, Technology and International Politics," *International Studies Quarterly* 21, no. 1 (March 1968): 1-15.

FRAIN, WILLIAM J. "Law and Order in the Technological Society," *Cross Currents* 12, no. 4 (Fall 1968): 459-70.

FRANK, JEROME. "The Place of the Expert in a Democratic Society," *Philosophy of Science* 16, no. 1 (January 1949): 3-24.

————. "Galloping Technology: a New Social Disease," *Etc.* 25, no. 1 (1968): 51-69. Because of the "interesting psychological point that our increasing power over nature has been accompanied by growing despair about ourselves," a psychologist argues for the need for man to learn to master himself.

FRANKE, HERBERT W. *Phänomen Technik.* Wiesbaden: F. A. Brockhaus, 1962. Pp. 95. A journalistic picture book.

FREED, J. ARTHUR. *Some Ethical and Social Problems of Science and Technology: a Bibliography of the Literature from 1955.* Report no. LAMS 3028. Los Alamos Scientific Laboratory, 1964. Distributed by the Office of Technical Services, U.S. Department of Commerce. Pp. 48. Covers period from 1955 to September 1, 1963. A good (although unannotated) bibliography of material which ranges from the professional ethics of engineers and scientists to their more general moral and social responsibilities.

FREEMAN, CHRISTOPHER. "Technology Assessment and Its Social Context," *Studium generale* 24, no. 9 (September 30, 1971): 1038-50. Places the rising concern with technology assessment in historical context—both philosophically and economically. Uses a few specific examples as well to demonstrate the weakness of laissez faire science and technology and the need for long-range cost analyses and public decisions about technological projects.

FREUND, PAUL A. "Organ Transplants: Ethical and Legal Problems," *Proceedings of the American Philosophical Society* 115, no. 4 (August 20, 1971): 276–81.

FRIEDMANN, GEORGES. *La crise du progrès: esquisse d'histoire des idées, 1895–1935.* Paris: Gallimard, 1936.

———. *Où va le travail humain?* Paris: Gallimard, 1950. Revised ed., 1967.

———. *Industrial Society: the Emergence of the Human Problems of Automation.* Translated by H. L. Sheppard. New York: Free Press, 1955. Pp. 436. From *Problèmes humains du machinisme industriel* (Paris: Gallimard, 1946).

———. "Technological Change and Human Relations," *Cross Currents* 10 (Winter 1960): 29–47.

———. *Anatomy of Work: Labor, Leisure, and the Implications of Automation.* Translated by W. Rawson. New York: Free Press, 1962. Pp. 203. From *Le travail en mietees — spécialisation et loisirs* (Paris: Gallimard, 1956).

———. *Sept études sur l'homme et la technique.* Paris: Gonthier, 1966. Pp. 214. "Technology is not neutral. Not only does it modify the natural milieu, but it also acts upon man, and profoundly transforms society." Discusses problems of life in the cities and country; of a technical milieu which is becoming more and more dense, omnipresent, and demanding; of work and leisure in a society of abundance; and finally those problems characteristic of the atomic age in which we now live.

FRISCH, ALFRED. "Technik und Geist," *Dokumente: Zeitschrift im Dienste übernationaler Zusammenarbeit* (Cologne) 15 (1959): 380–85.

———. "L'avenir des technocrates" [The future of technocrats], *Res publica* 9, no. 4 (1967): 649–60.

FROGER, DOM JACQUES. "The Electronic Machine at the Service of Humanistic Studies," *Diogenes* 52 (Winter 1965): 104–42.

FROMM, ERICH, ed. *Socialist Humanism: an International Symposium.* Garden City, N.Y.: Doubleday, 1965. Pp. 461. The book is divided into five sections: "On Humanism," "On Man," "On Freedom," "On Alienation," "On Practice." Contains a number of essays by European and East European authors confronting, at least implicitly, the problems of technology. See especially M. Neil, "The Phenomenon of Technology: Liberation or Alienation of Man?"

FROOMKIN, JOSEPH N. "Automation." Pp. 480–89 in David L. Sills, ed., *International Encyclopedia of Social Sciences.* Vol. 1. New York: Macmillan, and Free Press, 1968.

GABOR, DENNIS. *Inventing the Future.* New York: Knopf, 1964. Pp. 238. One of the more interesting pieces of futurology. There are sociological discussions of overpopulation, war, and other technological strains on modern society against the literary background of utopian writers such as Haldane, Huxley, Wells, Russell, etc. See also Gabor's *Innovations: Scientific, Technological, and Social* (New York: Oxford Univ. Press, 1970), and "Technology Autonomous," *New Scientist* 54 (May 25, 1972): 448–49.

GALBRAITH, JOHN K. *The New Industrial State.* New York: Houghton Mifflin, 1969. Pp. 583. "Technology means the systematic application of scientific or other organized knowledge to practical tasks" (p. 12). See also *The Affluent Society,* 2d ed. (New York: Houghton Mifflin, 1969). Galbraith describes this earlier book as foreshadowing the more complete theory.

GANZONI, WERNER. *Rheinau? 3 Aufsätze aus dem "Landboten"* [Rheinau? Three essays from the "Landboten"]. Winterthur: Kommissionsverlag W. Vogel, 1952. Pp. 30. First two essays use German theoretical criticisms of technology to evaluate the particular case of the building of a dam on the Rhine River near Rheinau. The third essay is an appreciation of the thinker Ludwig Klages.

GARCIA BACCA, JUAN DAVID. *Elogio de la technica* [Eulogy for technology]. Caracas: Monte Avila Editores, 1968. Pp. 181.

———. "Ciencia, tecnica, historia y

filosofía en la atmosfera cultural de nuestro tiempo" [Science, technology, history and philosophy in the cultural atmosphere of our time], *Cuadernos Americanos* (Mexico) 29, no. 3 (May–June 1970): 71–89.

GARDNER, JOHN W. "Toward a Humane Technology," *Science and Technology* (January 1968), pp. 56–60.

GARTMANN, HEINZ. *Man Unlimited: Technology's Challenge to Human Endurance.* Translated by Richard and Clara Winston. New York: Pantheon, 1957. Pp. 213. From *Stärker als die Technik* (Düsseldorf: Econ-Verlag, 1956). Concerned with physical fitness in relation to modern technology.

GATES, DAVID M. "Exploitation, Evolution and Ecology," *Technology Review* 71, no. 2 (December 1968): 34–43.

GERLACH, WALTHER. "Ortsbestimmung der Technik" [Assigning technology its place], *Deutsche Universitätszeitung* 10, no. 21 (1955): 7–11.

———. "Die Stellung der Technik zu Wissenschaft und Kultur" [The place of technology relative to science and culture], *VDI-Nachrichten* 9, no. 26 (1955): 11 ff. Also in *Allgemeine Wärmetechnik; Zeitschrift für Wärme- und Kältetechnik* (Dissen) 6 (1955): 207–15; and in *Zeitschrift für Bibliothekswesen und Bibliographie* (Frankfurt) 2, no. 3 (1955): 166–85.

———. "Forschung-Technik-Mensch" [Research-technology-man], *Uhr; Fachzeitschrift für das Uhrmacherhandwerk* (Cologne) 10, no. 15 (1956): 46–56.

———. "Der Mensch und die Technik — Gestern, Heute, Morgan" [Man and technology — yesterday, today, tomorrow], *Fernmelde-Praxis* (Wolfshagen-Scharbeutz) 34 (1957): 465–71.

———. "Zwei Übel unserer Zeit: Dämon Technik und Fetisch Wissenschaft" [Two evils of our time: demon technology and fetish science], *Deutsche Universitätszeitung* 19, no. 8 (1964): 5–9.

GERSON, ALFONS. "Freiheit und Verantwortung des Ingenieurs" [Freedom and responsibility of engineers], *Humanismus und Technik* 2, no. 2 (August 15, 1954): 89–94.

GIRARDEAU, EMILE FERNAND EUGÈNE. *Le progrès technique et la personnalité humaine* [Technical progress and the human personality]. Paris: Plon, 1955. Pp. 336. Attempts to popularize more technical and/or sociological studies of technology, evaluating them at the same time in terms of their importance to the human person. Remains superficial.

GOEDEKING, HANS. "Über den institutionellen Charakter der modernen Technik." Pp. 247–62 in Heinz-Dietrich Wendland, ed., *Sozialethik im Umbruch der Gesellschaft: Arbeiten aus dem Mitarbeiter und Freundeskreis des Instituts für Christliche Gesellschaftswissenschaften an der Universitat Münster.* Göttingen: Vandenhoeck & Ruprecht, 1969.

GOESHEL, HEINZ. "Die technologischen Entwicklungen und unsere Zukunft" [Technological developments and our future], *Universitas; Zeitschrift für Wissenschaft, Kunst und Literatur* 24, no. 8 (1969): 847–62.

GOLDBERG, JACOB PINHEIRO. *Éthica e technologica; fragmentos.* São Paulo: Ediçâo Fulgor, 1968. Pp. 148. Not a serious philosophical work.

GOLDBERG, MAXWELL, H. "The Technetronic Age." Chap. 4 in *Design in Liberal Learning.* San Francisco: Jossey-Bass, 1971.

GOLDSTEIN, MORITZ. "Zusammenbruch der Welt?" [Collapse of the world?], *Deutsche Rundschau* 83 (1957): 883–88.

GOMER, ROBERT. "The Tyranny of Progress," *Bulletin of the Atomic Scientists* 24, no. 2 (February 1968): 4–8. "To a much larger extent than we may realize or acknowledge, we are caught up in an evolutionary stream of our making but beyond our control. Even under the best of circumstances we will have to accept the fact that man must change to meet changes he has himself set off."

GONZALES RIOS, FRANCISCO. "Quelques conclusions sur l'aporie entre humanisme et technique." Pp. 78–85 in *Actes: IIe Congrès International de l'Union Internationale de Philosophie des Sciences.* Vol. 4. Neuchatel: Griffon, 1955.

GOODMAN, L. LANDON. *Man and Automa-*

tion. Baltimore: Penguin, 1957. Pp. 286. Part I outlines the technical aspects of automation. Part II is concerned with social and economic implications. A good general introduction.

GOODY, JOHN RANKINE. *Technology, Tradition and the State in Africa.* New York: Oxford Univ. Press, 1971. Pp. 88.

GORAN, MORRIS. "The Literati Revolt against Science," *Philosophy of Science* 7 (1940): 379-84.

GORKIN, J. *El hombre en la era cibernética y su filosofía* [Man in the cybernetic age and his philosophy]. Buenos Aires: Editorial Ciencia, 1967. Pp. 363.

GOULDNER, ALVIN W., and RICHARD A. PETERSON. *Notes on Technology and the Moral Order.* Indianapolis: Bobbs-Merrill, 1962. Pp. 96. "Moral order" is being used by the authors in an anthropological or sociological sense. Not really a philosophical study.

GRAEMIGER, SILVIO. "Wir Techniker und Goethe's Faust" [We technologists and Goethe's Faust], *Technische Rundschau* (Bern) 54, no. 51 (1962): 1-3.

GRANER, HANS. *Fluch und Segen der Technik; ein Beitrag zum neuen Beginn* [The curse and blessing of technology; a contribution toward a new beginning]. Stuttgart: Deutsche Verlags-Anstalt, 1946. Pp. 58. Of only historical value. Argues that technology is not an unalloyed good, that it can have both good and bad effects. His suggestions of possible bad effects. are mildly prophetic. But the fact that it was written right after World War II, before anybody knew what was going to happen to Europe, makes it extremely dated. – R. J. R.

GRASSI, ERNESTO. "Il tempo umano l'umanesimo contro la 'techne.' " Pp. 201-6 in *Umanesimo e scienza politica.* Milan: Marzorati, 1951.

GRAVIER, JEAN FRANÇOIS. *Décentralisation et progrès technique* [Decentralization and technological progress]. Paris: Portulan, 1953. Pp. 394. Primarily economics.

GRAZIA, SEBASTIAN DE. *Of Time, Work and Leisure.* New York: Twentieth Century Fund, 1962. Pp. 559. On the annihilation of leisure by time- and work-saving devices. Reviewed by Norman D. Kurland in *Technology and Culture* 4, no. 3 (Summer 1963): 370-72; and by Kenneth E. Boulding in *Scientific American* (January 1963), p. 157.

GREEN, HAROLD P. "The New Technological Era: a View from the Law," *Bulletin of the Atomic Scientists* 23, no. 9 (November 1967): 12-19. "Our national commitment to technological advance seems irresistable, irrevocable, and irreversible. Given, moreover, the tension-filled world in which we live, it is perhaps unthinkable that the U.S. would not continue its drive for continuing technological superiority. Protective measures must therefore be built into the development and practice of technology."

――――. "Technology Assessment and the Law: Introduction and Perspective," *George Washington Law Review* 36 (July 1968): 1033-43.

GREENBERG, DANIEL S. *The Politics of Pure Science.* New York: New American Library, 1968. Pp. 303.

――――. "The New Politics of Science," *Technology Review* 73, no. 4 (February 1971): 40-45.

GREENBERGER, MARTIN, ed. *Computers and the World of the Future.* Cambridge, Mass.: M.I.T. Press, 1962. Pp. 340. Originally published as *Management and the Computer of the Future* (Cambridge, Mass.: M.I.T. Press, 1962). Contains two articles relevant to ethical-political issues: C. P. Snow's "Scientists and Decision Making," with discussion by E. E. Morison and N. Wiener; and J. W. Forrester's "Managerial Decision Making," with discussion by C. C. Holt and R. A. Howard. Two other essays are more relevant to metaphysical-epistemological issues and are cited there. The other four articles are of marginal interest.

――――. *Computers, Communications, and the Public Interest.* Baltimore: John Hopkins Press, 1971. Pp. 315. Papers by J. Kemeny, H. A. Simon, A. Oettinger, J. Coleman, A. Westin, R. Brown, N. Johnson, and G. Wald.

GREGORY, R. L. "Social Implications of Intelligent Machines," *Machine Intelligence* 6 (1971): 3-13.

GREIFFENHAGEN, MARTIN. "Demokratie

und Technokratie," *Praxis* 3, no. 2 (1967): 214-27.

GREILING, WALTER. *Wie werden wir leben? ein Buch von den Aufgaben unserer Zeit* [How shall we live? A book on the problems of our time]. Düsseldorf: Econ-Verlag, 1954. Pp. 320. Intended for popular edification. More references to biological sciences than physical. Future painted in two colors: World War III or utopia. – R. J. R.

––––––. "Hat der technische Fortschritt Grenzen?" [Does technological progress have limits?], *Radius* 1 (1958): 39-44.

GRODZINS, MORTON, and EUGENE RABINOWITCH, eds. *The Atomic Age: Scientists in National and World Affairs.* New York: Basic, 1963. Pp. 616. A collection of articles from the *Bulletin of the Atomic Scientists,* 1945-62, which chronicles the birth and partial success of a political movement by scientists. Contains a large number of classical statements by scientific figures such as Leo Szilard, J. Robert Oppenheimer, Edward A. Shills, Albert Einstein, Hans Bethe, Eugene Rabinowitch, Ralph E. Lapp. Max Born, whose public consciences were aroused by the threat of nuclear war which they had helped to make possible; also a number of statements by governmental defenders such as Henry L. Stimson, Edward Teller. Cf. also Robert Jungk, *Brighter than a Thousand Suns; the Moral and Political History of the Atomic Scientists,* translated by J. Cleugh (New York; Harcourt Brace, 1958); Robert Gilpin, *American Scientists and Nuclear Weapons Policy* (Princeton, N.J.: Princeton Univ. Press, 1962); Joseph Robtlat, ed., *Pugwash: the First Ten Years: a History of the Conferences on Science and World Affairs* (New York: Humanities Press, 1967); Alice Kimball Smith, *A Peril and a Hope: the Scientists' Movement in America, 1945-1947* (Chicago: Univ. Chicago Press, 1965); and Donald A. Strickland, *Scientists in Politics: the Atomic Scientists' Movement, 1945-1946* (Lafayette, Ind.: Purdue Univ. Studies, 1968).

GROSS, FELIKS. "Consecuencias sociales del cambio tecnologico," *Revista Mex-icana de sociologia* (Mexico) 28, no. 2 (April-June 1966): 377-409.

GROSSMAN, HENRYK. "Die gesellschaftlichen Grundlagen der mechanistischen Philosophie und die Manufaktur," *Zeitschrift für Sozialforschung* 4 (1935): 161-231.

GROTJAHN, M. "The New Technology and Our Ageless Unconscious," *Psychoanalytical Forum* 1 (1966): 7-18.

GRUBER, ABRAHAM. "A Functional Definition of Primate Tool-Making," *Man* 4 (December 1969): 573-79.

GUILLERME, JACQUES. "Qu'est-ce qu'un technologue?" *La table ronde* (Paris) 214 (1965): 28-39.

GUNN, JAMES E., ed. *Man and the Future.* Lawrence: Univ. Kansas Press, 1968. Pp. 305. Lectures delivered at the Inter-Century Seminar at University of Kansas in 1967. Contributors such as R. B. Fuller, H. Clurman, and A. Clarke. Tends to be popular and eclectic.

GURVITCH, GEORGES, ed. *Industrialisation et technocratie.* Paris: Colin, 1949. Pp. 211. Contents: E. Mounier's "La machine en accusation"; E. C. Hughes's "Les recherches américaines sur les relations industrielles," with discussions by H. I. Laski, P. de Bie, G. Gurvitch, and P. Naville; Fourastié's "Technocratie et rendement économique"; G. Friedmann's "Les technocrates et la civilisation technicienne," with discussions by G. Le Bras, C. Bettelheim, G. Gurvitch, and R. Clémens; M. Bye's "Vers un quatrième pouvoir?" with discussions by G. Davy, H. Lefebvre, P. Naville, M. Prelot, and P. de Bie; Charles Bettelheim's "Les techniciens constituent-ils une classe sociale?" with discussions by G. Gurvitch, J. Lhomme, J. Hyppolite, P. Naville, and I. Meyerson; Jacques Vernant's "Evolution technique et structure sociale," with discussions by P. Gemaehling, P. de Bie, C. Bettelheim, and M. Barioux; H. Lefebve's "Les conditions sociales de l'industrialisation," with discussions by P. Gemaehling, G. Duveau, and P. Naville; H. I. Laski's "L'état, l'ouvrier et le technicien," with discussions by G. Le Bras, G. Gurvitch, E. C. Hughes, M. Leroy, G.

Duveau, J. Weiller, A. Varagnac, and P.-M. Schuhl; A. Veragnac's "Industrialisation et experimentation sociale," with discussions by A. Piganiol, G. Duveau, and H. Lefebvre, G. Gurvitch's "La technocratic est-elle un effet inévitable de l'industrialisation?"; concludes with general discussions by P. Francastel, E. C. Hughes, A. Pampu, M. Barioux, P. Kahn, P. de Bie, F. Bourricaud, P. Clemens, M. Pagès, P. Naville, E. Mounier, H. Lefebvre, H. I. Laski, and G. Gurvitch. Mounier's essay is available in English as "The Case against the Machine," in *Be Not Afraid* (London: Rockliff, 1951).

GUTIERREZ CASTRO, EDGAR. "La ciencia aplicada y la technologia en la politica del desarrollo" [Applied science and technology in the politics of evolution], *Revista de Planeacian y Desarrollo* (Bogota) 1, no. 3 (October 1969): 3-11.

GUZZO, AUGUSTO. "L'uomo, la macchina, la técnica," *Filosofia* (Turin) 17 (1966): 425-50. Comments and replies by: G. Calo, G. Calogero, C. Giacon, S. Caramella, V. Somenzi, V. Mathieu, A. Vasa, A. Guzzo (pp. 451-74).

HAHN, HERMANN H. "Der Ingenieur von Morgen" [The engineer of tomorrow], *VDI-Zeitschrift* 113, no. 4 (1971). 233-37. The engineer is the necessary mediator between the scientist and the politician.

HAIDANT, PAUL. *L'homme et la machine; la grande angoisse du temps présent.* Paris: Béranger, 1951. Pp. 31.

HALL, CAMERON, P. *Technology and People.* Valley Forge, Pa.: Judson, 1969. Pp. 159. "The basic issue lies not with technology but with Man. Will he match those mental and technical skills by which he develops technology with spiritual and ethical resources in the use to which he puts technology? To meet this test men must first become aware of technology as a pervasive fact in today's life and understand the nature of its impact upon people. The development of such awareness and understanding is a primary concern of this book" (p. 7). Basically a textbook

for adult education classes. Each chapter closes with study questions and supplementary reading suggestions; there is also a bibliography of audiovisual resources.

HALSBURY, EARL OF. "Automation— Verbal Fiction, Psychological Reality," *Impact of Science on Society* 7, no. 4 (1956): 179-201.

HAMMING, R. W. *Computers and Society.* New York: McGraw-Hill, 1972. Pp. 284. A popular introduction.

HAMMOND, JAMES. *Technical Talents and Cultural Needs.* Dimensions for Exploration Series. Oswego, N.Y.: Division of Industrial Arts and Technology, State Univ. College, 1965. Pp. 18. Man has specific technical aptitudes and capabilities that must be identified, evaluated, and developed within the context of "this whole education." To this end, it is necessary to apply the results of scientific research, i.e., scientific facts about the unchanging and measurable characteristics of human beings. Such research, "wisely applied will enable men to keep the ever more aggressive and impertinent mechanical robots of our age as our servants and not allow them to take over and become our masters."

HAMPSHIRE, STUART. "Suspect Sages," *New York Review of Books* 19, no. 4 (September 21, 1972): 12-13. A review of C. P. Snow's collection of essays entitled *Public Affairs* (New York: Scribner's, 1971) and Dennis Gabor's *The Mature Society* (New York: Praeger, 1972). Especially critical of Snow's worship of technology and Gabor's vague futurology.

HANDLIN, OSCAR. "Man and Magic: First Encounters with the Machine," *American Scholar* 33, no. 3 (Summer 1964): 408-19. A study of 19th-century American literary reactions. "The machine, which was a product of science, was also magic, understandable only in terms of *how* it worked, not of why. Hence the lack of comprehension or of control; hence also the mixture of dread and anticipation as in the past" (p. 418).

HARDIN, GARRETT. "An Evolutionist Looks at Computers," *Datamation* 15, no. 5 (May 1969): 98-109. A discussion of the effects of owning slaves (in the form of machines) on the slave owner. It tends to corrupt the owners and make them physically soft; as evolution has shown, the implementation of the knife meant the decline of the jaw of man. "Exosomatic" adaptations tend to bring about the degeneration of the "endosomatic" precursor. While knives do not present serious problems, other exosomatic inventions such as the heart pacemaker and artificial kidney do cause problems by preserving genetically deficient members of the species, thus making the species more vulnerable to accidents which we have no way of preventing.

HARRINGTON, MICHAEL. *The Accidental Century.* Baltimore: Penguin, 1965. Pp. 332. Argues that, while the technological and social revolution of the 20th century will continue, it must cease to be accidental; it must become conscious, planned, and democratic through the political intervention of man.

HARRISON, ANNETTE. *Bibliography on Automation and Technological Change and Studies of the Future.* Santa Monica, Calif.: RAND Corp., March 1968. This bibliography lists works associated with two ongoing RAND projects: "Automation and Technological Change" and a related informal study of future national and international problems. The appendix contains a list of private organizations currently engaged, or planning to engage in, future-oriented work.

HART, HORNELL. "Technological Acceleration and the Atom Bomb," *American Sociological Review,* vol. 11, no. 3 (June 1946).

HASKINS, C. P. "Technology, Science and American Foreign Policy," *Foreign Affairs* 40 (January 1962): 224-43.

HASTENFEL, PAUL, ed. *Markierungen; Beiträge zur Erziehung im Zeitalter der Technik* [Markings; essays on education in an age of technology]. Munich: Kösel-Verlag, 1964. Pp. 247. Contents: R. Schwarz's "Über 'das Psychische' im Bildungsproblem," A.

Petzelt's "Technik und Ethic," A. Wenzl's "Ur-Technik und Ur-Erziehung," J. Dolch's "Technik und Wirtschaft als Triebkräfte der Pädagogik der Völker," F. Schneider's "Prolegomena über den Einsatz technischer mittel im Bildungsprozess," J. Zielinski's "Das Problem der 'Doppelendigkeit' in der Pädagogik," M. Heitger's "Erziehung zum partiellen Verzicht," E. Weber's "Die Aufgaben der Medienpädogogik," E. Feldmann's "Ansichten der Jugend über den Einfluss des Fernsehens auf das Familienleben," F. Stückrath's "Zur Psychologie der niederen Publizistik," E. Dovifat's "Erfahrung und Information," K. Holzamer's "Theater, Werbung und Pädagogik," E. Wasem's "Die Pädagogische Film- und Fernsehforschung in München," W. Tröger's "Zur Entwicklung der Fernsehforschung," and F. Zieris's "Die Forschung zum Thema Jugend und Film der Bundes teufel."

HAUDRICOURT, ANDRÉ. "La technologie, science humaine," *La pensée: revue du rationalisme moderne,* new series, 15 (1964): 28-35.

HAUSEN, JOSEF. "Atome für den Frieden" [Atoms for freedom]. *Humanismus und Technik* 3, no. 2 (December 25, 1955): 72-90.

HAUSER, P. M. "Social Science and Social Engineering," *Philosophy of Science* 16, no. 3 (July 1949): 209-18.

HEALD, M. "Technology in American Culture." Pp. 103-17 in J. A. Hague, ed., *American Character and Culture.* De Land, Fla.: Everett Edwards, 1964.

HEINEMANN, FRITZ. "Können wir das Zeitalter der Technik überleben?" [Can we survive the age of technology?], *Universitas; Zeitschrift für Wissenschaft, Kunst und Literatur* 12 (1957): 673-82.

HELMER, OLAF. *Social Technology.* New York: Basic, 1969. Pp. 108. Concerned with "the gap between physical technology and socio-political progress." Examines new techniques of forecasting scientific and social innovation.

HELVEY, T. C. *The Age of Information; an Interdisciplinary Survey of Cybernetics.* Englewood Cliffs, N.J.: Educational

Technology Publications, 1971. Pp. 207. Begins with a rough conceptual analysis of cybernetics (defined as "the science of interaction") and then argues that cybernetics is involved in everything from sleep control in astronauts (chap. 5) to history (chap. 6), politics (chap. 9), marine sciences (chap. 11), and interpersonal relations (chap. 13). Also argues that information is more important to contemporary society than any other product. An optimistic appraisal.

HERSCHEL, WILHELM. "Regeln der Technik" [Rules of technology], *Neue juristische Wochenschrift* 21, no. 14 (1968): 617-23.

HESBURGH, THEODORE M. "Science and Technology in Modern Perspective," *Vital Speeches* (August 1, 1962), pp. 631-34.

―――. "Science Is Amoral: Need Scientists Be Amoral, Too?" *Way* (U.S.A.), 19 (April 1965): 20-25.

HETMAN, FRANÇOIS. "Le progrès technique, une illusion comptable?" *Analyse et prévision* (Paris) 9, no. 3 (1970): 155-75.

HEUSS, THEODOR. "Über die Bewertung der modernen Technik" [The evaluation of modern technology], *Physikalische Blätter* 12, no. 4 (1956): 145-50.

HEYKE, HANS-EBERHARD. "Über den Begriff des technischen Fortschritts," *Jahrbuch der Sozialwissenschaft und Bibliographie der Sozialwissenschaft* (Göttingen) 21, no. 2 (1970): 99-126.

HILDEBRAND, W. "Technik und Revolution; die Funktion der Technik in der Gesellschaftslehre des Marxismus-Leninismus und Sowjetkommunismus" [Technology and revolution; the function of technology in the sociology of Marxism-Leninism and Soviet communism], *Studium generale* 15 (1962): 334-55.

HILLER, EGMONT. *Automaten und Menschen*. Stuttgart: Deutsche Verlag-Anstalt, 1958. Pp. 104.

―――. *Humanismus und Technik*. Düsseldorf: Patmos Verlag, 1966. Pp. 108.

HILTON, MARY ALICE, ed. *The Evolving Society*. New York: Institute for Cybercultural Research Press, 1966. Pp. 410. This book has sections on "Basic Assumptions," "Computing Machines and Cybernated Systems," "The Evolving Society," "The Future Society—Concepts," "The Future Society—Reasons for Hope and Causes of Fear." Each section consists of papers and discussion from the First Annual Conference on the Cybercultural Revolution.

HINTON, LORD OF BANKSIDE. *Engineers and Engineering*. London: Oxford Univ. Press, 1970. Pp. 75. "It is not science but technology that is changing our lives and, of all the technologies, engineering is probably the most important."

HOFSTÄTTER, PETER R. "Der Einfluss der Technik auf die Psychologie" [The influence of technology on psychology], *VDI-Zeitschrift* 104, no. 26 (1962): 1322-26.

HOGGART, RICHARD. "Human Values and the Technological Society," *Technologist* 3 (1966): 40-42. There is a tendency to believe that technology can take the place of religion in telling us what to do. But within itself technology provides knowledge only about expediency and efficiency. Outside of this one enters the realm of value-judgments and choices. And there is no neutrality here: "Every choice is a moral choice." The grounds for judgment lie in reflecting upon "the quality of life being assumed and offered with a society." Argues for the preservation of human values in a technological society and insists on the importance of literature in developing an awareness of the complexity in human life. See also the author's "Two Ways of Looking," in *Speaking to Each Other* (New York: Oxford Univ. Press, 1970), 1:106-13.

HOLTHUSEN, HANS E. "Technik und Welteinheit" [Technology and world unity], *VDI-Zeitschrift* 99, no. 23 (1957): 1100-1106.

HOOS, IDA R. *Social Significance of Technological Advance*. Internal Working Paper no. 67. Berkeley: Univ. California, June 1967. A discussion of the effects of technology upon modern existence. Some points: Argues that reliance on the computer is so great "that the very problems to which society will turn a hand must be framed in quantitative terms amenable to electronic handling." In education, technology has depersonalized learning. Also contains an interesting analysis of "coin box morality," or people's tendency to think of the machine as their natural enemy so that they feel free to cheat technological devices. Concludes that technology must "be a part of instead of apart from man's moral commitment to the past, present, and future of all mankind."

HOSELITZ, BERT F., and WILBERT E. MOORE, eds. *Industrialization and Society.* Chicago, 1960; Paris: UNESCO, 1963. Pp. 437. Proceedings of the North American Conference on the Social Implications of Industrialization and Technological Change.

HOWE, GÜNTHER. "Technik und Freiheit" [Technology and freedom], *Eckart-Jahrbuch, 1961-1962,* pp. 209-27.

HOWLAND, W. E. "Engineering Education for Social Leadership," *Technology and Culture* 10, no. 1 (January 1969): 1-10. See also in this same journal J. G. Burke's "Comment: Let's Be Sure Technology Is for Man" (pp. 11-13), S. C. Florman's "Comment: Engineers and the End of Innocence" (pp. 14-16), and J. C. Wallace's "Comment: the Engineering Use of Human Beings" (pp. 17-19). See also W. E. Howland's " 'Technology for Man' Revisited," *Technology and Culture* 11, no. 2 (April 1970): 237-39; and S. C. Florman's "Reply" which follows (p. 240).

HUANT, ERNEST. *Masses, morale et machines, la morale devant l'hypertechnie et le conditionnement.* Paris: Éditions du Cèdre, 1967. Pp. 97.

HUBBLE, DOUGLAS. "Towards a Humanistic Technology: the Place of Medicine." Pp. 47-73 in Alan S. C. Ross, ed., *Arts v. Science.* London: Methuen, 1967. Modern medicine is the model for a humanistic technology.

HÜBNER, KURT. *Sinn und Aufgaben philosophischer Fakultäten an technischen Hochschulen* [The role of the faculty of humanities at a technical university]. Schriftenreihe des Stifterbandes für die Deutsche Wissenschaft, 1, no. 1. Essen Bredeney: Gemeinnützige Verwaltungsgesellschaft für Wissenschaftspflege G.m.b.H., Verlagsabteilung, 1966. Pp. 52.

HUTCHINS, EDWARD and ELIZABETH, eds. *Scientific Progress and Human Values.* New York: American Elsevier, 1967. Pp. 219. Proceedings of the conference celebrating the 75th anniversary of California Institute of Technology. Very marginal.

HUTCHINS, ROBERT M., et al. *Science, Scientists, and Politics.* Santa Barbara, Calif.: Center for the Study of Democratic Institutions, 1963. Pp. 16. Short papers by R. M. Hutchins, S. Buchanan, D. N. Michael, C. Sherwin, J. Real, and L. White, jr., presented at a conference on the role and responsibilities of science executives in the service of government. Several of the papers make reference to C. P. Snow and the "two cultures" debate.

HUXLEY, ALDOUS. *Ends and Means; an Inquiry into the Nature of Ideals and into the Methods Employed for Their Realization.* New York: Harper, 1937. Pp. 386. Despite some rather gross oversimplifications, this book, especially in its French translation, *La fin et les moyens* (Paris: Plon, 1939), has had a wide influence in Europe. The argument concerns the relation between the religious ideals of charity and nonattachment and modern industrialization and technology. After a brief theoretical analysis, the book gives "a kind of practical cookery book of reform," then concludes by asking, "What sort of world is this, in which man aspires to good and yet so frequently achieves evil?" Also translated into German, Spanish, Italian, and Danish. And for more material on the same theme, see the author's *Science, Liberty and Peace* (New York: Harper, 1946) as well as the novel *Island* (New York: Harper & Row, 1962).

————. *Tomorrow and Tomorrow and Tomorrow and Other Essays.* New York: Harper & Row, 1956. Pp. 301. Abridged paperback edition 1972. See especially "Liberty, Equality, Machinery."

HYNES, LEONARD. The Impact of Technology on Man and His World," *Chemistry and Industry* 26 (1967): 1112–16.

"The Impact of Technology." Part VII in A. K. Bierman and J. A. Gould, eds., *Philosophy for a New Generation.* New York: Macmillan, 1970. Contains Z. Brzezinski's "America in the Technetronic Age," K. Marx's "Estranged Labor," H. Marcuse's "Technology and the Control of Man's Freedom," and J. Pieper's "Work versus Leisure."

"L'idéologie technocratique et le teilhardisme," *Temps modernes* 22 (1966): 254–95. A long critical essay by an anonymous reviewer of L. Armand and M. Dranscourt's *Plaidoyer pour l'avenir* and F. Bloch-Laine's *Pour une reforme de l'entreprise.*

ISCHREYT, HEINZ. *Studien zum Verhältnis von Sprache und Technik* [Studies in the relation between language and technology]. Sprache und Gemeinschaft, vol. 4. Düsseldorf: Pädagogischen Verlag Schwann, 1965. Pp. 304. Concerned with the problem of standardizing terminology.

JANTSCH, ERICH. "Prevision technologique et recherche fondamentale," *Atomes* 246 (September 1967): 511–18. Discusses the relationship between basic and applied science, and that between science and technology, stating that today every important innovation is due to basic research (whereas, prior to the 20th century, science and technology evolved independently). Describes methods of technological forecasting and its importance in planning.

JEMOLO, ARTURO C., et al. *Il mondo moderno e la tecnica.* Milan: Tamburini Editore, 1967. Pp. 97. Several rather conventional discussions of the impact of science and technology on cultural and moral values, originally conceived for an academic conference, and expressive of a rather traditional Christian sense of "progress."—F. W.

JERVIS, GIOVANNI, and LETIZIA COMBA. "Contradictions du technicien et de la culture techniciste," *Temps modernes* 26 (April 1970): 1601–12.

JOHNSON, DAVID L., and ARTHUR L. KOB-

LER. "The Man-Computer Relationship," *Science* 138 (November 23, 1962): 873–79. As knowledge increases, man must resort to the use of the computer in order to cope with it. At the same time, "we find that the computer is being given responsibilities with which it is less able to cope than man is."

JOHNSON, PAUL. "A Morality for a Dynamic Society," *New Scientist* (December 4, 1969), pp. 506–7.

JOHNSTON, EDGAR G., ed. *Preserving Human Values in an Age of Technology.* Detroit: Wayne State Univ. Press, 1961. Pp. 132. Five Franklin Memorial Lectures: E. U. Condon's "The Challenge of Science to Human Values," H. S. Commager's "Human Values in the American Tradition," F. Biddle's "Freedom and the Preservation of Human Values," L. L. Mann's "The Clash of Ideologies in an Age of Technology," and E. R. Johnston's "Respect for the Individual: the Basic Human Value."

JOSHI, MADHU. "Law in Science and Science in Law," *Lex et scientia* 7, no. 2 (April–June 1970): 63–67. "Many a legal code today is tantamount to being the hand-book that comes along with a new spectometer, oscilloscope or what-have-you, which gives instructions on how to use it and how to maintain it."

JOUVENEL, BERTRAND DE. *The Art of Conjecture.* Translated by Nikita Lary. New York: Basic, 1967. Pp. 307. From *L'arte de la conjecture* (Paris: Editions du Rocher, 1964). "To be sure, an effort is being made to tie scientists to technological objectives . . . , while on the other hand the likelihood that empirical workers will make practical inventions is growing smaller. The once loose connection between progress in science and progress in technology may become close, but in social forecasting—the subject concerning us here—each of them matters to us separately and for quite different reasons: the progress of science because it is reflected strong if crudely in our current ideas . . . , and the progress of technology because it introduces new material facts" (p. 283). A theoretical discussion of futurology methodology by the director of the French Futuribles project.

JUNGK, ROBERT. *Tomorrow Is Already Here; Scenes from a Man-made World.* Translated by M. Waldman. London: Hart-Davis, 1954. Pp. 239.

―――. "Human Futures," *Futures* 1, no. 1 (September 1968): 34-39. Argues that man is not at the mercy of technology.

―――. "Altering the Direction of Technology," *Cross Currents*, vol. 20, no. 2 (Spring 1970). A concrete attempt to describe what a "humanized technology" might be like.

JUNGK, ROBERT, and JOHAN GALTUNG. *Mankind 2000.* London: Allen & Unwin, 1969. Pp. 368. R. Jungk is director of a German future studies project.

KAHLER, ERICH. "Nihilism and the Rule of Technics." In *Man the Measure.* New York: Pantheon, 1943. An inquiry into the meaning of fascism. See also the author's *The Tower and the Abyss* (New York: Braziller, 1957).

KAHN, HERMAN, and ANTHONY J. WIENER. *The Year 2000: a Framework for Speculation on the Next Thirty-three Years.* New York: Macmillan, 1967. Pp. 431.

KANTROWITZ, ARTHUR. "The Test," *Technology Review* 71, no. 7 (May 1969): 45-50. A new system for the democratic control of technology is needed in order to realize its potential.

KAPLAN, MAX, and PHILLIP BOSSERMAN, eds. *Technology, Human Values, and Leisure.* Nashville, Tenn.: Abingdon, 1971. Pp. 256. Proceedings of a conference on this topic held at the University of South Florida. Articles of note: M. Kaplan's "The Relevance of Leisure," R. Theobald's "Thinking about the Future," E. G. Mesthene's "Technology and Humanistic Values," H. Brown's "Technology and Where We Are."

KASPER, R. G., ed. *Technology Assessment.* New York: Praeger, 1971.

KATZ, AARON. "Toward High Information-Level Culture," *Cybernetica* 7, no. 3 (1964): 203-45. See also the author's "High Information-Level Culture—Further Aspects," *Cybernetica* 9, no. 1 (1966): 24-49.

KAUTZ, HEINRICH. "Weltgespräch um die Technik" [World discussion on technology], *Neue Ordnung* 11 (1957): 100-105.

KENISTON, KENNETH. "Does Human Nature Change in a Technological Revolution?" *New York Times* (January 6, 1969).

KERR, C., J. T. DUNLOP, F. H. HARBISON, and C. A. MYERS. *Industrialism and Industrial Man.* 2d ed. New York: Oxford Univ. Press, 1964. Pp. 263.

KIEFER, D. M. "Assessing Technology Assessment," *Futurist* 5, no. 6 (December 1971): 234-39. Technology assessment has expanded from a traditional concern with feasibility and economic profitability to include assessment of indirect effects and social impacts. Continuing questions: Who will perform technology assessments? How and where will they be made?

KING, ALEXANDER. "Management as a Technology," *Impact of Science on Society* 8, no. 2 (1957): 65-85.

KING, ROBERT W. "Whither the Technological State?" *Political Science Quarterly* 65, no. 1 (March 1950): 55-67. Definitely soft core. Technology leads to the well-run society. See also "Technology and Social Progress," *Political Science Quarterly* 76, no. 1 (March 1961): 3-10.

KINZEL, AUGUSTUS B. "Engineering, Civilization and Society," *Science* 156 (June 9, 1967): 1343-45.

KLAGES, HELMUT. *Technischer Humanismus.* Stuttgart: Enke, 1964. Pp. 191.

KLEIN, OTA. "Die wissenschaflich-technische Revolution und die Gestaltung des Lebensstils," *Futurum* 1 (1968): 225-49.

KNIEHAHN, WILHELM. "Vom technischen Geschehen unserer Zeit" [On the technological events of our time], *Humanismus und Technik* 5, no. 2 (April 7, 1958), 86-97.

KOELLE, HEINZ HERMANN. "Alternativen für die Zukunft der Technik," *VDI-Zeitschrift* 111, no. 18 (September 1969): 1241-47.

KOESTLER, ARTHUR. *The Ghost in the Ma-*

chine. New York: Macmillan, 1967. Pp. 384. A spychological and evolutionary study. Argues that "the explosive growth of the human brain resulted in a faulty coordination between ancient and recent brain structures, creating a pathological split between emotion and reason" which is at the heart of man's contemporary problems in the technological world. This is a sequel to Koestler's *The Sleepwalkers* (New York: Grosset & Dunlap, 1959) which portrays Galileo as the figure who first separated mensuration from metaphysics and thus opened the way for the scientific manipulation of nature.

KOHLER, J. "Abgrenzung zwischen Technik und Mensch" [Boundary between technology and man], *Naturwissenschaftliche Rundschau* (Stuttgart) 13, no. 12 (1960): 478 ff.

KÖNIG, RENÉ. "Der Einfluss der technischen Entwicklung auf Gesellschaft und Beruf" [The influence of technological development on society and profession], *VDI-Zeitschrift* 110 (1968): 381-84. Technology suffers from not being accorded as much respect as the humanities in education, as a result of a false contrast between applied science and mythical greatness. By a professor of sociology.

KORAC, VELJKO. "The Possibilities and Prospects of Freedom in Modern World," *Praxis* 4 (1968): 73-82.

KOSTA, JIRI. "Der Mensch und die Gesellschaft in der wissenschaftlich-technischen Revolution," *Futrum* 1 (1968): 167-72.

KOSTELANEZ, RICHARD, ed. *Beyond Left and Right; Radical Thought for Our Times.* New York: Morrow, 1968. Pp. 436. An anthology of writings on technology and the future of man. Favorably reviewed by W. D. Lewis, *Technology and Culture* 11, no. 1 (January 1970): 120-24, who especially praises the author's insightful and generally optimistic introduction. See also Kostelanetz, ed., *Human Alternatives* (New York: Morrow, 1971).

KRÄMER, ERWIN. "Die Serie und der präparierte Mensch" [The serial and programmed man], *Radius* 1 (1958): 25-33.

KRANING, L. ALAN. "Wanted: New Ethics for New Techniques," *Technology Review* 72, no. 5 (March 1970): 41-45. "The rights of privacy are not likely to be preserved in the absence of some clear definition of their nature. Here is an effort toward an agreed ethic drawn from principles laid down by Sartre."

KREITH, FRANK. "Science, Technology, and Moral Responsibility," *Colorado Quarterly* 19, no. 3 (Winter 1971): 235-45. A popular lecture arguing that we all need to accept responsibility for directing science and technology.

KROH, OSWALD. "Seelisches Leben im Zeitalter der Technik" [Spiritual life in the age of technology], *VDI-Zeitschrift* 96, no. 5 (1954): 131-37.

KRÜGER, KARL. *Raum, Volk, Technik* [Space, nation, technology]. Bonn: Bouvier, 1954. Pp. 95. The welfare of nations may be increased and the danger of overpopulation on earth avoided if man learns to think and act objectively, i.e., technologically. Basic thesis is similar to Buckminister Fuller's "Spaceship Earth" concept. Technology conceived as intelligent use of available space.

———. "Gesellschaftskrisen durch Technisierung" [Social crises through technization], *Studium generale* 15 (1962): 512-18.

———. "Was ist und was will die 'Regionaltechnik'?" [What is 'regional technology' and what is its goal?], *Humanismus und Technik* 8, no. 3 (1963): 105-9.

KRUTCH, JOSEPH WOOD. "Invention Is the Mother of Necessity." Pp. 12-18 in *And Even If You Do; Essays on Man, Manners and Machines.* New York: Morrow, 1967.

KUHNKE, HANS-HELMUT. "Technik und Begehrlichkeit" [Technology and greed], *Ganze Deutschland* 8, no. 23 (1956): 7.

KÜNG, EMIL. "Die moderne Technik und ihre sozialen Probleme," *Wirtschaft und Recht* (Zürich) 13 (1961): 213-23,

———. "Die menschliche Natur und die moderne Technik" [Human nature and modern technology], *Technische Rundschau* (Bern) 55, no. 2 (1963): 1 ff.

————. "Wettlauf um technischen Fortschritt" [The race about technological progress], *Technische Rundschau* (Bern) 55, no. 28 (1963): 57 ff.

————. "Technischer Fortschritt—menschliches Glück" [Technological progress—human good fortune], *Technische Rundschau* (Bern) 55, no. 32 (1963): 2 ff.

KÜNNETH, WALTER. "Die Technik als anthropologisch-ethisches Problem," *Studium Berolinense; Aufsätze und Beiträge zu Problemen der Wissenschaft und zur Geschichte der Friedrich-Wilhelms-Universität zu Berlin* (1960), pp. 21–33.

KUVACIC, IVAN. "Scientific and Technical Progress and Humanism," *Praxis* 5 (1969): 181–84.

KWANT, REMY C. *Philosophy of Labor*. Pittsburgh: Duquesne Univ. Press, 1960. Pp. 163. Contains some material on how modern technology has altered the nature of human labor. See also the same author's *Ontmoeting zam Wetanschap en Arbeid* [The meeting of science and labor] (Utrecht: Het Spectrum, 1958). For a good historico-philosophical discussion, see Adriano Tilgher, *Homo Faber: Work through the Ages*, translated by D. C. Fisher (Chicago: Regnery, 1958), which is a reprint of *Work: What It Has Meant to Men through the Ages* (New York: Harcourt, Brace & World, 1930). For more discussion relevant to this problem, see M.-D. Chenu, *The Theology of Work* (Chicago: Regnery, 1963); Thomas Garrigue Masaryk, *How to Work* (New York: Books for Libraries Press, 1969; first published, 1938); Yves Simon, *Work, Society, and Culture* (New York; Fordham Univ. Press, 1971); Frank Tannenbaum, *The True Society, a Philosophy of Labor* (New York: Knopf, 1951); and V. Vigiline, "Metaphysical Dimensions of Work," *Philosophy Today*, vol. 5, no. 2 (Summer 1961). For a more existential analysis, cf. Hannah Arendt, *The Human Condition*. The whole subject of the philosophy of work or labor deserves more study than it has been yet accorded.

LADEIRA, LUIS. "Do fenómeno technológico e suas ambiguidades," *Brotéria: Rivista contemporânea de cul-* *tura* (Lisbon) 81, no. 5 (1965): 450–58.

LALOIRE, M. "Progreso ténico, progreso humano," *Rivista internacional de sociología* 12 (1954): 205–18.

LAMBLE, J. HOSKIN, ed. *Men, Machines, and the Social Sciences*. Bath, Somersetshire: Bath Univ. Press, 1969. Pp. 98. Social and technological changes have transformed the background of industrial relations, exposing the need for new approaches in solving their problems. Papers on this topic from the fourth Bath Conference examine the use of the social sciences in industrial situations.

LA PORTE, TODD R. "Politics and Inventing the Future: Perspectives in Science and Government," *Public Administration Review* 27, no. 2 (June 1967): 117–27. Discusses science-technology and social change, political values, administrative organization, policy process, study of future.

LAPP, RALPH. *The New Priesthood; the Scientific Elite and the Uses of Power*. New York: Harper & Row, 1965. Pp. 244. See especially the chapter entitled "The Tyranny of Technology." Cf. also Don K. Price, *The Scientific Estate* (Cambridge, Mass.: Harvard Univ. Press, 1965), and R. E. Fitch, "Scientist as Priest and Savior," *Christian Century* 75 (March 26, 1958): 368-70, who argues that "the temptations that have perennially plagued the priesthood of religion now beset the new priesthood of science." Note also the author's later *Arms beyond Doubt; the Tyranny of Weapons Technology* (New York: Cowles, 1970).

LASCH, CHRISTOPHER. "Birth, Death and Technology: the Limits of Cultural Laissez-Faire," *Hastings Center Report* 2, no. 3 (June 1972): 1–4.

LATIL, PIERRE DE. *Ainsi vivrons'nous demain* [Thus we shall live tomorrow]. Paris: Centurion, 1958. Pp. 210. Popular sociology.

LAUE, THEODORE H. VON. "Modern Science and the Old Adam," *Bulletin of the Atomic Scientists* 19, no. 1 (January 1963): 2–5. Modern science was founded to escape from the "old Adam." But in fact "science too ... is caught in the senseless flux of events, contributing to human irrationality and caprice as much as alleviating it." See also the author's "The Subversive West," *Bulletin of the Atomic Scientists* 21 (May 1965): 25-28.

LEACH, GERALD. "Technophobia on the Left: Are British Intellectuals Anti-Science?" *New Statesman* 70 (1965): 286-87.

LEFÈBVRE, HENRI. *Position: contre les technocrates (en finir avec l'humanité-fiction).* Paris: Gonthier, 1967. Pp. 233.

LEIFER, WALTER, ed. *Man and Technology.* German Opinion on Problems of Today, no. 2. Munich: Hueber, 1963. Pp. 74. Contents: F. Klemm's "The Spiritual Basis of Technical Development," F. Tischler's "Technology—the Herald of Progress," W. Conze's "Historical Landmarks in a Technical Age," A. Hilckman's "Technology—Curse, Blessing or Our Responsibility?" A. Brunner's "The Danger of Technical Thinking," W. Schadewaldt's "Technology and Man," J. Bodamer's "Old People and Technology," A. Schoenknecht's "Space Travel and Belief in God," and P. Jordan's "Television from Mars and Venus?" The articles by Klemm, Hilckman, Brunner, and Schadewaldt are the most important.

LEON, BENJAMIN DE. "Is Science Morally Sterile?" *Bulletin of the Atomic Scientists* 24, no. 5 (May 1968): 54-55. Science has alienated itself from humanity and needs once again to consider the human goals for which it was originally intended.

LEOPOLD, ALDO. *A Sand County Almanac and Sketches Here and There.* New York: Oxford Univ. Press, 1949. Pp. 226. Enlarged ed. published in 1968 as *A Sand County Almanac, with Other Essays on Conservation from Round River.* The concluding essays are a powerful call for an antitechnological "land ethic" and an "ecological conscience."

LEOZ CENDOYA, SANTIAGO. *Ante la segunda revolucion tecnica: productividad, dignidad de la persona y relaciones humanas* [After the second technological revolution: productivity, dignity of the person, and human relations].

Madrid: Ediciones Studium, 1959. Pp. 394.

LERNER, MAX. "Big Technology and Neutral Technicians." Pp. 180-90 in H. Cohen, ed., *The American Culture.* New York: Houghton Mifflin, 1968. For an earlier version, see the author's "Universale Technologie und neutrale Techniker; Ausstrahlungen der voraussetzunglosen Maschine," *Perspektiven* 14 (1956): 138-41.

LERSCH, PHILIPP. *Der Mensch in der Gegenwart.* Basel: Reinhardt, 1955.

———. "Das Bild des Menschen in der Sicht der Gegenwart" [The image of man from the modern perspective], *Universitas; Zeitschrift für Wissenschaft, Kunst und Literatur* 13 (1958): 1-10.

LEVENSTEIN, AARON. "Technological Change, Work, and Human Values," *Social Science* 42, no. 2 (1967): 67-79.

LEWIS, WARREN K. "The Place of Engineering in Society and Civilization," *Technology Review* 71, no. 7 (May 1969): 18-23. How the engineer has given man his leisure and through it his spiritual as well as material wealth. The profit gained by society is the measure of the engineer's true success.

L'homme la technique et la nature. Paris: Rieder, 1938. Pp. 359. Contains, among others, the following articles from *Europe:* G. Friedmann's "Esquisse de quelques problèmes" (pp. 11-21); M. Bloch's "Technique et evolution sociale, réflexions d'un historien" (pp. 29-33); A. Varagnac's "L'Homme et les techniques pré-machinistes" (pp. 51-57); L. Zoretti's "Culture et métier" (pp. 58-66); L. Cheronnet's "Bilan de l'exposition" (pp. 67-96); LeCorbusier's "Espoir de la civilisation machiniste" (pp. 97-104); and A. Spire's "Pour la technique" (pp. 196-212).

LICHTHEIM, GEORGE. "Ideas of the Future," *Partisan Review* 33, no. 3 (Summer 1966): 396-410. Discusses the evolution of political institutions in modern times and the problem of applying a philosophy of history derived from this evolution to a prediction of the future.

———. "Technocrats vs. Humanists,"

New York Review of Books (October 9, 1969), pp. 51-56. An analysis (in the form of a review of seven recent French books) of the current political situation in France — with an emphasis upon the way in which technocrats have outmaneuvered the humanists of the left and the consequent problem both Marxists and liberals face in the technological society.

LINKE, WERNER. *Technik und Bildung; Möglichkeiten und Grenzen im Umgang mit der Technik* [Technology and education; possibilities and limits of education in association with technology]. Heidelberg: Quelle & Meyer, 1961. Pp. 209.

LIPHSHITZ, I. N. "The Social Roots of Technical Invention, a Critical Study of the Recent Literature," *Journal of the Patent Office Society* 17 (1935): 927-40.

LOESER, FRANZ. "Old and New Types of Futurology and Their Influence on Ethics." In John William Davis, ed., *Value and Valuation: Axiological Studies in Honor of Robert S. Hartman.* Knoxville: Univ. Tennessee Press, 1972.

LOUBET, EMILE. *L'honnête homme au siècle de l'ingenieur* [The honest man and the century of the engineer]. Paris: Editions Magnard, 1960. Pp. 31. A popular presentation which works from reflections on humanistic and technical culture, through a brief historical survey of the attitudes of society toward technology, to an argument which is summed up by the author in the following words: "Rabelais has already remarked that 'Science without conscience is the ruin of the soul'; but technique without conscience will be the ruin of humanity." — W. C.

LOVSKY, F. "La technique ou l'enjeu du siècle, "*Foi et vie* 53 (1955): 556-62. An article on Ellul's book of the same title, translated into English as *The Technological Society.*

LOWI, THEODORE J. *The End of Liberalism: Ideology, Policy, and the Crisis of Public Authority.* New York: Norton, 1969. Pp. 332. Contains a brief discussion of the evils of technology, arguing that they can be remedied by

the author's proposals for a "juridical democracy."

LUCHTENBERG, PAUL. "Vom Beitrag der Technik im Werden der Kultur" [On the contribution of technology to the evolution of culture], *VDI-Nachrichten* 12, no. 14 (1958): 5-6.

———. "Vom Anteil der Technik an den Wandlungen des Menschenbildes" [On the part of technology in the transformations of the image of man], *Convent; Akademische Monatsschrift* (Mannheim) 14 (1963): 162-63. Also in Ingrid Schindler, ed., *Pädagogisches Denken in Geschichte und Gegenwart* (Ratingen, 1964), pp. 315-30.

LUIJPEN, W. A. "Technocracy and Philosophy," *Science Counselor* 24 (March 1961): 6-7.

LYSTAD, MARY H. *Social Aspects of Alienation: an Annotated Bibliography.* Prepared for National Institute of Mental Health. Public Health Service Publication, no. 1978. Washington, D.C.: Government Printing Office, 1969. Pp. 92. A valuable survey of one type of socio-psychological problem in technological society. See especially section 2, "The Alienated Worker," for some in-depth annotations on important studies of alienation influenced by machines.

MACBRIDE, ROBERT O. *The Automated State; Computer Systems as a New Force in Society.* Philadelphia: Chilton, 1967. Pp. 407. Examines probable effects of the computer on jobs, business, and personal decisions. Argues that "the real future of the machine will be determined neither by technical potentials nor by its short-term profitability as a business tool, but rather by the degree to which its capabilities can be applied to the social and economic imperatives of the 60s and 70s." Four appendices on (1) the Ad Hoc Committee on the Triple Revolution; (2) report of the National Committee on Technology, Automation, and Economic Progress; (3) national data center proposals; and (4) a statement by Senator Edward Kennedy.

MCCARTHY, CALLUM. "Making the Best Use of Science," *New Scientist* 41 (February 27, 1969): 460-62. The interaction of science, technology, and society makes the practicing scientist into a technologist. Scientists are led astray by neglecting their social environment.

McCAULL, J. "The Politics of Technology," *Environment* 14, no. 2 (March 1972): 2-10. Discusses questions concerning the introduction of scientific information into social and political areas; examines the relevance of GNP as a measure of economic and social well-being and describes interactions among technology, GNP, and national policy.

MacDONALD, GORDON J. F. "The Modification of Planet Earth by Man," *Technology Review*, vol. 72, no. 1 (October-November 1969). "A summary of recent technological advances in understanding the planet earth leads to the following generalizations with regard to how man alters his environment: (1) Large-scale, man-made but frequently inadvertent, changes . . . are taking place in the physical environment. (2) The magnitude of the changes produced by man is of the same order of magnitude as that caused by nature. . . ."

McHALE, JOHN, *Future of the Future*. New York: Braziller, 1969. Pp. 332. Although philosophical questions are raised, this work is more concerned with the specifics of overpopulation, increased technological activity, and ecology.

MacKAY, DONALD M. "Machine and Societies." In Gordon E. W. Wolstenholme, ed., *Man and His Future; a Ciba Foundation Volume*. Boston: Little, Brown & Co., 1963.

MACKENSEN, LUTZ. *Sprache und Technik; Zwei Vorträge* [Language and technology; two essays]. Luneburg: Heiland-Verlag, 1954. Pp. 51. This is merely concerned with how to write good technical German. — J. V.

MADDOX, J. "Doomsday Men," *Encounter* 36 (January 1971): 64-68. Review article on A. Toffler's *Future Shock*, G. R. Taylor's *The Doomsday Book*, R. Dubos's *Reason Awake—Science for Man*, R. B. Fuller's *Utopia or Oblivion*,

S. Zuckerman's *Beyond the Ivory Tower*, and H. S. Rose's *Science and Society*.

"Man and Machine." Part VI in Patrick Gleeson, ed., *America, Changing . . .* (Columbus, Ohio: Merrill, 1968). Contains L. Mumford's "The First Megamachine," Handlin's "Man and Magic: First Encounters with the Machine," G. Piel's "Our Industrial Culture," H. Arendt's "Man's Conquest of Space," N. Wiener's "Some Moral and Technical Consequences of Automation," A. Samuel's "Some Moral and Technical Consequences of Automation: a Refutation," S. P. R. Charter's "The Computerized Intellectuals."

MANIERI, MARIA ROSARIA. *Umanesimo e civiltà neotecnica*. Saggi de *Il Protagora*, 2. Naples: Editons Glaux, 1969. Pp. 127.

MANSFIELD, EDWIN. *The Economics of Technological Change*. New York: Norton, 1968. Pp. 257. See also the companion volume, *Industrial Research and Technological Innovation; an Econometric Analysis* (New York: Norton, 1968).

MANSFIELD, EDWIN, JOHN RAPOPORT, JEROME SCHNEE, SAMUEL WAGNER, and MICHAEL HAMBURGER. *Research and Innovation in the Modern Corporation*. New York: Norton, 1971. Pp. 239. "New industries are born from the matching of new technologies to ever-expanding consumer wants and needs while older industries must adapt to changes in the marketplace. The mechanism and style of management also assume new characteristics, for the impact of technological change goes beyond the shop floor to the boardrooms of corporations large and small. For a decade and a half, economists have been engaged in an intensive effort to expand our knowledge of how new processes and products are invented, commercialized, and accepted. . . . The main focus of this book is on the individual firm and its attempt to develop and apply new technology. Utilizing their original studies—and placing particular emphasis on the drug industry—the authors consider the total cost and time involved in product innovation, the extent to which the largest firms tend to be innovators, the lag between invention and commercial application, and many other topics essential to effective management of the modern corporation." — Publisher's note.

MANSTEIN, BODO. *Im Würgegriff des Fortschritts* [In the choking grasp of progress]. Frankfurt: Europäische Verlagsanstalt, 1961. Pp. 500. Journalistic.

———. "Vom Terror der Technik." Pp. 438–94 in Karlheniz Deschner, ed., *Das Jahrhundert der Barbarei*. Munich, 1966.

"Man versus Technology." Part III in R. E. A. Shanab and G. J. Weinrath, eds., *Present Day Issues in Philosophy*. Dubuque, Iowa: Kendall Hunt, 1971. Includes K. Marx's "Machines and Labor," B. Russell's "Demonology and Science," N. Rescher's "The Allocation of Exotic Medical Lifesaving Therapy," B. Commoner's "Science and the Sense of Humanity," D. F. Sly's "Technology and Population Growth" (original), and J.-M. Levy-Leblond's "Address to the Academie des Sciences."

MARCSON, SIMON, ed. *Automation, Alienation and Anomie*. New York: Harper & Row, 1970. Pp. 479. "This volume is an interdisciplinary reader devoted to analyses of the technical and social problems generated by advanced technology and automation of the means of production. The selections are topically ordered under six headings: technology as an aspect of social change; principles of the technology of automation; organizational changes resulting from automation; impact of automation on industrial relations systems; impact of automation on management and occupational structure; industrialization, automation, alienation, and anomie." — W. H. Truitt, *Technology and Culture* 12, no. 1 (January 1971): 134. The review goes on to argue that, although of historical interest, the essays presented here do not discuss the problem of alienation in its most im-portant contemporary form. "The first manifestations of alienation are only now beginning to appear not in the work force, but among students and the "intelligentsia," not over issues of wages and labor, but over the quality of life, social priorities, etc."

MARINE, GENE. *America the Raped: the Engineering Mentality and the Devastation of a Continent*. New York: Simon & Schuster, 1969. Pp. 312.

MARKOVIĆ, MIHAILO. "Man and Technology," *Praxis* 2, no. 3 (1966): 343–52.

MARSAK, LEONARD MENDES, ed. *The Rise of Science in Relation to Society*. New York: Macmillan, 1964.

MARTINO, J. P. "Science and Society in Equilibrium," *Science* 165 (August 22, 1969): 769–72.

MARUYAMA, MAGOROH. "Toward Nonhierarchical Administration through Computers," *Cybernetica* 11, no. 2 (1968): 99–110. See also the author's "A Democracy Model Based on a Non-Western Epistemology," *Cybernetica* 12, no. 4 (1969): 214–20.

MASSIAH, GEORGES. "La technique: facteur déterminé par l'evolution de la société contemporaine," *Science* (Paris) 48 (1967): 44–47.

MAULNIER, THIERRY. "Dangers de la civilisation technique," *Revue de défense nationale* 24, no. 10 (1968): 1391–1405.

MAYZ VALLENILLA, ERNESTO. "Las categorías técnicas y la alienación" *Revista nacional de cultura* (Caracas) 30 (April–June 1969): 33–42.

MEAD, MARGARET, ed. *Cultural Patterns and Technical Change*. A Manual from the Tensions and Technology Series. New York: New American Library, 1955. Pp. 352. Argues for the study of the adaptation of contemporary primitives to new technologies to find models for the same process in Western civilization. Good bibliography. See also the author's *New Lives for Old: Cultural Transformation—Manus, 1928–1953* (New York: Morrow, 1956).

MEADOWS, PAUL. *The Culture of Industrial Man*. Lincoln: Univ. Nebraska Press, 1950. Pp. 216.

MEESSEN, M. "Mensch und Technik," *Neue Giesserei* 43 (1956): 615–19.

MEIER, HARRY KJELD. *I guder! Nogle studier over et palaeoteknisk kulturmonster i det tyvende arhundrede, umiddelbart for den endelige forfaldsperiode* [You, the gods! Some studies about a palaeontechnical culture-monster in the 20th century, just before the final period of decay]. Copenhagen: n.p., 1959. Pp. 157.

MEISSNER, FRITZ. *Revolution durch die Technik? Neue Formen durch die Technik* [Technological revolution? New forms through technology]. *Mensch und Arbeit,* supplement 1. Vienna, 1947. Pp. 78. Revolutions are political not technological. Technology does, however, have an impact on society; and Meissner considers the effect of the Industrial Revolution and automation on the working class and the position of technology within the framework of political and sociological radicalism and conservatism. Gives, in essence, a Marxist interpretation of the function of technology in and for society. — R. J. R.

MELMAN, SEYMOUR. "Who Decides Technology?" *Columbia University Forum* 11, no. 4 (Winter 1968): 13–19. Popular sociology.

MELVIN, BRUCE L. "Science and Man's Dilemma," *Science* 103 (March 1, 1946): 241–51. A pessimistic assessment with references to the use of nazism and fascism made of technology.

MENNICKEN, PETER. *Die Technik im Werden der Kultur* [Technology in the development of culture]. Wolfenbüttel: Wolfenbütteler Verlagsanstalt, 1947. Pp. 152.

MERCIER, ANDRE. "Science and Responsibility: Part I, The Theoretical Problem of Responsibility," *Studi internazionali di filosofia* 1 (1969): 5–76.

———. "Science and Responsibility: Part II, The Problem of Science and Responsibility in Practical Life," *Studi internazionali di filosofia* 2 (1970): 65–115. Argues that neither religion nor traditional morals can serve as an adequate guide to science and technology. To replace these outmoded concepts, he proposes the idea of authenticity, or living up to what it means to be human.

MERTON, ROBERT K. "A Note on Science and Technology in a Democratic Order," *Journal of Legal and Political Sociology* (October 1942), pp. 115–26.

———. "The Machine, the Worker, and the Engineer," *Science* 105 (January 3, 1947): 79–84.

MEYER, HERMANN J. *Die Technisierung der Welt; Herkunft, Wesen und Gefahren* [The technicization of the world; origin, essence and dangers]. Tubingen: Niemeyer, 1961. Pp. 300.

———. "Die Technik im Selbstverständnis des heutigen Menschen" [Technology in modern man's understanding of himself]. Pp. 750–82 in Richard Schwarz, ed., *Menschliche Existenz und moderne Welt: ein internationales Symposium zum Selbstverständnis des heutigen Menschen.* Vol. 1. Berlin: de Gruyter, 1967. Cf. also Bogdan Suchodolski's "Traditionen und Perspektiven des Bündnisses der Technik mit der Geisteswissenschaft," in vol. 2, pp. 66–86.

MICHAEL, DONALD N. "Speculations on the Relation of the Computer to Individual Freedom and the Right to Privacy," *George Washington Law Review* 33 (October 1964): 270–86.

———. *The Unprepared Society: Planning for a Precarious Future.* New York: Basic, 1968. Pp. 132. "We are almost certain to face disaster if we don't plan; we are almost certain to increase the likelihood of having a better world if we plan well. But we are also almost certain to be in deep trouble even with planning because our best plans will be developed and fostered by limited human beings picking and choosing among limited knowledge, very often ignorant of the extent of their own ignorance" (p. 91). "One of the first books of the future to hit squarely on the problem of 'values,' it remains the best and rests firmly on my own list of 'essential first books' on futuristics." — J. A. Dator, *Technology and Culture* 12, no. 1 (January 1971): 141. See also Donald N. Michael, ed., *The Future Society* (Chicago: Aldine, 1970), a collection of essays by M. M. Weber, R. A. Skedgell, W. G. Bennis, F. Davis, K. Keniston, H. S. Rowen, M. Pilisuk, M. Weidenbaum, L. J. Duhl, and H. P. Gouldner, which sees "the problem of future society as two fold: deriving an ethic appropriate for the imperatives to act in a situation of unprecedented complexity, or unprecedented technological potency, with the awareness that unanticipated secondary and tertiary consequences of our actions may very well overwhelm original intents; and developing means for evolving such an ethic in the presence of traditional moralities which for many years will guide the actions of many people."

―――. "Technology and the Human Environment," *Public Administration Review* 28, no. 1 (January–February 1968): 57–60. Argues that, "if we are to cope with the impact of technology, we must have major social inventions as well as hardware inventions, and the social inventions will . . . alter the very core of our way of life, values, beliefs, and aspirations over a fifty year perspective." Discusses what social technology or engineering is, its future growth, the coming pressures on it, the problems of forecasting studies, and the place of social technology in studies of the human environment.

MIGEON, HENRI. *Le monde après 150 ans de technique* [The world after 150 years of technology]. New ed. Paris: Editions d'Organisation, 1958. Pp. 138. Popular.

MILKEWITZ, HARRY. *El hombre y la técnica.* Montevideo: Editorial Alfa, 1964. Pp. 203.

MILLER, ARTHUR SELWYN. "The Rise of the Techno-Corporate State in America," *Bulletin of the Atomic Scientists* 25, no. 1 (January 1969): 14–19. Critical problems in the preservation of humanistic values have arisen due to the emergence of the techno-corporate synthesis.

MILLER, NATHAN. *Selected Readings in the Study of Technology and Society.* 2d ed. Pittsburgh: Carnegie Institute of Technology, 1941. Pp. 480. A text-book for engineers; still remains a valuable collection—with discussion questions, references, and notes. Articles by J. Dewey, J. D. Bernal, W. F. Ogburn, S. Chase, L. White, jr., C. Beard, H. Ford, W. N. Polakov, Berle and Means, L. Mumford, etc.

MIROSLAV, TOMS. "Der technische Fortschritt und das Wachstumsmodell," *Futurum* 1 (1968): 275–316.

MISHAN, E. J. *Technology and Growth: the Price We Pay.* New York: Praeger, 1969. Pp. 193. Published in England under the title *Growth: the Price We Pay,* this is a more popular version of the author's *The Costs of Economic Growth* (New York: Praeger, 1967). Argues that there is an indiscriminate pursuit of technological progress in the name of economic growth and human satisfaction. But human values are in jeopardy if some evaluation of technological advance is not made part of the innovation process. The consequent costs to human life, the environment, and social priorities are discussed within an economic framework. Cf. the negative review by R. W. Weiss of *The Costs of Economic Growth* in *Journal of Political Economy* 77, no. 1 (January–February 1969): 138–40; with the reply by W. A. Weisskopf, *Journal of Political Economy* 77, no. 6 (November–December 1969): 1036–39.

―――. "Futurism: and the Worst That Is Yet to Come," *Encounter* 36 (March 1971): 3–9.

MITRANI, NORA. "Réflexions sur l'opération technique, les techniciens et les technocrates," *Cahiers internationaux de sociologie* 19 (1955): 157–70.

―――. "Ambiguïté de la technocratie," *Cahiers internationaux de sociologie* 30 (1961): 101–14.

MITSUMOTO, YUASA. "The Scientific Revolution and the Age of Technology," *Journal of World History* 9, no. 2 (1965): 187–207.

Modern Technology and Patterns of Living: a Reference List. Automation Programme Document no. 9. Geneva: International Labour Office, 1969. Pp. 39. Good annotated bibliography on the social implications of modern technology. International in scope.

MONCRIEF, LEWIS. "The Cultural Basis for Our Environmental Crisis," *Science* 170 (October 25, 1970): 508–12.

MONZEL, NIKOLAUS. "Technik und Gemeinschaft" [Technology and community], *Jahres- und Tagungsbericht der Görres-Gesellschaft* (Cologne) (1954), pp. 19–35.

MOORE, J. A. *Science for Society: a Bibliography.* Prepared for Commission on Science Education, American Association for the Advancement of Science. 2d ed. Washington, D.C.: Education Department, American Association for the Advancement of Science, 1971. Pp. 76. Contains approximately 4,000 references, some briefly annotated to the literature. Entries classified in five major categories—General; Resources; Science; Technology; Society; Population; Health. One and three are most relevant to philosophy.

MOORE, WILBERT. *Industrial Relations and the Social Order.* Rev. ed. New York: Macmillan, 1951. Pp. 660. Reviews the literature and problems of industrial organization in different cultures.

———, ed. *Technology and Social Change.* Chicago: Quadrangle, 1972. A collection of articles from the *New York Times Magazine.*

MORRICE, HERBERT A. *The Chasm; the Protest of an Engineer.* London: Alliance, 1945. Pp. 121. "A civilisation that depends for its existence on the achievements of scientists and engineers is being flagrantly mismanaged by men whose training has not enabled them to understand what science and engineering could and should do" (p. 7). The "chasm" equals the divorce of theory from practice; its remedy is socialism.

MORRISON, ROBERT S. "Science and Social Attitudes," *Science* 165 (July 11, 1969): 150–56. A brief but insightful discussion of recent and growing disenchantment with science and technology. Calls for scientists to reevaluate their own attitudes toward their profession and its relationship to society.

MUCK, H. "Die unbewältigte Technik" [Uncontrolled technology], *Grosse Entschluss* 22 (March 1967): 264–67.

MUCKERMANN, FREDERICK. *Der Mensch im Zeitalter der Technik.* Lucerne: Stocker, 1943. Pp. 342.

MUCKERMANN, HERMANN. *Anthropolgie und Sozialethik im Zeitalter der Technik* [Anthropology and social ethics in the age of technology]. Berlin: Morus-Verlag, 1950. Pp. 29.

———. "Der Humanismus des Ingenieurs," *Humanismus und Technik* 1, no. 1 (1953): 25–34.

———. "Lenkende Humanitas— dienende Technik" [Humanity in control—technology in its service], *Humanismus und Technik* 1, nos. 3–4 (1953): 127–34.

MUELLER, H. F. "Technik als Schicksal" [Technology as fate], *Praktische Energiekunde* (Karlsruhe) 9, no. 5 (1961): 97–107.

MUKERJI, D. P. "Mahatma Gandhi's Views on Machines and Technology." Pp. 63–75 in Jean Meynaud, ed., *Social Change and Economic Development.* Paris: UNESCO, 1963. First printed in *International Social Science Bulletin* 6, no. 3 (1954): 411–24.

MÜLLER, M. "Filosofía-ciencia-técnica o la filosofía en la edad de las ciencias" [Philosophy-science-technology or philosophy in the age of the sciences], *Convivium* (Barcelona), vol. 8 (July–December 1959).

MUÑOZ SOLER, RAMÓN PASCUAL. *Gérmenes del futuro en el hombre; hacia una individualidad expansiva y participante* [Seeds of the future of man; toward the expansion and participation of individuals]. Buenos Aires: Ediciones Arayú, 1966. Pp. 197.

MURARO, ROSE MARIE. *A automaçao e o futuro do homem* [Automation and the future of man]. Petropolis: Ediçâo Vozes, 1968. Pp. 157.

MURPHY, EARL FINBAR. *Governing Nature.* Chicago: Quadrangle, 1967. Pp. 333. A knowledgeable and multidimensional account of man's attempt to manage natural resources with a central message: that mankind is destroying the world in which he lives because he is provided with no legal, economic, or institutional reasons for doing otherwise. Rather, the motives are all on the other side. Question: Can legal, economic, or institutional brakes be applied simply out of necessity without positive or philosophical reasons?

MUTHESIUS, EHRENFRIED. *Der letzte Fussgänger; oder, die Verwandlung unserer Welt* [The last pedestrian; or, the transformation of our world]. Munich: Beck, 1960. Pp. 207. Popular.

NEGROPONTE, NICHOLAS P. "Toward a Humanism through Machines," *Technology Review* 71, no. 6 (April 1969): 44–54.

NELSON, RICHARD, MERTON PECK, and EDWARD KALACHEK. *Technology, Economic Growth, and Public Policy.* Washington, D.C.: Brookings Inst., 1967. Pp. 238. "This book explores the relations among research, development, innovation, and economic growth; considers the manner in which the economy adapts to technical change and the problems encountered in the processes of adaptation; and recommends several policy changes designed to encourage technological change consistent with other public policy objectives." See also R. Nelson, "The Economics of Invention: a Survey of the Literature," *Journal of Business* 32 (1959): 101–27.

NEUHÄUSLER, ANTON. "Das Ende des technischen Zeitalters" [The end of the technological age], *Hochland* 56 (1963–64): 345–54.

NICHOLS, W. A. "Technocrats and Society," *Technologist* 2, no. 1 (Winter 1964–65): 78–88. A good bibliographical review of the literature, including: Burnham's *The Managerial Revolution;* Whyte's *The Organization Man;* Berel and Mean's *The Modern Corporation and Private Property;* and Florence's *Ownership, Control, and Success of Large Companies.*

NIEBURG, HAROLD L. *In The Name of Science.* Chicago: Quadrangle, 1966. Pp. 431. Basically politics of science.

NIEDEN, ERNST ZUR. —*Und trotzdem Zeit* [—And in spite of this time]. Baden-Baden: Lutzeyer, 1960. Pp. 64. Popular.

NONENBRUCH, FRITZ. *Politik, Technik und Geist.* Munich: Hoheneichen-Verlag, 1939. Pp. 327. The author states that this is "a personal work" of a man "conscious of his life's task, to grow into National Socialism." —R. J. R.

"Non-Scientists Dissect Science," *Impact of Science on Society,* special issue, vol. 19, no. 4 (October–December 1969). Contents: "Flawed Science, Damaged Human Life" (an interview with Robert Graves), T. J. Mboya's "Technology in the Development of Africa," E. S. Muskie's "The Challenge of the Technological Revolution," "The Creative Spirit in Art and Science" (an interview with J. Miró), M. A. Asturias's "A Dream of Waters Glittering with Stars," G. T. Pecson's "The Good and Bad of Science and Technology," P. Emmanuel's "A Poet's View of Science," C. H. Malik's "Ten Limitations of Natural Science," S. Gerasimov's "Science, Morality, and the Humanities," and M. Lubis's "The Frustrations of Science and Technology." Cf. "The Scientists Riposte," *Impact of Science on Society,* vol. 20, no. 2 (April–June 1970), which contains A. Koestler's "Humility and Duty in Science" (an interview); H. A. Foecke's "Engineering in the Humanistic Tradition," C. H. Waddington, N. Ginsburg, T. F. Malone, K. Mellanby, and J. S. Weiner's "Five Scientists View the Impacts of Technology"; M. Holub's "Science in the Unity of Culture"; and D. Bohm's "Fragmentation" in Science and in Society."

NORDHAUS, WILLIAM D. *Invention, Growth, and Welfare: a Theoretical Treatment of Technology.* Cambridge, Mass.: M.I.T. Press, 1969. Pp. 168. "This book represents the most systematic attempt to date to assimilate a theory of invention into the framework of received economic theory." —C. Freeman, *Technology and Culture* 12, no. 2 (April 1971): 375. Basically a book on theoretical economics, however.

NUNES, A. SEDAS. "Técnica e sociedade," *Brotéria: revista contemporânea de cultura* (Libson) 80, no. 2 (1965): 155–66.

O'DEA, THOMAS. "Technology and Social Change: East and West," *Western Humanities Review*, vol. 13 (1959).

OFFE, CLAUS. "Technik und Eidimensionalität," in J. Habermas, ed., *Antworten auf Herbert Marcuse*. Frankfurt: Suhrkamp, 1968.

OFTINGER, K. "Konfrontation der Technik mit dem Recht" [Confrontation of technology with the law], *Schweizerische technische Zeitschrift* 64, no. 47 (1967): 997–1005.

OGBURN, WILLIAM F. *Technology and International Relations*. Chicago: Univ. Chicago Press, 1949. Pp. 202. Contents: W. F. Ogburn's "Introductory Ideas on Inventions and the State," "The Process of Adjustment to New Inventions," and "Aviation and International Relations"; H. Hart's "Technology and the Growth of Political Areas"; A. P. Usher's "The Steam and Steel Complex and International Relations"; W. T. R. Fox's "Atomic Energy and International Relations"; R. Leigh's "The Mass-Communications Inventions and International Relations"; B. Brodie's "New Techniques of War and National Policies"; and Q. Wright's "Modern Technology and the World Order."

O'NEILL, JOHN JOSEPH. *Engineering the New Age*. New York: Washburn, 1949. Pp. 320. Calls for the development of a social consciousness by science and technology. Conceives of technology as conscious imitation of the processes of nature; engineering is man's way of living in harmony with a mechanistic cosmos. See especially chaps. 1, "Partnership in the Cosmos"; 11, "Techniques of Control"; and 12, "Origin of Engineering Techniques."

ORLANS, HAROLD. "Science and the Polity, or How Much Knowledge Does a Nation Need?" *Proceedings of the American Philosophical Society* 115, no. 1 (February 17, 1971): 4–9.

ORLEANS, L. A., and R. P. SUTTMEIR. "The Mao Ethic and Environmental Quality," *Science* 170 (December 11, 1970): 1173–76. As an ethic of frugality, of "doing more with less," Maoism may have something to offer citizens of Western industrial societies beset with ecological problems.

OTTO, MAX C. *Science and the Moral Life*. New York: New American Library, 1949. Pp. 192. This selection of the writings of Otto on such topics as "Realistic Idealism" and "Scientific Humanism" is essentially an argument for that typically American attitude toward the world generally known as pragmatism. Although now over twenty years old, these essays still represent what must be called the prevailing attitude, even in academic institutions devoted to the study of technology and society.

PAESCHKE, HANS. "Moralia der Technik," *Merkur* 11, no. 9 (1957): 871–85.

PAGGIARO, LUIGI. "C'e um progresso nella technologia contemporania?" *Sapienza* 23 (1970): 465–68.

PANIKKAR, RAYMOND. *Técnica y tiempo*. Buenos Aires: Columba, 1967. Pp. 71.

PAPPENHEIM, FRITZ. "Technology and Alienation." Pp. 37–44 in *The Alienation of Modern Man; an Interpretation Based on Marx and Tönnies*. New York: Monthly Review Press, 1959. Simply a summary of the contemporary indictments of technology. Chap. 5, "Retrospect and Outlook," contains a reply arguing that technology is neutral (pp. 105–9).

PARSONS, HOWARD L. "Technology and Humanism," *Praxis* 5, nos. 1–2 (1969): 164–80. An analysis of the relation between technology (as the theory and practice of technique) and ideas of man's generic fulfillment in Fromm, Montagu, Morris, Sheldon, Marx, Veblen, Marcuse, and Lenin. Conclusion: For man's fulfillment, technology must be humanistically, democratically controlled.

PARSONS, TALCOTT. "The Impact of Technology on Culture and Emerging New Modes of Behavior," *International Social Science Journal* 12, no. 4 (1970): 607–27. Argues that, despite the critics, technology does increase freedom.

PAUL, LESLIE. "The Effect of Technology on Man," *Hibbert Journal* 55 (October 1956): 20–29. Technology is necessary to man; therefore the criticisms of Ruskin, Morris, Bernanos, etc., are foolish. Yet the full development of the technological attitude (or "method") does create some problems.

PAVESE, ROBERTO. "Valori umani e progresso tecnico, oggi," *Sapienza* 18 (1965): 100-104.

PENDERSON, OLAF. *Mennesket og teknikken; Kommentarer til Atomalderens Problemer* [Mankind and technology: a commentary on the problems of the atomic age]. Copenhagen: Frost-Hansen, 1950. Pp. 172.

PEPERZAK, A. *Techniek en dialoog* [Technology and dialogue]. Delft: Waltman, 1967. Pp. 16.

PERRUCCI, ROBERT, and JOEL E. GERSTL, eds. *The Engineers and the Social System.* New York: Wiley, 1969. Pp. 334. Original essays concerned with the engineering profession in a social context. Includes bibliography.

PERTICONE, GIACOMO. "La sfida technologica" [The challenge of technology], *Rivista internazionale di filosofia del diritto,* Series 4, 46, no. 1 (1969): 95.

PETER, WALTER G., III. "Ethical Perspectives in the Use of Genetic Knowledge," *Bio-Science* (November 15, 1971), pp. 1133-37.

PFENDER, MAX. "Der Mensch und seine Technik," *Humanismus und Technik* 4, no. 1 (1956-57): 16-27. Articles of the same title in *Technische Rundschau* (Bern) 48, no. 25 (1956): 1-3; and *VDI-Zeitschrift* 98, no. 23 (1956): 1315-21. By a German engineer.

———. "Die Technik," *Architekt BDA* (Essen) 9 (1960): 19 ff. Also in *Bauwelt* (Berlin) 51 (1960): 41-44.

———. "Die Technik, Verrat oder Dienst an der Menschheit" [Technology, betrayal or service to mankind], *Physikalische Blätter* 19 (1963): 289-95.

PFLICKE, G. "Die Rolle des sozialistischen Rechts bei der Förderung der technologischen Entwicklung" [The role of socialist justice in the promotion of technolgical development], *Staat und Recht* 10, nos. 11-12 (1961): 2078-97.

PHILIPSON, MORRIS, ed. *Automation: Implications for the Future.* New York: Vintage, 1962. Pp. 456.

"Philosophy and the Future of Man." Proceedings of the American Catholic Philosophical Association, vol. 42. Washington, D.C.: Catholic Univ.

America Press, 1968. Includes: M. J. Adler's "The Challenge to the Computer," J. K. Feibleman's "The Human Future from Scientific Findings," W. G. Pollard's "The Key to the Twentieth Century," P. T. Durbin's "Philosophy and the Futurists," and L. K. Dupre's "Secular Man and His Religion."

PICHT, GEORG. *Technik und Überlieferung; die Überlieferung der Technik, die Autonomie der Vernunft und die Freiheit des Menschen* [Technology and tradition; the tradition of technology, the autonomy of reason and the freedom of man]. Hamburg: Furche-Verlag, 1959. Pp. 26.

———. *Prognose, Utopie, Planung; die Situation des Menschen in der Zukunft der technischen Welt* [Prognosis, utopia, planning; the situation of man in the future of a technical world]. Stuttgart: E. Klett, 1967. Pp. 62. See also the author's *Mut zur Utopie: die grossen Zukunftsaufgaben* (Munich: Piper, 1969).

PIEL, GERARD. *Science in the Cause of Man.* New York: Knopf, 1961. Pp. 298. Assorted popular essays. Five comprise the Walgreen Lectures in 1955 and trace the historical development of the connection between American democracy and the growth of science.

———. "Technology and Democracy." Pp. 45-60 in Edward Reed, ed., *Challenges to Democracy: the Next Ten Years.* New York: Praeger (for Center for the Study of Democratic Institutions), 1963. Popular sociology. Piel is the editor of *Scientific American.*

PIERCE, J. R. "Technology and Freedom," *New Scientist* 25 (March 11, 1965): 650-51. There are two conflicting tendencies in technology: enhancing the power of central authorities, the other giving more freedom to the individual.

PIETSCH, MAX. "Der Kampf um die technische Führung in der Welt" [The struggle over technological leadership in the world], *Universitas; Zeitschrift für Wissenschaft, Kunst und Literatur* 13 (1958): 1257-65.

———. *Die Industrielle Revolution: Von Watts Dampfmaschine zu Automation und*

Atomkernspaltung. Freiburg: Herder, 1961. Pp. 187.

PILLEY, JOHN. "The Humanities and the Technologist," *Universities Quarterly* 11, no. 2 (1957): 127-38.

"Planning and Designing for the Future: the Breakthrough of the Systems Approach," *Futures* 1, no. 5 (September 1969): 440-44. Describes a common approach to the future found in R. Dubos's *So Human an Animal* (New York: Scribner's, 1968), J. Forrester's *Urban Dynamics* (Cambridge, Mass.: M.I.T. Press, 1969), and A. Peccei's *The Chasm Ahead* (New York: Macmillan, 1969). This involves three basic ideas: (1) The future of man and society are to be dealt with in terms of a system which links them to the environment shaped by nature, technology, or social development. (2) These systems form complex dynamic wholes, which implies that they are high-order, multiple-loop, nonlinear, feedback structures. (3) Actively shaping the future implies changing the structures of these systems, not just the variables—that is, it involves systems engineering.

PLATE A. *Het diepste punt, een cultuurbeechouwing* [The deepest point, reflections on culture]. Leiden: Stenfert Kroese, 1952. Pp. 88.

PLESSNER, HELMUTH. "Technik und Gesellschaft in Geganwart und Zukunft" [Technology and society in the present and the future], *Universitas; Zeitschrift für Wissenschaft, Kunst und Literatur* 24 (1969): 1241-47.

POHL, BRUNO. "Das Rätsel der Technik" [The riddle of technology], *Reformatio* 9 (1960): 632-40.

POHLE, WOLFGANG. "Technik und Politik," *VDI-Zeitschrift* 99, no. 23 (1957): 1130-39.

POLAKOV, WALTER N. *The Power Age; Its Quest and Challenge.* New York: Covici Friede, 1933. Pp. 247. Argues that the Machine Age has given way to the Power Age. "In the Machine Age, the rigid forms and definite connections of machine parts had their parallel in static, dogmatic forms of ethical standards. The fixed character of machines and supposedly unchangeable human nature was not questioned any more than were 'absolute' moral values. On the other hand, the dynamic characteristics of the Power Age, activated by electric current devoid of bulk, permeated with the concept of 'field' manifested in the rapid rate of changes and accelerated flow of production, find their expression in the changed views on spatio-temporal ethical values and the demonstrated actual rapid changes in 'human nature.' "

PONIATOWSKI, MICHEL. *Les choix de l'espoir.* Paris: Grasset, 1970. Pp. 260. Discusses the agricultural, industrial, and scientific revolutions in relation to mankind and projects the political and social developments needed by the end of the 20th century to counteract the dehumanizing effects of these revolutions.

PORSTMANN, WALTER. *Technische Andachten* [Technical meditations]. Hamburg: Fabriknorm, 1956. Pp. 132. Written entirely in lower-case type, which is defended as being most in accord with technology. The primary concern of the essays in this book is the question of standardization, and as such it deals with a subject not often isolated in other books on technology. Might be an informative monograph for someone interested specifically in this subject. The essays are really more like notes of 1-2 pages.—R. J. R.

POSTLEY, JOHN A. *Computers and People.* New York: McGraw-Hill, 1960. Pp. 246. A general book on the history of the development of computers and on their use by management.

POZZO, GIANNI M. "Il tecnicismo e le sue aporie," *Sapienza* 23 (1970): 460-64.

PRAT, HENRI. *Métamorphose explosive de l'humanité.* Paris: Société d'Edition d'Enseignement Supérieur, 1960-61. Pp. 343. Reprinted in abridged form (Paris: Editions Planète, 1966). Pp. 255.

PREHUDA, ROBERT W. *Designing the Future; the Role of Technological Forecasting.* Philadelphia: Chilton, 1967. Pp. 310. Contains annotated bibliography, pp. 287-91. See also the author's *Suspended Animation: the Re-*

search Possibilities That May Allow Man to Conquer the Limiting Chains of Time (New York: Chilton, 1969).

PRICE, DON K. "Purists and Politicians," *Science* 163 (January 3, 1969): 25-31. Argues that the scientific mode of thought is not a fundamental threat to human values. The threat is from "the very pragmatic reductionism which assumes that applications of advanced technology are automatically beneficial, or that we are always justified in granting special concentrations of technological and industrial power freedom from central political authority."

———. "Science and Technology in a Democratic Society: Educating for the Scientific Age." Pp. 21-36 in Kingman Brewster et al., *Educating for the Twenty-first Century.* Urbana: Univ. Illinois Press, 1969. ". . . we cannot think of a value system that is apart from our science, which we can either teach potential scientists independently at the outset of their training, or set up as the governing political standard apart from and superior to science." See also the author's *The Scientific Estate* (Cambridge, Mass.: Harvard Univ. Press, 1965).

———. "Knowledge and Power." Pp. 135-42 in Paul J. Piccard, ed., *Science and Policy Issues: Lectures in Government and Science.* Itasca, Ill.: Peacock, 1969.

Proceedings: "XXI° Congresso Nazionale di Filosofia," *Filosofia* (Turin), vol. 18, no. 1 (January 1967). Contains: V. Somenzi's "Uomini e macchine," U. Spriito's "Cybernetica e biologia," E. Paci's "Obiezioni a 'uomini e macchine,'" A. Vasa's "Osservazioni alla relazione Somenzi," F. Lombardi's "Osservazioni," G. Galogero's "Lettera a Vittorio Somenzi; l'uomo, l'automa, lo schiavo," A. Guzzo's "Le profezie degli scienziati e la loro legittimata," and V. Somenzi's 'Rispote alle osservazioni."

PYKE, MAGNUS. *Slaves Unaware? A Mid-Century View of Applied Science.* London: Murray, 1959. Pp. 208. Ambitious but superficial.

———. *The Science Myth.* New York: Macmillan, 1962. Pp. 179. Points to need for ecological approach to problems of environment that are caused or intensified by technological developments.

QUARG, GÖTZ. *Wider den technischen Kulturpessimissmus* [Against technical-cultural pessimism]. Düsseldorf: Deutscher Ingenieur-Verlag, 1949. Pp. 55. Popular defense of technology. Quarg finds some basis for a negative reaction to technology in postwar Germany. But he defends technology with an appeal to man's striving nature. Scientific technology was historically the result of a failure of "transcendental" reason. Nevertheless, one must be aware that technology brings dangers as well as benefits. Concludes with a religious appeal; faith in the salvation of mankind by technology has its roots in Christian faith.

———. "Leistungen und Aufgaben des Abendlands für die Technik der Menschheit" [Achievements and tasks of the Occident for the technology of humanity], *Studium generale* 15 (1962): 551-60. For other articles pre-1956, see Dessauer's bibliography.

QUINN, FRANCIS X., S.J., ed. *The Ethical Aftermath of Automation.* Westminster, Md.: Newman, 1962. Pp. 270. The real concern of the articles in this volume is not so much ethical problems as problems of economic or social justice—unemployment, job training, etc.

RABINOWITCH, EUGENE. "Scientific Revolution: Man's New Outlook," *Bulletin of the Atomic Scientists* 19, no. 7 (September 1963): 15-18.

———. "Scientific Revolution: the New Content of Politics," *Bulletin of the Atomic Scientists* 19, no. 8 (October 1963): 11-16.

———. "Scientific Revolution: the End of History," *Bulletin of the Atomic Scientists* 19, no. 9 (November 1963): 9-12.

———. "Scientific Revolution: the Beginnings of the World Community," *Bulletin of the Atomic Scientists* 19, no. 10 (December 1963): 14-17. The previous four articles were originally lectures concentrating first on the cosmological and then on the practical and technological implications of modern science. A Spinozistic metaphysics based on Bohr's theory of complementarity is joined to a messianic liberalism based on necessity. A good essay because it reveals what a prominent scientist thinks about the social and political meaning of science-technology.

———. "Science Popularization in the Atomic Age," *Impact of Science on Society* 17, no. 2 (1967): 107–13.

———. "The Mounting Tide of Unreason," *Bulletin of the Atomic Scientists* 27, no. 5 (May 1971): 4–9. "A new system of values and a new code of behavior is needed for human survival and progress in the age of abundant production capacity."

———. "Thoughts for 1972—Living Dangerously in the Age of Science," *Bulletin of the Atomic Scientists* 28, no. 1 (January 1972): 5–8. Argues that mankind must change its sense of values and philosophy to cope with "the new technological apocalypse," whose "horsemen" are the threat of an all-destroying nuclear war, population explosion, and destruction of our natural habitat; presents a strong case for countering the bad side effects of technology with yet more utilization of our technological capabilities to overcome them.

RADHAKRISHNAN, SARVEPALLI. "Die Technik—Diener, nicht Herr" [Technology—servant, not master], *Physikalische Blätter* 18 (1962): 441–44.

RAMO, SIMON. *Century of Mismatch.* New York: McKay, 1970. Pp. 204. How advanced technology can bring a richer, more comfortable, more democratic, and more fulfilling existence to a society that knows how to master its machines.

RANDALL, JOHN HERMAN, JR. *Our Changing Civilization: How Science and the Machine Are Reconstructing Modern Life.* New York: Stokes, 1929. Pp. 362.

RAS, NORBERTO. "Sociedad, technología y desarrollo" [Society, technology and evolution], *Trimestre económico* (Mexico) 34 (April–June 1967): 267–85.

RASMUSSEN, J. P., ed. *The New American Revolution: the Dawning of the Technetronic Era.* New York: Wiley, 1972.

RAUSCH, JÜRGEN. "Die Technik als Provokation des Menschen," *Jahresring;*

ein Querschnitt durch die deutsche Literatur und Kunst der Gegenwart (Stuttgart) 1 (1954): 25–37.

———. *Europa im Zeitalter der unbewältigten Technik* [Europe in the age of uncontrolled technology]. Bremen: Angelsachsen-Verlag, 1956. Pp. 22.

———. *Der Mensch als Märtyrer und Monstrum; Essays.* Stuttgart: Deutsch Verlags Anstalt, 1957. Pp. 253.

RAVETZ, ALISON. "Modern Technology and an Ancient Occupation: Housework in Present-Day Society," *Technology and Culture* 7, no. 2 (Spring 1965): 256–60. "Social factors may inhibit the spread of modern technology to certain occupations, even in societies that widely adopt it in other spheres.... A blatant example is housework."

RAVETZ, JEROME R. *Scientific Knowledge and Its Social Problems.* Oxford: Clarendon, 1971. Pp. 449.

REINHARDT, KURT F. "Streit um die Technik," *Erasmus speculum scientiarum* 10 (1957): 577–80.

REINICKE, ADOLF. "Technik und Menschenführung" [Technology and human leadership], *Taktik Truppenpraxis; Technik und Ausbildung für Offiziere aller Truppen* (Darmstadt), supplement (1959), pp. 669–71. Cf. also Karl Bauer's "Aber Technik und Menschenführung sind nicht dasselbe," ibid., pp. 185–86.

REINTANZ, GERHARD. "Wissenschaftliche-technische Revolution und einige Probleme des Völkerrechts" [Scientific-technolgical revolution and some problems of popular rights], *Staat und Recht* 18, no. 7 (1969): 1074–84.

REISCHOCK, WOLFGANG. *Die Bewältigung der Zukunft: Technische Revolution, polytechnische Bildung* [Mastering the future; technical revolution, polytechnical education]. Berlin: Volk & Wissen, 1966. Pp. 178.

RESEK, ROBERT W. "Neutrality of Technical Progress," *Review of Economics and Statistics* 45 (1963): 55–63.

Revolution der Roboter; Untersuchungen über Probleme der Automatisierung; eine Vortragsreihe [Revolution of robots; inquiries concerning problems of automation; a lecture series]. Munich: Isar Verlag, 1956. Pp. 198. Lectures deliv-

ered at the Arbeitsgemeinschaft So-
zialdemokratischer Akademiker.

RICH, ARTHUR. "Auftrag und Grenze
der Technik" [Mission and limit of
technology], *Schweizerische Bauzeitung*
76, no. 18 (1958): 274-77.

RICHARDS, I. A. "The Technological
Crisis." In *So Much Nearer; Essays toward a World English.* New York: Harcourt, Brace & World, 1968.

RICHTA, RADOVAN. *Člověk a technika v
revoluci našich dnu.* Prague: Čs. společ-
nost pro šiřeni politických a
vědeckých znalosti, 1963. Pp. 85.

———, ed. *Civilisation at the Crossroads.*
Prague: Institute of Philosophy,
Czech Academy of Sciences, 1967.
Translated by Marian Šlingová. 3d
enlarged ed. White Plains, N.Y.: International Arts and Sciences Press,
1969. Pp. 372. From *Civilizace na roz-
cesti* (Prague: Svoboda, 1967). General
argument of this collection of essays is
for a more deliberate political control of science-technology for human
ends. Presents the East European
perspective. See a review of the
French translation by Tom Bottomore, *New York Review of Books* 17,
no. 7 (November 4, 1971): 12- 19. Cf.
also the article on Czechoslovakian
thought on this subject by Nigel Calder, *New Statesman* (August 30, 1968).

———. "Die wissenschaftlich-technische
Revolution und die Alternativen der
modernen Zivilisation," *Futurum* 1
(1968): 173-204. See also the author's
"Technology and the Human Situation" (in Czech), *Filosofický časopis,*
vol. 16, no. 5 (1968), as well as R.
Richta and O. Sulc, "Forecasting and
the Scientific and Technological Revolution," *International Social Science
Journal,* vol. 21, no. 4 (1969).

———. "The Scientific and Technological Revolution and Development of Man" (in Russian), *Voprosy
filosofii* 24, no. 1 (1970): 68-79. A
critique of the conditions of life in
advanced capitalist society followed
by a vision of how technology can be
integrated with socialist society. Argues that science-technology has
made the expansion of all man's abili-
ties a factor in the progress of production; today individual creative de-

velopment is becoming as important
as labor and capital were earlier.

RICHTER, J. *Die Damonie der Technik und
ihre Uberwindung* [The evil of technology and how to vanquish it].
Frankfurt: Klein-schriften-Verlag,
1951. Pp. 39.

RICKMAN, HANS PETER. *Living with Technology.* London: Holder & Stoughton,
in association with Hilary Rubinstein,
1967. Pp. 128.

RIDEAU, E. "Technique et avènement de
l'homme," *Revue de l'action populaire*
(1951), pp. 293-302.

RIEGER, LADISLAU. "Le travail comme
base de la civilisation." Pp. 1048-50
in *Proceedings of the Xth International
Congress of Philosophy; Amsterdam,
1948.* Amsterdam: North-Holland,
1949.

RIENOW, ROBERT, and LEONA RIE-
NOW. *Moment in the Sun; a Report on
the Deteriorating Quality of the American
Environment.* New York: Dial, 1967.
Paperback reprint, 1970. Pp. 286. A
comprehensive documentation which
argues for a reorientation of political
values. The notes contain a good deal
of useful bibliographical material. See
the author's "Conservation for Survival," *Nation* (August 26, 1968), pp.
138-42, for a summary statement of
their position. Cf. also C. R. Harris,
"Model for a New Radicalism," *Nation*
(November 10, 1969), pp. 496-500,
which argues that the issue of ecology
transcends class differences.

RIHA, JEANNE. "Technology and Man,"
Humanist 26, no. 3 (May-June 1966):
88- 89. Journalistic report on a conference at the Center for the Study of
Democratic Institutions.

RIU, FEDERRIO. "Una extraña in-
terpretación de la técnica," *Revista nacional de cultura* (Caracas) 30
(April-June 1969): 57-63.

RIVIÈRE, MARC. *Economie bourgeoise et
pensée technocratique.* Paris: Editions sociales, 1966. Pp. 225.

ROBERTSON, C. G., et al. *Humanism and
Technology and Other Essays.* London:
Humphrey Milford, 1924; Oxford
Univ. Press, 1925. Pp. 91. Other au-
thors are T. H. Holland, C. H. Desch,
H. Fowler, F. W. Burstall, W. Cramp.

ROGERS, EVERETT M. *Diffusion of In-*

novations. New York: Free Press, 1962. This is an exhaustive review of studies of diffusion and adoption gathered from almost 600 different publications. The author then presents a paradigm of innovation diffusion and adoption, elements of which have been used by others investigating technological diffusion and technology transfer.

ROHRMOSER, GUNTER. "Humanität und Technologie," *Studium generale* 22 (1969): 771–82. Does modern technology bring about the end of humanity? Or does it make humanity possible for the first time in history? Or is the relationship between humanity and technology essentially irrational? Argues that these three views have their historical place in the fundamental principles of Marxist anthropology and, as such, stand in the way of a true development of humanity.

ROIG, GIRONELLA, JUAN S. I. "Humanismo y técnica," *Espíritu* (Barcelona) 16 (1967): 43–68.

ROMANO, BRUNO. *Tecnica e giustizia nel pensiero di Martin Heidegger* [Technology and justice in the thought of Martin Heidegger]. Milan: Giuffre, 1969. Pp. 233.

RÖPKE, WILHELM. "Die Technik in der Gesellschaftskrisis der Gegenwart" [Technology in the social crisis of our time], *Universitas; Zeitschrift für Wissenschaft, Kunst und Literatur* 7 (1952): 673–79.

──────. *A Humane Economy; the Social Framework of the Free Market.* Translated by E. Henderson. Chicago: Regnery, 1960. Pp. 312. From *Jenseits von Angebot und Nachfrage* (Erlenback-Zurich: Rentsch, 1958). A work on the social implications of technology by an internationally known economist.

ROSE, STEPHEN, and HILARY ROSE. "Democracy and Science," *New Scientist* 44 (November 20, 1969): 397–400. "At a time of rapid technological change, the decision-making processes need to be opened up at all levels." The lack of communication and accessibility is the root problem of contemporary science and its social relations. How to avoid this danger is the subject of this article: a reshaping of goals toward "an open, accessible and Man-centered science." See also the authors' *Science and Society* (London: Penguin, 1969).

──────. "Social Responsibility: the Myth of the Neutrality of Science," *Impact of Science on Society* 21, no. 2 (April–June 1971): 137–50.

ROSEN, S. MCKEE, and LAURA ROSEN. *Technology and Society.* New York: Macmillan, 1941. Pp. 474. Tries to present "a balanced picture which might aid the student and the citizen to gain some perspective of the kind of world which they are called upon and will be called upon to face." Contains description of the "technological base" of manufacture, transportation, agriculture, construction, and science, followed by chapters discussing the economic, social, and political effects of modern technology. Introduction by William F. Ogburn on "National Policy and Technology."

ROSSEM, ARNOLD VAN. *Leven in een riskante technische wonderwereld; de invloed van de techniek op het menselijk bestaan* [Life in a hazardous technical wonder-world; the influence of technology on human existence]. Amsterdam: Wereldbibliotheek, 1968. Pp. 224. Popular.

ROTH, PAUL. "Mensch und Technik," *Stimmen der Zeit* 149 (1951): 1–10.

ROUGEMENT, DENIS DE. "Man versus Technics," *Encounter* 10 (January 1958): 43–52.

──────. "La technique, facteur de paix," *Bulletin technique de la suisse romande* 91, no. 5 (1965): 53–57.

ROWE, A. P. "From Scientific Idea to Practical Use," *Minerva* 2, no. 3 (Spring 1964): 303–19. Argues for the need to study how scientific research can be made more practically effective.

RUGG, HAROLD O. *The Great Technology; Social Chaos and the Public Mind.* New York: Day, 1933. Pp. 308.

SACKMAN, HAROLD. *Computers, System Science, and Evolving Society: the Challenge of Man-Machine Digital Systems.* New York: Wiley, 1967. Pp. 638.

SALIZZONI, ANGELO. "L'uomo, la tecnica,

la civilta," *Incontri cultural* (Rome) 1 (1968): 395-406.

SALOMON, JEAN JACQUES. "L'Amerique s'interroge sur la science," *Atomes* 248 (November 1967): 658-61. Discusses some of the literature questioning the aims, methods, institutionalization, and power of science in the United States, concentrating on the work of D. K. Price, R. Lapp, H. L. Neiburg, Nelson-Peck-Kalachek, E. S. Skolnikoff, and S. A. Lakoff.

———. "Science Policy and Its Myths," *Diogenes* 70 (Summer 1970): 1-26.

SALTER, JAMES ARTHUR. *Modern Mechanization and Its Effects on the Structure of Society.* London: Oxford Univ. Press, 1933. Pp. 42.

SANTESMASES, J. GARCÍA. "A Few Aspects of the Impact of Automation on Society," *Impact of Science on Society* 11, no. 2 (1961): 107-26.

SANTOS, DELFIM. "A técnica como fundamento da cultura," *Revista brasileira de filosofia* (São Paulo) 16 (1966): 471-90.

SAUERBRUCH, FERDINAND. "Mensch und Technik," *Erziehung* 14 (1938): 65-78.

SAUVY, ALFRED. "The Information of Machines and of Men: Wizards and Technocrats," *Diogenes* 62 (Summer 1968): 1-24. Machines have always inspired the emotions of hope and/or anguish. But it is wrong to criticize innovation; criticism should be against misuse of the machine and enslavement to it.

SAYRE, W. S., and B. L. R. SMITH. *Government, Technology, and Social Problems.* New York: Institute for the Study of Science in Human Affairs, Columbia Univ. 1969. Pp. 33. The prospects and problems of applying science and technology to the solution of social problems are discussed and assessed.

SCHELSKY, HELMUT. *Die sozialen Folgen der Automatisierung* [The social consequences of automation]. Düsseldorf: Diederich, 1957. Pp. 46. Contains two essays: "Zukunftsaspekte der industriellen Gesellschaft" (1953), and "Die sozialen Folgen der Automatisierung" (1957).

———. *Der Mensch in der wissenschaftlichen Zivilisation* (Man in scientific civ-

ilization]. Arbeitsgemeinschaft für Forschung des Landes Nordrhein-Westfalen; Veröffentlichungen Geisteswissenschaften, vol. 96. Cologne: Westdeutscher Verlag, 1961. Pp. 68.

———. "Demokratischer Staat und moderne Technik," *Atomzeitalter* 5 (1961): 99-102.

———. "Die Zukunft des Menschen in der industriellen Arbeitswelt," *Soziale Welt* 13 (1962): 97-108.

———. *Schule und Erziehung in der industriellen Gesellschaft* [School and education in the industrial society]. Weltbild und Erziehung, no. 20. Würzburg: Werkbund-Verlag, 1965. Pp. 82. Consists of two long essays: (1) "Soziologische Bemerkungen zur Rolle der Schule in unserer Gesellschaftsverfassung," and (2) "Beruf und Freizeit als Erziehungsziele in der modernen Gesellschaft."

SCHENK, GUSTAV. *Die Grundlagen des 21. Jahrhunderts; über die Zukunft der technischen Welt* [The foundations of the 21st century; on the future of the technical world]. Berlin: Safari-Verlag, 1963. Pp. 420.

SCHERER, GEORG. *Absurdes Dasein und Sinnerfahrung; über die Situation Menschen in der technischen Welt* [Absurd existence and sense experience; on the situation of man in the technological world]. Essen: Driewer, 1963. Pp. 118.

SCHERL, AUGUST. *Technik formt unsere Welt; eine enzyklopädische Darstellung* [Technology forms our world; an encyclopedic description]. Bonn: Athenäum-Verlag, 1957. Pp. 600.

SCHICK, ALLEN. "The Cybernetic State," *Transaction*, vol. 7, no. 4 (February 1970). Traces the evolution of the state from the political to the administrative to the bureaucratic and finally to the cybernetic state. Primarily concerned with the cybernetic state and the extent to which government will possibly control the lives of the individuals it proposes to govern.

SCHIFFERS, NORBERT. "Ethik für Techniker," *Orientierung; Katholische Blätter für weltanschauliche Information* (Zurich) 30, no. 21 (1966): 233-35.

SCHILPP, PAUL A. "A Challenge to Philosophers in the Atomic Age," *Philoso-*

phy 24 (1949): 56–58. Conference paper; a weak and unspecific call on philosophers to give some attention to the historical problems with which they live. See also "Does Philosophy Have Anything to Say to Our Atomic Age?" in *Proceedings of the XIIth International Congress of Philosophy; Venice and Padua, Sept. 12–18, 1958* (Florence: Sansoni, 1958–61), 8:239–45.

SCHIMANK, H. "Technik als formende Kraft unserer Zivilisation" [Technology as forming power of our civilization], *Physikalische Blätter* 10, no. 12 (1954): 533–43.

SCHLESINGER, ARTHUR M. "An American Historian Looks at Science and Technology." Pp. 67–75 in *Nothing Stands Still.* Cambridge, Mass.: Harvard Univ. Press, 1969.

SCHLESINGER, JAMES R. "Systems Analysis and the Political Process," *Journal of Law and Economics* 11 (October 1968): 281–98.

SCHMANDT, JUERGEN. "Wissenschaft und Wirtschaft: das Beispiel der Kentststoffindustrie," *Wirtschaft und Wissenschaft,* vol. 13, no. 1 (January 1965).

———. "Le 'Scientific Statesman,' " *Etudes philosophiques,* new series, 21, no. 2 (April–June 1966): 165–86.

———. "Machines et pedagogie," *Atomes* 244 (June 1967): 377–80. Reprinted as "Technologie et enseignement," in *Coucours medical* 89, no. 22 (June 3, 1967): 4449–56.

———. "Technik und Sozialer Fortschritt," Pp. 90–102 in Robert Jungk, ed., *Menschen im Jahr 2000.* Frankfurt: Umschau Verlag, 1969.

———. "Technology and Education." Pp. 76–97 in Arthur M. Kroll, ed., *Issues in American Education: Commentary on the Current Scene.* New York: Oxford Univ. Press, 1970.

———. "Wissenschaft, Technik und Gesellschaft," *Umschau in Wissenschaft und Technik* 70, no. 5 (February 26, 1970): 131–37.

———. "Technology and Social Purpose." Pp. 1–17 in Paul W. DeVore and William J. Smith, eds., *Education in a Technological Society.* Morgantown: West Virginia Univ., Office of Research and Development, Appalachian Center, 1970. Most modern

critiques are ideological whether romantic or socialist, and do not lead to understanding social problems induced by technological problems. Examining these problems reveals a connecting thread: We must match our scientific and technological progress with the creation of new social and political institutions.

SCHMÉLZER, HORST. "Soziale Voraussetzungen des technischen Fortschritts" [Social preconditions of technological progress], *Die neue Ordnung in Kirche, Staat, Gesellschaft und Kultur* 23, no. 4 (1969): 266–79.

SCHMID, KARL. *Mensch und Technik; die sozialen und kulturellen Probleme im Zeitalter der 2. industriellen Revolution* [Man and technology; social and cultural problems in the period of the second industrial revolution]. Bonn: Vorstand der SPD, 1956. Pp. 26. A political party tract.

SCHMID, REINHARD, ed. *An der Schwelle des dritten Jahrtausends* [On the threshold of the third millennium]. Stuttgart: Spectrum Verlag, 1967. Pp. 152.

SCHMOOKLER, J. *Invention and Economic Growth.* Cambridge, Mass.: Harvard Univ. Press, 1966. Pp. 332. Analysis of the effects of economic growth on technological changes and inventions. Covers patents, innovations in various industries, the role of the intellectual stimulus, productivity advance, and the extent of the market. Comprehensive statistics and list of inventions (pp. 217–328).

SCHNEIDER, JULIUS. *Für oder Gegen die Technik* [For and against technology]. Munich: Verlag Braun & Schneider, 1934. Pp. 108. Cover of the book shows a man in a business suit shaking hands with a robot. Although the author openly expresses prewar Nazi sympathies and is often naïve, the book as a whole contains some intriguing historical material. – R. J. R.

SCHNEIDER, P. K. "Science, cybernétique et conscience," *Archives de philosophie,* vol. 30, no. 2 (April–June 1967).

SCHOOLER, DEAN, JR. *Science, Scientists, and Public Policy.* New York: Free Press, 1971. Pp. 339. Includes bibliography.

SCHOUTEN, I. F. "Die Anpassung der

Technik an den Menschen" [The adaptation of technology to man], *VDI-Zeitschrift* 104 (1962): 698–99.

SCHREITERER, MANFRED. "Naturwissenschaft und Politik," *VDI-Zeitschrift* 111, no. 24 (December 1969): 1702. Politics is now impossible without a close interrelationship of natural sciences, engineering sciences, and politics.

SCHUMACHER, EDGAR. "Wunscherfüllende Technik" [Wishfulfilling technology], *Schweizerische Technische Zeitschrift* 52, nos. 23–24 (1955): 422–25.

SCHUMACHER, FRITZ. *Schöpferwille und Mechanisierung; Fortsetzung der Schrift: der "Fluch der Technik"* [Creative will and mechanization; sequel to "Fluch der Technik"]. Hamburg: Verlag Boysen & Maasch, 1933. Pp. 31. "The domination of spirit which is embodied in technology leaves more room to the maker to unfold soulish powers than pessimists admit." – W. C.

SCHWABE, G. "Über Rückwirkungen der technischen Zivilisation auf den Menschen" [On the repercussions of technological civlization upon man], *Studium generale* 15, no. 8 (1962): 495–512.

SCHWARZE, BRUNO C. *Die Technik, ihre Problematik und Praxis; eine Einführung für Nichttechniker, insbesondere für nichttechnische Studierende aller Fakultäten* [Technology, its problematics and practice; an introduction for the nontechnician, especially for nontechnical students of all faculties]. Berlin: Schmidt Vorwort, 1949. Pp. 224.

SCHWEINITZ, KARL DE, JR. *Industrialization and Democracy; Economic Necessities and Political Possibilities.* New York: Free Press, 1964. Pp. 309. Devoted to the arguments for and against the view that democracy is intimately, even causally, related to industrialization.

Science and Freedom. Boston: Beacon, 1955. Pp. 295. Proceedings of a conference convened by the Congress for Cultural Freedom, Hamburg, July 23–26, 1953. Contains opening papers by J. Pieper and M. Hartmann, followed by five sessions of papers, with discussions: (1) The Organization of Science contains M. Polanyi's "Pure and Applied Science and Their Appropriate Forms of Organisation," J. R. Baker's "Freedom and Authority in Scientific Publication"; (2) Science and the State contains S. K. Allison's "Loyalty, Security and Scientific Research in the United States," J. D. Millett's "Universities, the State and Freedom," J. Thibaud's "The Implications of State-financed Research," L. Raiser's "State Support of Universities, and Academic Freedom," T. Komai's "Freedom and the Universities in Japan"; (3) Science and Its Method contains H. Mehlberg's "The Method of Science, Its Ranges and Limits," B. Snell's "Science and Dogma," R. Aron's "The Concepts of 'Class Truth' and 'National Truth' in the Social Sciences," H. Plessner's "Ideological Tendencies among Academic Thinkers"; (4) Science in Chains contains S. Hook's "Science and Dialectical Materialism," S. Landshut's "The Theory of Dialectical Materialism," V. Gitermann's "The Study of History in the Soviet Union," Th. Dobzhansky's "The Fate of Biological Science in Russia"; (5) Science and Citizen contains W. R. Niblett's "Neutrality or Profession of Faith?" R. Apéry's "The Problem of Neutralism," H. Thirring's "Faith and Objectivity," A. Jores's "Science and Moral Responsibility," and T. Litt's "Science and Objectivity."

"Science and Technology in Contemporary Society," *Daedalus,* vol. 91, no. 2 (Spring 1962). Contents: R. Aron's "The Education of the Citizen in Industrial Society," E. Ashby's "The Administration: Bottleneck or Pump?" A. Huxley's "Education on the Nonverbal Level," L. S. Kubie's "The Fostering of Creative Scientific Productivity," W. A. Lewis's "Education for Scientific Professions in the Poor Countries," R. J. Morison's "The University and Technical Assistance," and G. S. Brown's "New Horizons in Engineering Education."

Science and Technology for Development. Report of the United Nations Conference on the Application of Science and Technology for the Benefit of the Less Developed Areas. New York, United Nations, 1963. Full report contains eight volumes: (1) "World of

Opportunity"; (2) "Natural Resources"; (3) "Agriculture"; (4) "Industry"; (5) "People and Living—Population, Health, Nutrition, Rural Development, Urbanization"; (6) "Education and Training"; (7) "Science and Planning"; (8) "Plenary Proceedings, List of Papers, Index." Volume 1, pp. 30–36, with sections on "Science as a Common Heritage," and "Science and Society" disclose what might be called the internationalist's attitude toward technology.

"Science for Mankind," *Science Journal,* special issue, vol. 5A, no. 4 (October 1969). A collection of articles dealing with almost every aspect of modern society affected by modern science and technology. Silver's "Science and Society," E. G. Mesthene's "Technology and Human Values," K. E. Boulding's "Research for Peace," P. Leyhausen's "The Dilemma of Social Man," B. Commoner's "Evaluating the Biosphere," R. Dubos's 'The Human Environment," A. D. C. Peterson's "Real Goals for Education," H. Miller's "Real Goals for Medicine," Llewelyn-Davies and P. Cowan's "Science and the City," and C. H. Waddington's "Assessing the Priorities."

Science, Technology, and American Diplomacy: a Selected, Annotated Bibliography of Articles, Books, Documents, Periodicals, and Reference Guides. Prepared for Subcommittee on National Security Policy and Scientific Developments, Foreign Affairs Committee, U.S. Congress, House. Washington, D.C.: Government Printing Office, 1970. Pp. 69. Emphasis is on materials of a general nature from 1965 to 1969, although several basic works published before 1965 are listed. Items are ordered alphabetically in four categories: (1) articles, (2) books and documents, (3) basic periodicals, and (4) bibliographic tools.

Science, Technology, and American Diplomacy; toward a New Diplomacy in a Scientific Age. Prepared for Subcommittee on National Security Policy and Scientific Developments, Foreign Affairs Committee, U.S. Congress, House. Washington, D.C.: Government Printing Office, April 1970. Pp. 28.

"Science, Technology and the Law," *Impact of Science on Society,* vol. 21, no. 3 (July–September 1971). Contains: H. Pedersen's "Life, Machines and Law" (an interview), L. H. Tribe's "Towards a New Technological Ethic: the role of Legal Liability," D. W. Meyers's "Organ Transplantation and the Law," G. P. Zhukov's "Space Law: the New Extra-terrestrial Jurisprudence," Eugène Pépin's "Space Law: Legal Aspects of Direct Broadcasting by Satellite," D. P. O'Connell's "Legal Problems of the Exploitation of the Ocean Floor," E. Szabady's "The Legalizing of Contraceptives and Abortions," and S. M. Sheniti's "Multiple Reproduction Processes and Authors' Rights."

SEABORG, GLENN T. *Science, Technology and the Citizen.* Washington, D.C.: Government Printing Office, 1969. Pp. 16. Remarks by the chairman of the Atomic Energy Commission at a Nobel symposium on the place of values in a world of facts, Stockholm, Sweden, September 17, 1969.

———. *Science and Technology Rededicated.* Washington, D.C.: Government Printing Office, 1970. Pp. 8. Commencement address at Michigan Technological University, June 13, 1970.

SEDLMAYR, HANS. "Gefahr und Hoffnung des technischen Zeitalters" [Peril and hope of the technological age], *Österreichische Ingenieur-Zeitschrift* (Vienna) 1 (1958): 11–17.

———. "Technik und Natur; zur geistigen Situation unserer Gegenwart" [Technology and nature; on the spiritual situation of the present time], *Die Bautechnik* (Berlin) 37 (1960): 245.

SELSAM, HOWARD. *Ethics and Progress: New Values in a Revolutionary World.* New York: International Publishers, 1965. Pp. 126. A Marxist argues that moral progress can be usefully defined in terms analogous to scientific and technological progress, that is, in terms of man's increasing mastery of the physical world. Reviewed in D. C. Hodges's " 'Moral Progress' from Philosophy to Technology," *Philosophy and Phenomenological Research* 28, no. 3 (March 1968): 430–36.

SEQUENZ, H., ed. *Der Geist der Technik.* Vienna and New York: Springer-Verlag, 1966. Pp. 112. Speeches and essays on the 150th anniversary of the Technischen Hochschule in Vienna (November 8–13, 1965). Much like the German Engineers' Society yearbooks, that is, filled with a good deal of mutual praise and favorable comments of one technologist on another. But it presents an oddly typical Austrian position on the dangers of technology. Having been involuntarily annexed and involved in World War II, they seem to feel irresponsible for any of the horrors of technological war. In some ways much more American than German in their feeling about technology. —R. J. R.

SHANKS, MICHAEL. *The Innovators; the Economics of Technology.* Baltimore: Penguin, 1967. Pp. 294. Popular introduction to the field and problems of the management of innovation.

SHEEHAN, J. C. "On Applied Science." Pp. 95–98 in J. D. Summerfield and L. Thatcher, eds., *The Creative Mind and Method.* Austin: Univ. Texas Press, 1960.

SIEBERS, GEORG. "Die vier Epochen in der Weltgeschichte der Technik" [The four epochs in the world history of technology], *Saeculum; Jahrbuch für Universalgeschichte* (Freiburg-Breisgau) 13 (1962): 401–18.

———. *Das Ende des technischen Zeitalters* [The end of the technical age]. Freiburg: Alber, 1963. Pp. 238.

SIEGFRIED, ANDRÉ. *America Comes of Age; a French Analysis.* Translated by H. H. and D. Hemming. New York: Harcourt, Brace & World, 1927. Pp. 358. From *Les Etats-Unis d'aujourd'hui* (Paris: Colin, 1927).

———. *Technique et culture dans la civilisation moderne.* Paris: Centre de perfectionnement technique et administratif, 1948. Pp. 17.

———. *Über die Idee des Fortschritts* [Concerning the idea of progress]. Vienna: Amandus-Editions, 1948. Pp. 35. Considers the relation between technological and moral progress. Man can develop spiritually without the benefit of technology; and in fact too much technological progress can stop moral progress. Recalls Rousseau's *Discourse on the Arts and Sciences* —R. J. R.

———. "La technique et la culture dans une civilisation moderne," *Bulletin technique de la Suisse romande* 81, no. 2 (1955): 37–42.

SIMON, ERNST LUDWIG. "Das Wesen der Technik und unsere Verantwortung" [The essence of technology and our responsibility], *Convent; Akademische Monatsschrift* (Mannheim) 11 (1960): 213–19.

SIMONDON, GILBERT. "Culture et technique," *Morale et enseignement* (Brussels) 14, nos. 3–4 (1965): 3–16.

SIMONET, ROGER. *Les derniers progrès de la technique* [The final progress of technology]. Paris: Calmann-Levy, 1950.

SINGER, RICHARD E. "Man and Technology," *Saturday Review* (December 26, 1959), pp. 65–66. "Man's dilemma is not that he must come to terms with the things he has created. His great challenge is to come to terms wih himself." By a rabbi.

SKLAIR, LESLIE. "The Revolt against the Machine: Some Twentieth Century Criticisms of Scientific Progress," *Journal of World History* 7 no. 3 (1970): 479–87. A brief survey of criticisms of scientific progress from different thinkers throughout the 20th century, including L. P. Jacks, Gina Lombroso, M. Tailledet, Simon Kuznets, Julius Wolf, Gabriel Marcel.

———. "Sociology of the Opposition to Science and Technology: with Special Reference to the Work of Jacques Ellul," *Comparative Studies in Society and History* 13 (April 1971): 217–39. There is a reply by R. L. Meier.

SKOLNIKOFF, EUGENE B. *Science, Technology and American Foreign Policy.* Cambridge, Mass.: M.I.T. Press, 1967. Pp. 316. A general discussion of the relationship between science and foreign policy with a specific analysis of policy-making processes. Shows how uncertainties inherent in judgments about science and technology are affected by political factors and, in turn, how political factors depend on scientific and technological judgments.

——. "Technology and the Future Growth of International Organizations," *Technology Review* 73, no. 8 (June 1971): 38–47.

SLAGTER, S. *De mens in Maatschappij, Techniek en Cultuur* [Man in society, technology, and culture]. Ziest: de Haan, 1959. Pp. 164.

SLICHTER, SUMNER H. *Technology and the Great American Experiment.* Madison: Univ. Wisconsin Press, 1957. Pp. 18. 1957 Engineer's Day address at University of Wisconsin. "Technology today is a principal influence (probably the most powerful single influence) keeping strong and vigorous the philosophy of democracy that has led the United States to embark upon one of the most magnificent experiments in human history . . . the attempt to make available to all members of the community . . . the opportunity to lead the good life and to acquire the physical comforts of life" (p. 2). Also in J. T. Dunlop, ed., *Potentials of the American Economy; Selected Essays of Sumner H. Slichter* (Cambridge, Mass.: Harvard Univ. Press, 1961), pp. 5–14.

SMIT, J. "Techniek en cultuur; over de aard van hun relatie thans," *Streven* (Bruges) 19 (1965–66): 342–48.

SNELL, J. R. "Fight for Ethics in Engineering," *Journal of Engineering Education*, vol. 46 (November 1955). Part of a wide ranging debate on professional and moral responsibilities. Cf. also: P. L. Alger's "Unfinished Business of Engineering Ethics," and "Growing Importance of Ethics to Engineers," *Electrical Engineering*, vol. 75 (June 1956), and *Electrical Engineering*, vol. 77 (January 1958); "Candid Answers on Engineering Ethics," *Electrical World*, vol. 145, no. 62 (February 27, 1956); and R. W. Clouses's "Engineering Ethics: Can It Be Defined?" *Civil Engineering*, vol. 26 (February 1956). For a more complete listing of materials in this area, see J. Arthur Freed's *Some Ethical and Social Problems of Science and Technology: a Bibliography of the Literature from 1955* (Los Alamos, N.M.: Los

Alamos Scientific Laboratory of the Univ. Calif., 1964). See also *Engineering Issues,* vol. 98, no. PP1 (January 1972), for other articles by engineers on the ecological crisis. Other issues of this journal contain relevant articles as well.

SNYDER, E. E. *Please Stop Killing Me!* New York: New American Library, 1971.

"The Social Consequences of Technical Progress." *International Social Science Bulletin* 4, no. 2 (Summer 1952): 243–346. A symposium which includes the following: G. Friedmann's "Introduction," S. H. Frankel's "Some Conceptual Aspects of Technical Change," W. F. Ogburn's "Social Effects of Technology in Industrialized Societies," W. E. Moore's "Social Consequences of Technical Change from the Sociological Standpoint," J. Fourastie's "Economics and the Social Consequences of Technical Progress," E. Thorsrud's "The Social Consequences of Technical Change from the Psychological Standpoint," R. Savatier's "Law and the Progress of Techniques," A. P. Elkin's "Western Technology and the Australian Aborigines," H. D. Laswell's "Appraising the Effects of Technology," and a "Tentative Bibliography."

"Social Implications of Modern Science." *Annals of the American Academy of Political and Social Science* 249 (January 1947): 225. A symposium edited by H. M. Dorr, which includes: A. H. Compton's "The Atomic Crusade and Its Social Implications," B. Brodie's "The Atomic Dilemma," R. E. Cushman's "Civil Liberties in the Atomic Age," A. W. Bromage's "Total War and the Preservation of Democracy," D. G. Marquis's "Psychology of Social Change," E. B. Stason's "Law and Atomic Energy," Slichter's "Some Economic Consequences of Science," Thompson's "Impact of Science on Population Growth," H. B. Lewis's "Nutrition," G. W. Anderson's "The Political Impact of Modern Science on Public Health," L. Wirth's "Responsibility of Social Science," W. Haber's "Security, Freedom and Modern Technology," H. Keniston's "The

Humanities in a Scientific World," and G. Shuster's "Good, Evil, and Beyond."

SOMERS, GERALD G., EDWARD L. CUSHMAN, and NAT WEINBERG, eds. *Adjusting to Technological Change.* New York: Harper & Row, 1963. Pp. 230. Concerned with labor-management relations. "Technological change is accepted as an industrial fact, associated with gains and costs, calling for understanding, adjustment, accommodation and aid."

SPAGNOLI, H. "The Challenge of Technology." In C. Argoff, ed., *The Humanities in the Age of Science; Essays in Honor of Peter Sammartho.* Cranbury, N.J.: Farleigh Dickinson Univ. Press, 1968.

SPECHT, K. G. "Der Einfluss der technischen Entwicklung auf die Struktur der Gesellschaft" [The influence of technological development on the structure of society], *VDI-Nachrichten* 12, no. 21 (1958): 5-7.

SPEER, ALBERT. *Inside the Third Reich.* Translated by R. and C. Winston. New York: Macmillan, 1970. Pp. 600. The memoirs of the supreme technocrat in Hitler's Germany who lived to have grave doubts about technology.

SPICER, EDWARD H., ed. *Human Problems in Technological Change.* New York: Wiley, 1952. Pp. 301. A collection of anthropological case studies of the influence of technological innovations on various primitive cultures.

SPIRITO, UGO. "La crisi dell umanita e la technologia," *Giornale critico della filosofia italiana* 11 (January-March 1971): 151-53.

STACHOWIAK, HERBERT. "Zur Bedeutung des Rundfunks in der Kulturgestaltung der Gegenwart" [On the meaning of the radio in the cultural framework of the present], *Humanismus und Technik* 3, no. 2 (December 25, 1955): 96-110.

———. "Bedeutung des Fernsehens in der Kulturgestaltung der Gegenwart" [On the meaning of television in the cultural framework of the present], *Humanismus und Technik* 4, no. 2 (February 1, 1957): 99-116.

STAHL, R. "Die Technik verändert die Sprache" [Technology changes language], *Welt der Schule* 16, no. 2 (1963): 64-67.

STANTON, FRANK, GERARD PIEL, and OSCAR HANDLIN. *Three Lectures: the Heritage of Mind in a Civilization of Machines.* Waterville, Me.: Colby College, 1963. Pp. 60. The only essay of real value is by Handlin, and a later version of this can be found under the title "Man and Magic: First Encounter with the Machines," *American Scholar* 33, no. 3 (Summer 1966): 408-19.

STAUFENBIEL, FREDERICK. "Révolution technique et révolution culturelle," *La nouvelle revue internationale* 9, no. 11 (1966): 54-63.

STEVENS, CHANDLER H. "Citizen Feedback and Societal Systems," *Technology Review* 73, no. 3 (January 1971): 38-45. New methods of communication have made it easy for leaders to communicate to their constituencies while weakening the voices of the people in reaching their governments. But cf. also Thomas B. Sheridan, "Citizen Feedback: New Technology for Social Choice," *Technology Review* 73, no. 3 (January 1971): 46-51, which argues that technology can be used more effectively to help people communicate their values to each other and to reach a consensus on decisions.

STILLFRIED, ALFONS VON. "Ursprung und Sinn der technischen Zivilisation" [Origin and meaning of technological civilization], *Natur und Kultur* 54 (1962): 201-4.

STOLZE, DIETHER. *Den Göttern gleich; unser Leben von morgen* [Like the gods; our life of tomorrow]. Vienna: Desch, 1959. Pp. 326. Popular futurology. — R. J. R.

STRAKOSCH, ALEXANDER. *Leben und Lernen im Zeitalter der Technik; Erfahrungen und Erlebnisse* [Living and learning in an era of technology; experiences and adventures]. Basel: Zbinden, 1955. Pp. 73.

STREHL, ROLF. *The Robots Are among Us.* London and New York: Arco, 1955. Pp. 316. From *Die Roboter sind unter uns, ein Tatsachenbericht* (Oldenbourg: Stalling, 1952). A popular warning against the dangers of "The Invasion of Brain Machines" and "The Dictatorship of the Automaton." Though somewhat dated by events, it contains a good deal of interesting historical material, especially on the computer and popular reactions to it.

STRUNTZ, F. "Technik als Kulturproblem; zur Kritik der Grundlagen des technischen Zeitalters" [Technology as a cultural problem; contribution to a critique of the foundations of the technological age], *Scientia* 60 (1936): 162-68.

Study of Technology Assessment. Report of Committee on Public Engineering Policy, National Academy of Engineering, July 1969. Washington, D.C.: Government Printing Office, 1969. Pp. 208. Includes bibliography.

SUPEK, RUDI. "Karl Marx et l'epoque de l'automation," *Praxis* 3, no. 1 (1967): 31-38.

——. "Der technokratische szientismus und der sozialistische humanismus," *Praxis* 3, no. 2 (1967): 155-75.

SUSINOS, FRANCISCO. "Técnica y humanismo," *Rivista de filosfia* (Madrid) 19 (April-September 1960): 213-29.

SWANN, W. F. G. "Engineering and Pure Science," *Annual Report of the Board of Regents of the Smithsonian Institution* (1952), 201-16.

SWERDLOFF, PETER. "Take That, You Soulless Son of a Bitch!" *Playboy* 19, no. 7 (July 1972): 130 ff. A popular report on the Neo-Luddite reaction against machines.

SZENT-GYORGYI, A. *Science, Ethics, and Politics.* New York: Vantage, 1963. Pp. 91. Expansion of an article of the same title in *Science* 125 (February 8, 1957): 225-26. See also the author's *The Crazy Ape* (New York: Philosophical Library, 1970); and "The Dual World of Man: Reflections on Science and Government," *Bulletin of the Atomic Scientists* 25, no. 8 (October 1969): 33-34 and 37.

SZILVASSY, ARPAD. "Democracia e progresso technologico," *Sociologia* (São Paulo) 24, no. 3 (September 1962): 209-21. Includes bibliography.

TAVISS, IRENE. "The Technological Society: Some Challenges for Social Science," *Social Research* 35, no. 3 (Autumn 1968): 521-39. "Brief outline

of the need for new conceptual structures in the social sciences necessitated by "post-industrial society."

——. "Futurology and the Problem of Values," *International Social Science Journal* 11, no. 4 (1969): 574-84.

——, ed. *The Computer Impact.* Englewood Cliffs, N.J.: Prentice-Hall, 1970. Pp. 297. A (1) "General Introduction" followed by a collection of pieces of articles on (2) "The Computer Potential," (3) "The Economy," (4) "The Polity," and (5) "Culture," by H. Borko, H. Kahn, P. Armer, H. A. Simon, A. Westin, J. Diebold, D. N. Michael, A. G. Oettinger, B. Mazlish, etc. A helpful introduction to the large number of social issues raised by computer technology.

Technical Information for Congress. Prepared by Science Policy Research Division, Legislative Reference Service, Library of Congress for Subcommittee on Science, Research, and Development, Science and Astronaut's Committee, U.S. Congress, House, Washington, D.C.: Government Printing Office, April 25, 1969. Pp. 521. Contains (p. 11) an attempt to distinguish between basic research (pursues the discovery of facts about nature), applied research (employs knowledge gained through basic research to realize some social goal or create new technological options), and technology (the exploitation of the option supplied by applied research).

"The Technocratic Society." Part XIII in Paula R. Struhl and Karsten J. Struhl, eds., *Philosophy Now: an Introductory Reader.* New York: Random House, 1972. Contains selections from A. Toffler, J. Ellul, and E. Fromm. Part II also contains M. Minsky's "Machines That Think."

Technology Assessment; Annotated Bibliography and Inventory of Congressional Committee Organization for Science and Technology. Prepared for Subcommittee on Science, Research and Development, Science and Astronautics Committee, U.S. Congress, House. Washington, D.C.: Government Printing Office, July 15, 1970. Pp. 92. A bibliography (pp. 1-52) is a good guide to literature of the period 1966–70, especially to governmental documents and conference papers.

"Technology for Man," Symposium in *Technology and Culture* 10, no. 1 (January 1969): 1–19. Contents: W. E. Howland's "Engineering Education for Social Leadership," J. G. Burke's "Let's Be Sure Technology Is for Man," S. C. Florman's "Engineers and the End of Innocence," J. C. Wallace's "The Engineering Use of Human Beings."

TESCONI, C. A., and V. C. MORRIS. *The Anti-Man Culture*. Urbana: Univ. Illinois Press, 1972.

THENBERG, ANNE MARIE. *Teknik och manniskovarde* [Technology and human value]. Stockholm: Natur och kultur, 1958. Pp. 125.

THEOBALD, ROBERT. "Cybernation, Unemployment and Freedom." Pp. 49–67 in R. M. Hutchins and M. J. Adler, eds., *The Great Ideas Today 1965*. Chicago: Encyclopaedia Britannica, 1965.

———, ed. *Dialogue on Technology*. Indianapolis: Bobbs-Merrill, 1967. Pp. 109. First published as a special issue of *Motive* magazine on "Technology and Values" (March–April 1967). Contents: E. McIrvine's "The Admiration of Technique," W. R. Cozart's "Human Imagination in the Age of Space," R. Kean's "The Dialogue Community: the University in a Cybernated Era," C. R. DeCarlo's "Educational Technology and Value Systems," M. B. Bloy, Jr.'s "Technology and Theology," J. Landau's "Yes–No: Art–Technology," and R. Theobald's "Afterword."

THEUNISSEN, GERT H. "Die Komik in den Maschinen; eine heitere Philosophie der Technik" [The comic in the machine; a gay philosophy of technology], *Rheinischer Merkur; Wochenzeitung für Politik, Kultur und Wirtschaft* (Cologne-Koblenz) 12, no. 3 (1957): 7.

THRING, MEREDITH W. "The Design of Robot Slaves for the Creative Society," *Cybernetica* 13, no. 1 (1970): 55–67.

TODD, ARTHUR J. "The Impact of Industry on the Orient." Pp. 51–155 in

Three Wise Men of the East. Minneapolis: Univ. Minnesota Press, 1927. An early analysis of attitudes and practice in Japan, China, and India.

TOFFLER, ALVIN. *Future Shock*. New York: Random House, 1970. Pp. 505. Discusses what is happening to individuals and groups who are overwhelmed by technological change, with suggestions on how to cope. Chap. 19, "Taming Technology," contains some interesting prescriptions, especially for facing the problems of "technological backlash." Includes bibliography.

TOPCHIEV, A. V. "Interdependence of Science and Society," *Bulletin of the Atomic Scientists* 19, no. 3 (March 1963): 7–11. A Soviet academician's views on science, technology, and society—and on the Soviet society as creating the best conditions for programmed progress in science.

TOURAINE, ALAIN. *The Post-Industrial Society—Tomorrow's Social History: Classes, Conflicts, and Culture in the Programmed Society*. Translated by L. Mayhew. New York: Random House, 1971.

TOUSCOZ, JEAN. "La recherche scientifique el le droit international," *Progrès scientifique* 111 (August 1967): 31–42. Discusses problems of international law associated with scientific research (basic, applied, and mission oriented). Analyzes international diplomatic practices in this area and the transformation of international law occasioned by the rapid development of scientific research activity.

TOYNBEE, ARNOLD J. "The Impact of Technology on Life." Part IV in *Change and Habit: the Challenge of Our Time* (New York and London: Oxford Univ. Press, 1966), pp. 119–227. A brief exploration of the problems of population, urbanization, congestion, mechanization, boredom, affluence, and leisure, arguing that the humane use of technology requires a voluntary renunciation of national sovereignty. See also *Civilization on Trial* (New York: Oxford Univ. Press, 1948) and the monumental *A Study of History* (London: Oxford

Univ. Press, 1934-61). Note, as well, the author's *Surviving the Future* (London: Oxford Univ. Press, 1971), a series of conversations with Prof. K. Wakaizumi of Japan which touches on these topics.

TREUE, WILHELM. "Gesellschaftliche Voraussetzungen der Industrialisierung" [Sociological preconditions of industrialization], *VDI-Zeitschrift* 110, no. 1 (January 1968): 1-5.

TRIBE, LAURENCE H. "Legal Frameworks for the Assessment and Control of Technology," *Minerva* 9, no. 2 (April 1971): 243-55.

———. "Towards a New Technological Ethic: the Role of Legal Liablity," *Impact of Science on Society* 21, no. 3 (July-September 1971): 215-22. Considers the social and ethical significance of the idea that the producers of technology should be legally responsible for any harmful consequences whether or not such consequences were foreseeable.

The Triple Revolution. Santa Barbara, Calif.: Ad Hoc Committee on the Triple Revolution, 1964. A manifesto also published in *Liberation* 9, no. 2 (April 1964): 2-8. The "triple revolution" refers to revolutions in cybernation, weaponry, and human rights that must be met by political changes. See also Robert Theobald, ed., *The Guaranteed Income* (Garden City, N.Y.: Doubleday, 1966).

"L'uomo e la macchina." In *Atti del XXI Congresso nazionale de filosofia; Pisa, April 22-25, 1967.* Vol. 1. Turin: Edizioni de Filosofia, 1967.

VALLÉE, ROBERT. "Cybernetics and the Future of Man," *Impact of Science on Society* 3, no. 3 (Autumn 1952): 171-80. "We are now at the beginning of a second industrial revolution. The first resulted from new methods for obtaining energy; the second arises from man's new capacity for performing certain mental processes by automatic means—in other words, it is born of cybernetics. The 'devaluation' of muscular energy is now being followed by the 'devaluation' of the brain."

VATTER, HAROLD G., et al. "The March of Technology Calls for a New Philos-

ophy of Poverty," *Contemporary Review* (London) 213 (1968): 169-77.

VAUX, KENNETH, ed. *Who Shall Live? Medicine, Technology, Ethics.* Philadelphia: Fortress, 1970. Pp. 199. Contains M. Mead's "The Cutural Shaping of the Ethical Situation," E. G. Mesthene's "Technology and Values," R. Drinan's "Abortion and the Law," P. Ramsey's "The Ethics of Genetic Control," J. Fletcher's "Technological Devices in Medical Care," and H. Thielicke's "Ethics in Modern Medicine." Brief bibliography.

VERGEZ, ANDRÉ. *Marcuse.* Paris: Presses Universitaires de France, 1970. Pp. 103.

VERRES, DANIEL. "Le réformisme technocratique, espoir suprême du capitalisme," *Temps modernes* 25 (November 1969): 729-45.

VORESS, H. E. *Science and Society: a Bibliography.* Prepared for U.S. Atomic Energy Commission. U.S. AEC. Report no. WASH-1182. Washington, D.C.: Government Printing Office, July 1971. Pp. 21. An alphabetical listing by author of 410 references, to literature from 1968 to March 1971; includes author index.

WADDINGTON, C. H., et al. "Five Scientists View the Impacts of Technology," *Impact of Science on Society* 20, no. 2 (April-June 1970): 137-50. Technology is being questioned because the world is concerned about the destructive effects that its rapid and undisciplined growth are having on man and his environment—upon, in other words, the *quality* of human life. C. H. Waddington, N. Ginsburg, T. F. Malone, K. Mellanby, and J. S. Weiner discuss some of the aspects of this problem: the gaps in our knowledge of man-environment-technology relationships, the social involvement of scientists, environmental pollutants, and how population growth multiplies the difficulties.

WAELDER, ROBERT. "Assets and Liabilities of the Modern Movement." Pp. 54-83 in *Progress and Revolution.* New York: International Universities Press, 1967. Modern progress produces value conflicts, the most outstanding of which "appears between adjustment to reality by means of changing oneself so as to fit in with reality and adjustment by means of changing reality to make it fit one's needs."

WAFFENSCHMIDT, WALTER. "Technik und Menschheit," *Studium generale* 15, no. 7 (1962): 479–90.

WALKER, CHARLES R. *Technology in an Age of Reflection.* Dimensions for Exploration Series. Oswego, N.Y.: Division of Industrial Arts and Technology, State Univ. College, 1963. Pp. 14. The best method for "solving the problems of technological change and for utilizing technology for human ends, are the sciences of Man, including under that term all disciplines which deal with the human body, the human mind and society."

———, ed. *Technology, Industry, and Man: the Age of Acceleration.* New York: McGraw-Hill, 1968. Pp. 362. Revised edition of *Modern Technology and Civilization* (New York: McGraw-Hill, 1962). Text designed for a class on modern technology and society. Contains a large number of thematically ordered short readings by L. Mumford, B. Gille, C. Singer, H. Ford, M. Mead, F. Taylor, L. White, D. Bell, A. Huxley, D. Price, etc.

WALLACE, WILLIAM A. "A Thomist Looks at Teaching Machines," *Dominican Educational Bulletin* 4 (1963): 13–23.

WALLIA, C. S., ed. *Toward Century 21: Technology, Society, and Human Values.* New York: Basic, 1970. Questions approached: How, in the face of an ever-expanding technology, are man's ethical and human values to be preserved? For what goals shall he use his increasing mastery of the environment? Do scientists, as professionals, have a social responsibility? Which aspects of man's cultural heritage are still valid for the world of today and tomorrow? Among the 24 contributors are: C. Bay, K. E. Boulding, P. R. Ehrlich, W. H. Ferry, and A. Watts.

WALTERS, GERALD. "The Third Dimension: a Political Philosophy for Science," *Technology and Society* 4, no. 1 (November 1, 1967): 20–29. Science

and government as well as science and society must establish more communication to maintain an advanced technological state's momentum. Discusses D. Price, H. Brooks, C. P. Snow, R. Gilpin, R. Dubos, etc.

WARD, JOHN WILLIAM. "The Organization Society," *University* 5 (Summer 1960): 8–11. Organization is a social extension of machine rationality.

WASHBURN, SHERWOOD L. "Tools and Human Evolution," *Scientific American* 203, no. 3 (September 1960): 63–75. Introductory article for a special issue on the human species, with special reference to how tools made man. See also Herbert Butterfield's "The Scientific Revolution," ibid., pp. 173–92, on the interaction of science and technology from the Renaissance to the present.

WASMUTH, E. "Der Mensch und die zweite technische Revolution," *Wissenschaft und Weltbild* 8 (1955): 248–57.

WEAVER, WARREN. *Science and Imagination.* New York: Basic, 1967. Pp. 295. Selected papers on probability and information theory, implications of space technology and cybernetics, and the role of science and religion in modern life. Marginal.

WEBER, EDWARD. "Vom Gleichgewicht der sittlichen und technischen Werte" [The balance of moral and technological values], *Makromolekulare Chemie* (Heidelberg) 46, no. 21 (1955): 979–82. Also in *Die Schweizer Uhr; Fachorgan der gesamten Uhrenindustrie*, vol. 29, no. 5 (1956).

WEISS, DONALD H. " 'Cybernated Society' and Human Dignity," *Southwestern Journal of Philosophy* 1, no. 2 (1968): 143–51. Rapid technological changes in the means of production and distribution of goods is eliminating the need for massive labor forces, and this means that "we must reshape the concept of human dignity and moral worth by restating the nature of human work in terms of the standard definitions of physical science."

WEISSKOPF, W. "Repression and the Dialectics of Industrial Civilization," *Review of Social Economy* 23 (September 1965): 116–26.

WEIZENBAUM, JOSEPH. "The Two Cultures of the Computer Age," *Technology Review* 71, no. 6 (April 1969): 54–57.

WELLS, WARNER. "Our Technological Dilemma, or an Appraisal of Man as a Species Bent on Self-Destruction," *Bulletin of the Atomic Scientists* 16, no. 9 (November 1960): 362–65.

WESTIN, ALAN F. *Privacy and Freedom.* New York: Atheneum, 1967. Pp. 487. Comprehensive study of psychological, social, political, and legal aspects of privacy. Examines threats to privacy from the technology of surveillance, then analyzes ethical and legal implications of these new developments. Recommends federal action to safeguard against continuing abrogation of privacy.

——. *Information Technology in a Democracy.* Cambridge, Mass.: Harvard Univ. Press, 1970. Pp. 499. A reader on the effects of information systems, systems analysis, and new techniques of decision making on the organization of American government.

WHEELER, GEORGE SHAW. "Technologie et caractère de la production," *La nouvelle revue internationale* 10, no. 1 (1967): 186–99.

WHEELER, HARVEY. *Democracy in a Revolutionary Era: the Political Order Today.* Santa Barbara, Calif.: Center for the Study of Democratic Institutions, 1970. Pp. 216. Examines political changes in modern society as it is developing under the reins of science and technology. Among the specific topics considered are: democracy and the politics of cultural development; the rise of bureaucratic cultures; the scientific revolution; ideology; balance-of-power politics, old and new; and world order. See also *The Political Implications of the Scientific Revolution* (Heidelberg: Studiengruppe für Systemforschung, n.d.).

WHITNEY, W. R. "Technology and Material Progress," *Proceedings of the American Philosophical Society* 70 (1931): 255–62.

WILENSKY, HAROLD L., and C. N. LEBEAUX. *Industrial Society and Social Welfare; the Impact of Industrialization on the Supply and Organization of Social Welfare Services in the United States.*

New York: Russell Sage Foundation, 1958. Pp. 401.

WILHELMSEN, FREDERICK D., and JANE BRET. *The War in Man; Media and Machines.* Athens: Univ. Georgia Press, 1970. Pp. 122. See also Wilhelmsen and Bret, *Telepolitics* (Montreal and New York, 1972).

WILLER, HORST. "Zum Begriff des technischen Fortschritts" [On the concept of technological progress], *Berichte über Landwirtschaft,* new series, 45, no. 4 (1967): 673–86.

WILLIAMS, ELGIN. "The Technological Way of Life," *Inter-American Economic Affairs* 6 (Summer 1952): 71–82. Discusses Ortega, A. Reyes, and N. Wiener.

WILLIAMS, HUBERT, ed. *Man and the Machine.* London: Routledge, 1935. Pp. 219.

WILLIAMSON, HUGH P. "Technology, a Threat to Democracy," *American Journal of Economics and Sociology* 16 (1956–57): 281–90.

WILPERT, PAUL. "Das Phänomen Technik," *VDI-Zeitschrift* 104 (1962): 684–88.

WILSON, J. C. "Technology and Society," *Proceedings of the Academy of Political Science* 30 (May 1970): 158–65.

WINTHROP, HENRY. "Science, Technology and Social Order," *Darshana International* (Moradabad, India) 6 (July 1966): 13–24.

WIRTANEN, ATOS KASIMIR. *Tekniken, manniskan, kulturen; teknokratisk historie-tolkning* [Technology, man, culture; a technocratic interpretation of history]. Uppsala: Bokgillet, 1959. Pp. 193.

WOHLSTETTER, ALBERT. "Technology, Prediction, and Disorder," *Bulletin of the Atomic Scientists* 20, no. 8 (October 1964): 11–15. The consequences of technological action are increasingly difficult to predict as a result of increased complexity and multiplied opportunities for chance interactions.

WOLFF, MICHAEL F. *Technology and the Curriculum.* Dimensions for Exploration Series. Oswego, N.Y.: Division of Industrial Arts and Technology, State Univ., 1967. Pp. 17. An outline of some of the technological changes that will shape our world for the remainder of the century and the effect these will have on the educational curriculum. The educational emphasis will be on teaching "technical concepts," models and their use for analysis of prediction in engineering, and on "the idea of a system itself and the disciplines by which we analyse a system."

WOLIN, SELDON, S., and JOHN H. SCHARR. "Education and the Technological Society," *New York Review of Books* 13, no. 6 (October 9, 1969): 3-6. Included along with five other essays on the revolt of the young beginning with the Berkeley Free Speech Movement and their future under Nixon's technocracy, in *Berkeley and Beyond: Essays on Politics and Education in the Technological Society* (New York: New York Review of Books, 1970).

WOLLAN, MICHAEL. "Controlling the Potential Hazards of Government-sponsored Technology," *George Washington Law Review* 36 (July 1968): 1105-37.

WOODWARD, H. N. *The Human Dilemma.* New York: Brookdale, 1971.

ZARCO DE GEA, JUAN. "Reflexiones sobre la técnica," *Boletín cultural y bibliográfico* (Bogota) 5, no. 10 (1962): 1250-56.

ZBINDEN, HANS. *Technik und Geisteskultur* [Technology and spiritual culture]. Munich: Oldenbourg, 1933. Pp. 130.

———. *Von der Axt zum Atomwerk – über Macht und Not der Menschen in der Technik* [From the axe to the atomic power plant – concerning human power and need in the technical era]. Zurich: Artemis-Verlag, 1955. Pp. 136.

———. "Aviso acerca de la técnica," *Folia humanística* (Barcelona) 2 (1964): 5-25.

———. "Problemas culturales de la era técnica," *Folia humanística* (Barcelona) 4 (1966): 981-87.

———. "Die Technik im Licht kultureller Verantwortung" [Technology in light of cultural responsibility], *Bulletin des Schweizerischen Elektrotechnischen Vereins* 58, no. 26 (1967): 1249-59.

———. "Der Mensch im Spannungsfeld: der modernen Technik," *Deutsches*

Museum, *Abhandlungen und Berichte* 38, Heft 1 (1970): 5-60. A bibliography of Zbinden's writings on the problems of technology and civilization (p. 60).

ZIMAN, JOHN. "Social Responsibility: the Impact of Social Responsibility on Science," *Impact of Science on Society* 21, no. 2 (April-June 1971): 113-22. Cf. also Milton Leitenberg, "Social Responsibility: the Classical Scientific Ethic and Strategic-Weapons Development," ibid., pp. 123-36; and Steven Rose and Hilary Rose, "Social Responsibility: the Myth of the Neutrality of Science," ibid., pp. 137-50.

ZIMMERMAN, BILL, LEN RADINSKY, MEL ROTHENBERG, and BART MEYERS. "Science for the People," *Liberation* 16, no. 10 (March 1972): 20-31.

ZUCKERMAN, SOLLY. "Society and Technology," *Journal of the Royal Society of Arts* 117 (1969): 617-31.

———. "Technological Choice: the Social Cost," *New Scientist* 48 (December 3, 1970): 389-91. "The more technologists and scientists understand how events move in the political sphere, the more they are able to participate in political decisions, the more society benefits from technology, and the less there is to fear."

ZVEGINTZOV, M. "Management in a Modern Scientific and Technological Age," *Impact of Science on Society* 11, no. 1 (1961): 53-73.

C. APPENDIX: SOVIET AND EAST EUROPEAN MATERIALS

[GENERAL NOTE: Works in this section are primarily from the U.S.S.R. and the German Democratic Republic. With regard to the U.S.S.R.: Good surveys of Soviet attitudes toward cybernetics (the special technology about which philosophical questions are most often raised in Russia) can be found in Loren R. Graham, "Cybernetics in the Soviet Union," in Walter Z. Laqueur and Leopold Labedz, eds., *The State of Soviet Science* (Cambridge, Mass.: M.I.T. Press, 1965); Loren R. Graham, "Cybernetics," first published in *Survey* (London) (July 1964), but also available in George Fischer, ed., *Science and Ideolo-*

gy in Soviet Society (New York: Atherton, 1967), pp. 83–106, and in Loren R. Graham, *Science and Philosophy in the Soviet Union* (New York: Knopf, 1972); Lee R. Kerschner, *Cybernetics in Soviet Russia* (Washington, D.C.: Public Affairs Press, 1965); Lee R. Kerschner, "Cybernetics and Soviet Philosophy," *International Philosophical Quarterly* 6, no. 2 (June 1966): 270–85. Each of these works contains good bibliographic references. But the best bibliography to date is that prepared by Donald Dinsmore Comey, "Soviet Publications on Cybernetics," *Studies in Soviet Thought* 6, no. 2 (June 1964): 142–61, with the supplemental information on "Western Translations of Soviet Publications" by Lee R. Kerschner, ibid., pp. 162–77. The present bibliography makes no attempt to duplicate the references in this work; items from the Comey-Kerschner bibliography are included here only on a very selective basis. (N. B. There are at least two major different systems for transliterating Russian into English. We have endeavored to use the Library of Congress system as much as possible.)

With regard to the German Democratic Republic: As already mentioned in the Introduction, the best bibliography of East German literature on the philosophy of technology is Erwin Herlitzius's "Technik und Philosophie," *Informationsdienst Geschichte der Technik* (Dresden) 5, no. 5 (1965): 1–36. Again, works from this bibliography are relisted here only on a very selective basis. Also, although there are a number of German translations of Russian works (and vice versa), these have not been listed. In the case of both Russia and East Germany, we have concentrated on works published since the Comey-Kerschner and Herlitzius bibliographies were prepared.

For more general surveys of East European attitudes toward technology, see the following: Maxim W. Mikulak, "Cybernetics and Marxism-Leninism," in Charles R. Dechert, ed., *The Social Impact of Cybernetics* (New York: Simon & Schuster, 1967);

and Jiri Zeman, "Cybernetics and Philosophy in Eastern Europe," in Raymond Klibansky, ed., *Contemporary Philosophy; a Survey*, Vol. 2: *Philosophy of Science* (Florence: La Nuova Italia Editrice, 1968), pp. 407–13. For more specialized studies in this same area, see: Donald P. Bakker, *The Philosophical Debate in the U.S.S.R. on the Nature of Information* (M.A. thesis, Columbia Univ., 1966); Michael Csizmos, "Cybernetics—Marxism—Jurispuridence," *Studies in Soviet Thought* 11, no. 2 (June 1971): 90–108; Helmut Dahm, "Die Interpretation naturaler und sozialer Regelkreise in der marxistischen Philosophie," *Studies in Soviet Thought* 9, no. 1 (March 1969): 1–26; Lee R. Kerschner, "Cybernetics, Matter and the Dialectic," *Cybernetica* 11, no. 4 (1968): 235–52; Peter Kirschenmann, "On the Kinship of Cybernetics to Dialectical Materialism," *Studies in Soviet Thought* 6, no. 1 (March 1966): 37–41; Peter Kirschenmann, "Problems of Information in Dialectical Materialism," *Studies in Soviet Thought* 8, nos. 2–3 (June–September 1968): 105–21; Peter Kirschenmann, *Information and Reflection; on Some Problems of Cybernetics and How Contemporary Dialectical Materialism Copes with Them* (Dordrecht: Riedel, 1970); and Thomas W. Robinson, "Game Theory and Politics: Recent Soviet Views, *Studies in Soviet Thought* 10, no. 4 (December 1970): 291–315.]

AFANASYEV, V. *The Scientific Management of Society*. Moscow: Progress Publishers, 1970. Pp. 300. Discusses the use of cybernetics in social administration and argues that cybernetics alone cannot solve social problems. A Marxist-Leninist social analysis is needed in which cybernetic methods play a secondary role.

ÁGOSTON, L. "Zum wissenschaftlichen Begriff der Technik und über die Diskussion bezüglich seiner Definition" [On the scientific concept of technology and on the discussion relative to its definition], *Periodica polytechnica: Chemical Engineering–Chemisches Ingenieurwesen* 8, no. 4 (1964): 295–313.

AKCHURIN, I. A. "Development of Cybernetics and the Dialectic" (in Russian), *Voprosy filosofii* 19, no. 7 (1965): 22-30. English summary (p. 184).

AKHIEZER, A. S. "The Scientific and Technological Revolution and the Problem of Governing the Development of Society" (in Russian), *Voprosy filosofii* 22, no. 8 (1968): 12-23. English summary (p. 187). The ability of man to make labor the object of his own development is transformed by Communist society into an ability to make the development of society as a whole the object of his practice. And governing this process of development requires that productive knowledge be converted into a specific type of labor. This is what the scientific and technological revolution accomplishes.

AKHIEZER, A. S., L. B. KOGAN, and L. B. IANITSKY. "Urbanization, Society and the Scientific-Technological Revolution" (in Russian), *Voprosy filosofii* 23, no. 2 (1969): 43-53. English summary (pp. 187-88).

ALBERT, JOHANNES. "Bemerkungen zu Problemen der technischen Revolution" [Remarks on problems of technical revolution], *Wissenschaftliche Zeitschrift der Technischen Universität Dresden* 15, no. 4 (1966): 860-61.

ALBRECHT, ERHARD. *Einführung in die Philosophie*, Part 1: *Philosophie und Einzelwissenschaften* Greifswald: Pressestelle der Ernst-Moritz-Arndt-Universität, n.d. Pp. 236. This book is composed of three essays (two by Albrecht, one by H. Quitzsch) with an outline of problems and relevant bibliographic material. The first article, by Albrecht, is entitled "Logik/ Erkenntnistheorie und Kybernetik als methodologische Grundlagen der Einzelwissenschaften (Mit Erläuterung logischer und erkenntnistheoretischer Grundbegriffe)." But see also the sections of the outline and bibliography on "Philosophische Fragen der Kybernetik" and "Philosophische und soziologische Fragen der Technik."

ALBRING, WERNER. "Das Zusammenwirken von Grundlagenwissen und Technik" [The cooperation of basic knowledge and technology], *Grundlagenforschung und Technik; Sitzungsbe-richte der Deutschen Akademie der Wissenschaften zu Berlin, Klasse für Mathematik, Physik und Technik*, vol. 1 (1967).

ANISIMOV, S. F. *Cybernetics and Computers in the U.S.S.R.; Man and Machine*. Philosophical Problems of Cybernetics. Washington, D.C.: Joint Publications Research Service, Government Printing Office, November 24, 1959. Pp. 53. From *Chelovek i mashina; filosofski problemy kibernitiki* (Moscow: "Znanie," 1959).

ANISIMOV, S. F., and A. VISLOBOKOV. "Nekotorye filosofskie voprosy kibernetiki" [Some philosophical problems of cybernetics], *Kommunist* 2 (1960): 108-18.

ANISIMOV, S. F., and A. WISLOBOKOW. "Einige philosophische Fragen der Kybernetik," *Technik* 15, no. 9 (1960): 577-84.

ANTONOV, N. P., and A. N. KOCHERGIN. "Priroda myshleniia i problema ego modelirovaniia" [The nature of thought and the problem of its modeling]," *Filosofski nauki*, vol. 6 (1963).

APOSTOL, GH. P. "Technological Progress under Socialism and Its Economic and Social Consequences" (in Rumanian), *Cercetari filozofice*, vol. 6, no. 6 (1959).

———. "Technological Progress in the Course of Achieving the Construction of Socialism in the People's Republic of Rumania and the Struggle to Dominate the Forces of Nature" (in Rumanian), *Cercetari filozofice*, vol. 9, no. 5 (1962).

———. "The Application to the Production of Goods of Science and Modern Technology—an Essential Factor in the Development of the Technological and Material Base of Socialism" (in Rumanian), *Cercetari filozofice*, vol. 10, no. 6 (1963).

BABSKII, E. B., and E. S. GELLER. "Cybernetics and Life," *Soviet Studies in Philosophy* 8, no. 4 (Spring 1970): 354-70. From a discussion in *Voprosy filosofii* 23, no. 9 (1969): 124-32, of recent Soviet applications of cybernetics to biology and medicine.

BAREŠ, GUSTAV. *Zrození atomového věku; obavy a naděje* [Birth of the atomic age; fears and hopes]. Prague: Mladá fronto, 1961. Pp. 454.

BAVKOBA, V. G. "Leisure Time and the Improvement of the Technological Level of Engineers and Other Technical Workers" (in Russian), *Voprosy filosofii* 19, no. 4 (1965): 69–74.

BECHER, J., and FRIEDRICH. "Staatlich-rechtliche Probleme der technischen Revolution im Lichte der Subjekt-Objekt-Beziehungen" [Problems of public law of the technological revolution in the light of subject-object relations], *Deutsche Zeitschrift für Philosophie* 15, no. 1 (1967): 26–39.

BECKER, J. "Zur Dialektik von Technik, Ökonomie, Staat und Recht in der wissenschaftlich-technischen Revolution" [On the dialectic of technology, economy, state, and law in the scientific-technological revolution], *Technik* 23, no. 5 (1968): 295–97.

BERG, A. I. *Cybernetics in Service of Communism; Theory of Information, Computer Technology, Semiotics.* Washington, D.C.: Joint Publications Research Service, Government Printing Office, June 29, 1967. From A. I. Berg and E. Kolman, eds., *Kibernetiku–na sluzhbu kommunizmu* (Moscow: "Gosenergoizdat," 1961). Includes bibliographies. A widely circulated essay by the director of a Cybernetic Council established by the U.S.S.R. Academy of Sciences. For earlier versions of this article and translations, see the Comey-Kerschner bibliography.

———. "On Certain Problems concerning Cybernetics," *Soviet Studies in Philosophy* 1, no. 1 (Summer 1962): 57–65. From *Voprosy filosofii*, vol. 14, no. 5 (1960).

BERG, A. I., and IU. I. CHERNIAK. *Informatsiia i upravlenie* [Information and control]. Moscow: "Èdonomika," 1966.

BERG, A. I., and KOLMAN, eds. *Vozmozhnoe i nevozmozhnoe v kibernetike* [Possibilities and impossibilities in cybernetics]. Moscow: "Nauka," 1964. Pp. 221. Essays on the applications of cybernetics to philosophy.

BERG, A. I., et al., eds. *Kibernetika, myshlenie, zhizn'* [Cybernetics, thought, and life]. Moscow: "Mysl'," 1964. For

other articles by Berg, see the Comey-Kerschner bibliography.

BERNICKE, ERHARD. "Technologie und wissenschaftlich-technische Revolution," *Einheit* 22, no. 9 (1967): 1107–15.

BIRIUKOV, B. V. *Discussion of Philosophical Problems of Cybernetics.* Washington, D.C.: Joint Publications Research Service, Government Printing Office, September 15, 1965. Pp. 4.

BIRIUKOV, B. V., and A. G. SPIRKIN. "The Philosophical Problems of Cybernetics" (in Russian), *Voprosy filosofii* 18, no. 9 (1964): 111–19. Included in A. I. Berg, ed., *Kibernetika, myshlenie, zhizn'* (Moscow: "Mysl'," 1964). For other works by Biriukov, see the Comey-Kerschner bibliography. See especially *Conference on the Philosophical Problems of Cybernetics* (Washington, D.C.: Joint Publications Research Service, Government Printing Office, February 21, 1963).

BLUME, P. "Zur allseitigen Durchsetzung der Gruppentechnologie" [On the all-round establishment of group technology], *Technische Gemeinschaft; Zeitschrift für Theorie und Praxis der freiwilligen Gemeinschaftsarbeit der deutschen technischen Intelligenz* (Berlin) 10 (1962): 333–36.

BOBER, J. "Cybernetics and Its Approach to Reality" (in Czech), *Filosofický časopis,* vol. 15, no. 4 (1967).

BOBNEVA, M. I. "Technology and Man" (in Russian), *Voprosy filosofii* 15, no. 10 (1961): 70–81. English summary (pp. 185–86).

———. "On the Problems of the Present Day Scientific and Technological Revolution" (in Russian), *Voprosy filosofii* 18, no. 9 (1964): 154–57.

BOHRING, GÜNTER. "Menschheit und Technik," *Technik* 21, no. 5 (1966): 293–97.

———. "Wesen und Erscheinungsformen der burgerlichen 'Philosophie der Technik' in der Gegenwart," *Wissenschaftliche Zeitschrift der Technishen Hochschule für Chemie Leuna-Merseburg* (Halle-Saale) 11, no. 3 (1969): 200–209.

BORSHCHEV, V. B., F. Z. ROKHLIN, et al. *Some Philosophical Problems of Cybernetics.* Washington, D.C.: Joint Publications Research Service, Government Printing Office, March 20, 1961. Pp. 10.

BUCHHEIM, GISELA. "Zur Revolutionsbegriff in der Entwicklung der Technik" [The notion of revolution in the development of technology], *Wissenschaftlich Zeitschrift der Technischen Universität Dresden* 16, no. 3 (1967): 731-36.

BUDKO, N. S. "Can the Information-processing Equipments Be Considered as Machines?" (in Russian), *Voprosy filosofii* 20, no. 11 (1966): 75-80.

BULOCHNIKOVA, L. A. "The Scientific and Technological Revolution in Agriculture and Its Social Consequences" (in Russian), *Voprosy filosofii* 23, no. 1 (1969): 24-34. English summary (p. 186).

CERNEY, J. "The Philosophical Problem of Practice and Its Contemporary False Interpretations" (in Czech), *Filosofický časopis*, vol. 18, no. 1 (1970).

CIZEK, F. "On the Methodological Significance of Cybernetics" (in Czech), *Filosofický časopis*, vol. 5 (1963).

DESTOUCHES, JEAN-LOUIS. "Technological Development and the Capitalist Society" (in Russian), *Voprosy filosofii* 22, no. 12 (1968): 28-38. English summary (pp. 185-86).

DRĂGHICI, ION. "Structure and Function in Cybernetic-Systems" (in Rumanian), *Analele Universitătii Bucureşti; Seria Ştinţe Sociale, Filozofie* 17 (1968): 34-43.

DRAKIN, V. I., and G. L. SMOLIAN. "All-round Automation: Human Factors" (in Russian), *Voprosy filosofii* 21, no. 2 (1967): 38-48. English summary (p. 185).

DUMITRESCU, D. "Mathematics and Technological Progress" (in Rumanian), *Cercetari filozofice*, vol. 7, no. 5 (1960).

EFIMOV, ANATOLII NIKOLAEVICH. *Rol' tekhniki v razvitii sotsialisticheskogo proizvodstva; Stenogramma lektsii prochitannoi v lektoriakh Sverdlovska* [The role of technology in the development of socialist production; a report on the lectures given in the auditoriums of Sverdlovsk]. Vsesoiuznoe obshchestvo po rasprostranenii politicheskikh nauchnikh znanii, series II, no. 36. Moscow: "Znanie," 1955. Pp. 39.

ELEK, TIBOR. "Über die Wechselbeziehungen zwischen technischen Wissenschaften, Naturwissenschaften und Mathematik" [On the mutual relations between technical science, natural science and mathematics], *Freiburger Forschungsheft*, Series D: *Kultur und Technik* (Leipzig) 53 (1967): 27-40.

ENGLER, B. "Die Bedeutung der Technologie unter den Bedingungen der wissenschaftlich-technischen Revolution," *Fachschule: Zeitschrift für das Direkt-, Fern- und Abendstudium an den Fachschulen der DDR* (Leipzig) 16, no. 4 (1968): 102-5.

EPISKOPOSOV, G. L. *Tekhnicheskii progress i problema sokhraneniia mira; kritika burzhuaznoi filosofii i sotsiologii* [Technological progress and the problem of preservation of peace; criticism of bourgeois philosophy and sociology]. Moscow: Press of Institute for International Relations, 1960. Pp. 58. An interesting study by someone well versed in the most influential Western literature concerned with modern technology (especially nuclear weapons) and its impact on international affairs. Chapter titles: "Technology and Politics," "The Myth of World Peace," "Under the Flag of World Peace—Striving toward the Peaceful Rule of Imperialism," "Peace on the Basis of the Bomb," "Bombs and Ethics," "The Universal and Full Disarmament—the Lasting Guarantee of the Preservation of Peace."

_____. *Tekhnika i sotsiologiia* [Technology and sociology]. Moscow: "Vyshaia shkola," 1967. Pp. 288.

FILIPEC, JINDŘICH, PŘEMYSL MAYDL, and RADOVAN RICHTA. "Zur theoretischen Analyse der wissenschaftlich-technischen Revolution," *Deutsche Zeitschrift für Philosophie* 18, no. 8 (1970): 947-59.

FINEBURG, Z. I. "Contemporary Society and Science Fiction" (in Russian), *Voprosy filosofii* 21, no. 6 (1967): 32-43.

_____. "The Prospects of the Scientific-Technological Revolution and Development of the Individual" (in

Russian), *Voprosy filosofii* 23, no. 2 (1969): 32–42.

———. "Contemporary Stage of Scientific and Technological Revolution and Social Planning" (in Russian), *Voprosy filosofii* 25, no. 10 (1971): 28–38. English summary (p. 186).

FRIEDT, HEINZ. "Die Triebkräfte und die gesellschaftliche Rolle des technischen Fortschritts im Sozialismus und in Kapitalismus" [The driving forces and the social role of technological progress in socialism and capitalism], *Technik* 15, no. 6 (1960): 385–88. For other works by this author see the Herlitzius bibliography.

GEIST, RUDOLF VON, and ERWIN HERLITZIUS. "Der Mensch in technischen Systemen," *Deutsche Zeitschrift für Philosophie* 16, no. 6 (1968): 659–73.

GEORGIEV, F. I., and G. F. KHRUSTOV. Certain Preconditions and Specific Features of Consciousness," *Voprosy filosofii* 19, no. 10 (1965): 14–21. English summary (p. 18). English translation entitled "On the Preconditions and Essential Elements of Consciousness," *Soviet Studies in Philosophy* 4, no. 4 (Spring 1966): 42–48.

GIMEL'SHTEIB, E. KH. "Cybernetics and the Problem of Goals," *Soviet Studies in Philosophy* 4, no. 4 (Spring 1966): 49–55. From *Filosofskie nauki*, vol. 3 (1965).

GLUSHKOV, VIKTOR MIKHAĬLOVICH. "Thinking and Cybernetics" (in Russian), *Voprosy filosofii* 17, no. 1 (1963): 36–48. English summary (p. 182). English translation: *Thought and Cybernetics* (Washington, D.C.: Joint Publications Research Service, Government Printing Office, March 22, 1963); and "Thinking and Cybernetics," *Soviet Studies in Philosophy* 11, no. 4 (Spring 1964): 3–13.

———. *Introduction to Cybernetics*. Translated by Scripta Technica, Inc. Edited by George M. Kranc. New York: Academic Press, 1966. Pp. 322.

GRAFE, H. "Technik und Recht" [Technology and law], *Wissenschaftliche Zeitschrift der Technischen Hochschule Dresden* 4 (1954–55): 141–44.

GROSS, FELIKS. *Druga rewolucja przemyslowa; zagadnienia współczesnego socjalizmu* [The second industrial revolution; problems of contemporary socialism]. Paris: Swiatlo, 1958. Pp. 271.

HASELOFF, F. "Zu einigen Problemen des biologisch-technologischen Prozesses: ein Beitrag zur allgemeinen Technologie," *Fachschule: Zeitschrift für das Direkt-, Fern- und Abendstudium an den Fachschulen der DDR* (Leipzig) 16, no. 1 (1968): 4–10.

HENNEBERG, H. "Die wissenschaft in der technischen Revolution," *Die technische Gemeinschaft; Zeitschrift für Theorie und Praxis der freiwilligen Gemeinschaftsarbeit der deutschen technischen Intelligenz* (Berlin) 14, no. 1 (1966): 8–9.

HEPPENER, SIEGLINDE. "Marxistisch-leninistiche Produktivkrafttheorie und weltanschaulich-theoretische Probleme der wissenschaftlich-technischen Revolution" [Marxist-Leninist theory of productive power and ideological-theoretical problems of the scientific-technical revolution), *Deutsche Zeitschrift für Philosophie* 19, no. 4 (1971): 449–67.

HERING, DIETRICH. "Didaktische Merkmale der wissenschaftlich-technischen Revolution" [Didactic characteristics of the scientific-technological revolution]. *Wissenschaftliche Zeitschrift der Technischen Universität Dresden* 15, no. 4 (1966): 817–20.

———. "Probleme des sozialistischen Bildungswesen in den Zusammenhängen der wissenschaftlich-technischen Revolution" [Problems of socialistic education in connection with the scientific-technological revolution], *Wissenschaftliche Zeitschrift der Technischen Universität Dresden* 16, no. 2 (1967): 455–58.

HERLITZIUS, ERWIN. "Wissenschaftlich-technische Revolution und sozialistisches Schöpfertum" [Scientific-technological revolution and socialist production], *Die technische Gemeinschaft; Zeitschrift für Theorie und Praxis der freiwilligen Gemeinschaftarbeit der deutschen technischen Intelligenz* (Berlin) 15, no. 1 (1967): 8–10.

———. "Philosophische Probleme der technischen Revolution," *Wissenschaftliche Zeitschrift der Technischen*

Universität Dresden 16, no. 2 (1967): 463-68.

HOFMAN, F. "Zur Problematik der Begriffsbestimmung Technologie als Wissenschaft" [On the problematic of defining technology as a science]. *Technik* 20, no. 11 (1965): 727-33.

HOFMANN, G. "Grundlagen und Entwicklung der Technologie als Wissenschaft" [Foundations and development of technology as science], *Fachschule: Zeitschrift für das Direkt-, Fern- und Abendstudium an den Fachschulen der DDR* (Leipzig) 13, no. 11 (1965): 323-30.

HOLLITSCHER, WALTER. *Wissenschaft und Sozialismus heute and morgen* [Science and socialism, today and tomorrow]. Vienna: Kommunistische Partei Österreichs, 1958. Pp. 30.

HÖNTZSCH, R. "Technologie—eine Lebensfrage" [Technology—a vital question], *Die technische Gemeinschaft; Zeitschrift für Theorie und Praxis der freiwilligen Gemeinschaftsarbeit der deutschen technischen Intelligenz* (Berlin) 15, no. 12 (1967): 13-15.

Human Personality and Modern Technology. Washington, D.C.: Joint Publications Research Service, Government Printing Office, March 11, 1969. Pp. 33. Articles from *Nauka i Tekhnika*, vol. 12 (1968).

IL'IN, V. A. "The Congress on Cybernetics Held in Namur (Belgium): Brief Survey," in *Philosophical Problems of Cybernetics* (Washington, D.C.: Joint Publications Research Service, Government Printing Office, May 15, 1961). See also I. Ia. Aksenov, Iu. Ia. Bazilevskiĭ, and R. R. Vasil'ev, "O II mezhdunarodnom kongresse po kibernetike" [On the Second International Congress on Cybernetics]. Pp. 311-19 in A. A. Liapunov, ed., *Problemy kibernetiki*, vol. 2 (Moscow: Fizmatgiz, 1959), which is also available in English as *Problems of Cybernetics* (New York: Pergamon).

——, ed. *Filosofskie voprosy kibernetiki* [Philosophical problems of cybernetics]. Moscow: Socĕkgiz, 1961. Pp. 392. English translation entitled *Philosophical Problems of Cybernetics* (Washington, D.C.: Joint Publications Re-search Service, Government Printing Office, May 15, 1961). For other works by Il'in see the Comey-Kerschner bibliography.

ITELSON, L. B. "Contemporary Technique and Man's Psychological Potentialities" (in Russian), *Voprosy filosofii* 15, no. 4 (1961): 60-70. English summary (p. 183). For other articles by this author, see the Comey-Kerschner bibliography.

JACOB, HORST. "Philosophische Diskussion zum Thema 'Mensch und Technik,' " *Deutsche Zeitschrift fü Philosophie* 7, no. 1 (1959): 143-45.

JANTSCH, RUDOLF. "Die Veränderung der Technologie in der wissenschaftlich-technischen Revolution" [The change of technology in the scientific-technological revolution], *Deutsche Zeitschrift für Philosophie* 15, no. 12 (1967): 1418-30.

JAROSZEWSKI, T. M. "Rewolucja techniczna a humanizm," *Nowe Drogi* 14, no. 2 (1960): 57-70.

JOBST, EBERHARD. "Philosophische Probleme des Wechselverhältnisses von technischer Wissenschaft und Naturwissenschaft" [Philosophical problems of the relationship between technical science and natural science], *Wissenschaftliche Zeitschrift der Technischen Hochschule Karl-Marx-Stadt* 9, nos. 1-2 (1967): 81-92.

——. "Spezifische Merkmale der technischen Wissenschaft in ihrem Wechselverhältnis zur Naturwissenschaft" [Specific characteristics of technological science in its interrelation with natural science], *Deutsche Zeitschrift für Philosophie* 16, no. 8 (1968): 928-35.

JOHN, ERHARD. "Wissenschaftlich-technische Revolution—sozialistisches Menschenbild—sozialistisch realistische Kunst" [Scientific-technical revolution—socialist image of man—socialist realistic art], *Deutsche Zeitschrift für Philosophie* 14, no. 10 (1966): 1255-66.

KAGAN, M. S. "Art as a Systems Object and the Systems Analysis of Art" (in Russian), *Voprosy filosofii* 23, no. 11 (1969): 109-16.

KALLABIS, HEINZ VON. "Der technische

Fortschritt und die Entwickling der Gruppenbeziehungen und -interaktionen im sozialistischen Industriebetrieb" [Technical progress and the development of group relations and interactions in the socialist enterprise], *Deutsche Zeitschrift für Philosophie* 13, no. 12 (1965): 1465-73.

————. "Technischer Fortschritt und das Problem realer Demokratie im Wirtschaftsgeschehen" [Technical progress and the problem of real democracy in economic matters], *Deutsche Zeitschrift für Philosophie* 14, no. 6 (1966): 644-53.

KALMÄR, LÁSZLÓ. "Einige philosophische Probleme der Kybernetic." In *Naturwissenschaft und Philosophie; Beiträge zum internationalen Symposium über Naturwissenschaft und Philosophie anlässlich der 550-Jahr-Feier der Karl-Marx-Universität*, Berlin, 1960.

KANTOR, K. M. "Correlation of Social Organization and Individual under Conditions of the Scientific and Technological Revolution" (in Russian), *Voprosy filosofii* 25, no. 10 (1971): 39-51. English summary (p. 186). A critique of the theories of R. Garaudy.

————. "Scientific and Technological Revolution and the Problem of Changes in the Structure and Historical Role of the Working Class" (in Russian), *Voprosy filosofii* 25, no. 12 (1971): 30-42. English summary (p. 186). Continues the author's analysis of the theories of R. Garaudy and H. Marcuse.

KEDROV, B. M. "The Growing Role of Science in Our Time" (in Russian), *Voprosy filosofii* 21, no. 5 (1967): 15-26. English summary (p. 183).

KELLNER, EVA, and REINHARD MOCEK. "Naturwissenschaft und Ethik; Weltanschauliche Probleme der wissenschaftlich-technischen Revolution in der Auseinandersetzung zwischen Sozialismus und Kapitalismus" [Natural science and ethics; philosophical problems of the scientific-technical revolution in the struggle between socialism and capitalism], *Deutsche Zeitschrift für Philosophie* 17, no. 10 (1969): 1157-79.

KHARLAMOV, ALEKSANDR PAVLOVICH. *V*

tupike burzhuaznogo teknnitsizma [The impasse of bourgeois technicism]. Moscow: "Mycl'," 1967. Pp. 76.

KHURSIN, L. A. "On the Mechanism of Development of Science at the Present Stage of the Scientific and Technological Revolution" (in Russian), *Voprosy filosofii* 25, no. 6 (1971): 34-44. English summary (p. 185).

KIRCHGASSNER,W. "Auswirkungen der wissenschaftlich-technischen Revolution auf die Sprachentwicklung" [The effects of scientific-technological revolution upon linguistic development], *Wissenshaftliche Zeitschrift der Ernst Moritz Arndt-Universität; Gesellschafts- und Sprachwissenschaftliche Reihe* (Greifwald) 15, no. 2 (1966): 161-64.

KLAUDY, P. "Gedanken zur Bedeutung der Technik" [Thoughts on the significance of technology], *Österreichische Ingenieur-Zeitschrift; Zeitschrift des Österreichischen Ingenieur- und Architekten-Vereines* (Vienna) 10, no. 1 (1967): 1-7.

KLAUS, GEORG. *Kybernetik in philosophischer Sicht*. Berlin: Dietz, 1961. Pp. 491. An influencial East German philosopher who applies cybernetics to the problems of dialectical and historical materialsm. For reviews and critical discussions of this work, see J. Cibulka's "O dialektické pojeti kvality a kvalitativnich zmen," *Otazky marxistickej filosofie* 19 (1964): 327-36; V. Stoljarow's "G. Klaus: Kybernetik in philosophischer Sicht," *Deutsche Zeitschrift für Philosophie* 11 (1963): 250-53; H. Metzler's "Information in kybernetischer und philosophischer Sicht," *Deutsche Zeitschrift für Philsophie* 10 (1962): 621-38; W. Thimm's "Zum Verhältnis von Bewusstein und Information," *Deutsche Zeitschrift für Philosophie* 11 (1963): 851-64; H. Vogel's "Materie und Information," *Deutsche Zeitschrift für Philosophie* 11 (1963): 314-23; and B. S. Griaznov's "Cybernetics in the Light of Philosophy" (in Russian), *Voprosy filosofii* 19, no. 3 (1965): 161-65.

————. *Kybernetik und Gesellschaft*. Berlin: VEB Deutscher Verlag der Wissenschaften, 1963. 2d ed. 1965. Pp. 372.

————. "Kybernetische Reflexionen zum

subjektiven Moment der Erkenntnis" [Cybernetic reflections on the subjective impetus of cognition], *Deutsche Zeitschrift für Philosophie* 13, nos. 10-11 (1965): 1293-1302.

———. *Kybernetik und Erkenntnistheorie* [Cybernetics and the theory of knowledge]. Berlin: VEB Deutscher Verlag der Wissenschaften, 1966. Pp. 411.

KLIMENKO, K. "Tekhnicheskii progress v period razvernutogo stroitel'stva kommunizma" [Technological progress in the period of developed building of communism], *Voprosy èkonomiki; ezhemesiachnyi zhurnal* (Moscow) 6 (1959): 3-13.

KOENNE, N. "Zur Frage nach dem Phänomen 'Technik,'" *Österreichische Ingenieur-Zeitschrift Zeitschrift; des Österreichischen Ingenieur- und Architekten-Vereines* (Vienna) 8, no. 11 (1965): 366-71.

KOLMAN, ERNST. "Na obranu kybernetiky" [In defense of cybernetics], *Filosofický časopis* 8 (1960): 515-29. Argues against Pavlov that cybernetics is significant for science and philosophy.

———. "Work, Cybernetics, and Communism" (in Czech), *Filosofický časopis*, vol. 13, no. 1 (1965).

———. "Some Cybernetical Aspects of Cosmology in the Light of Philosophy" (in Russian), *Voprosy filosofii* 23, no. 7 (1969): 96-102. English summary (p. 187). For other works by this author see the Comey-Kerschner bibliography. A select list of translations: "What Is Cybernetics?" *Behavioral Science* 2 (1959): 132-46; *Cybernetics* (Washington, D.C.: Joint Publications Research Service, Government Printing Office, July 6, 1960); "On the Philosophical and Social Problems of Cybernetics," in V. A. Il'in, ed., *Philosophical Problems of Cybernetics* (Washington, D.C.: Joint Publications Research Service, Government Printing Office, December 15, 1961); and *Philosophy and Cybernetics* (Washington, D.C.: Joint Publications Research Service, Government Printing Office, May 2, 1962).

KORIUKIN, V. I., and IU. P. LOBASTOV. "Living Beings, Artificial Creations, and Cybernetics," *Soviet Studies* in *Philosophy* 3, no. 4 (Spring 1965): 32-39. From *Filosofskie nauki*, vol. 1 (1963).

KRAJEWSKI, W. "Computers and Thought; Some Remarks on Methodology" (in Polish), *Studia Filozoficzne*, vol. 1 (1963). English summary.

KRAVCHENKO, I. I., and V. S. MARKOV. "Scientific and Technological Progress and Problems of Development of the Individual under Socialism" (in Russian), *Voprosy filosofii* 25, no. 9 (1971): 26-37. English summary (p. 186).

KRESCHNAK, HORST, and HAROLD ZIMMER. "Gegen die Entstellung des Verhältnisses der marxistischen Philosophie zur Kybernetik" [Against the distortion of the relation of Marxist philosophy to cybernetics], *Deutsche Zeitschrift für Philosophie* 17, no. 1 (1969): 101-5.

KUDRIASHOV, ANATOLII PETROVICH. *Sovremennaia nauchno-tekhnicheskaia revolutsiia i ee osobennosti* [The contemporary scientific-technological revolution and its features]. Moscow: "Mysl'," 1965. Pp. 174. Major chapter titles: "Concerning the Character and Categories of the Contemporary Scientific-Technological Revolution," "The Present Scientific-Technological Revolution and the Intensification of the Rotting of Capitalism," and "The Significance of the Present Scientific-Technological Revolution for the Decisive Objective of Constructing Communism." As these titles indicate, the work is rather ideological. —M. A. S.

KUGEL, S. A. "The Changes Occurring in the Social Structure of Socialist Society under the Impact of Scientific and Technological Revolution" (in Russian), *Voprosy filosofii* 23, no. 3 (1969): 13-22. English summary (p. 186).

KURAKOV, I. G. "The Development of Technology on the Basis of Socialism" (in Russian), *Voprosy filosofii*, vol. 13, no. 1 (1959).

———. *Science, Technology and Communism; Some Questions of Development.* Translated by C. Dedijer. Oxford: Pergamon, 1966. Pp. 126. From *Nauka, Tekknika i voprosy stroitel'stva kom-*

munizma (Moscow: Soc-Econ. Lit., 1963). A Soviet economist's analysis. Using Strumylin's definition of science as "a direct productive force," Kurakov deals with the problem of how to take into account the various kinds of research work inputs in cost-benefit analysis and calculations at the micro- and macroeconomic levels. There is a brief bibliography of Soviet literature.

————. "Forecasting Scientific and Technological Progress" (in Russian), *Voprosy filosofii* 22, no. 10 (1968): 21–35. English summary (pp. 185–86). This as well as some other articles from this volume have been translated and published under the title *Philosophical Problems of Forecasting and Organizing U.S.S.R. Science* (Washington, D.C.: Joint Publications Research Service, Government Printing Office, January 10, 1969).

KUZIN, ALEKSANDR AVRAMIEVICH. *K. Marks i problemy tekhniki* [K. Marx and problems of technology]. Moscow: "Nauka" (U.S.S.R. Academy of Sciences, Institute of History of Natural Science and Technology), 1968. Pp. 112. Major topics covered: I, "The Teaching of Karl Marx concerning the Methods of Material Production"; II, "Capitalistic Industrial Production"; III, "Communism and Technology"; IV, "Technology and Public Awareness"; V, "K. Marx on the Evolution of Technology"; Conclusion; with an appendix collecting the basic definitions given by Marx on the categories of technology and a bibliography of Russian literature. A valuable study. — M. A. S. Reviewed by A. Parry, *Technology and Culture* 11, no. 2 (April 1970): 326–27.

LADA, IGOR' VASIL'EVICH. *Kontury griadushchego* [Contours of the future]. Moscow: "Znanie," 1965. Pp. 379.

LAPORTE, H., and E. PADELT. "Saubere Sprache in der Technik" [Fine language in technology], *Die technische Gemeinschaft; Zeitschrift für Theorie und Praxis der freiwilligen Gemeinschaftsarbeit der deutschen technischen Intelligenz* (Berlin) 12, no. 12 (1964): 542–43.

LASSOW, EKKHARD. "Probleme der Pro-

duktivkrafttheorie in der Periode des umfassenden Aufbaus des Sozialismus und der technisch-wissenschaftlichen Revolution" [Problems of the productive forces theory in the period of the all-round construction of socialism and the technological and scientific revolution], *Deutsche Zeitschrift für Philosophie* 15, no. 4 (1967): 373–98.

LETSCHE, H. "Über das Verhältnis von Technik und Kunst in technizistischer und klerikaler Sicht" [On the relationship between technology and art from technical and clerical points of view], *Deutsche Zeitschrift für Philosophie* 15, no. 2 (1967): 192–201.

LEVI, E. I. *Contemporary Problems of Bionics and Their Philosophical Significance*. Washington, D.C.: Joint Publications Research Service, Government Printing Office, December 12, 1966. From *Filosofskie nauki*, vol. 5 (1966).

LIEBSCHER, HEINZ. "Kybernetik und Methodik methodologischer Forschung" [Cybernetics and methodics of methodological research], *Deutsche Zeitschrift für Philosophie* 15, no. 7 (1967): 821–27.

LIAPUNOV, A. A., ed. *Problemy Kibernitiki*. Moscow: Fizmatgiz, 1958-. A yearly collection of essays. Published in English as *Problems of Cybernetics* (New York: Pergamon, 1960-).

LIAPUNOV, A. A., and A. I. KITOV. "Cybernetics in Technology and Economics" (in Russian), *Voprosy filosofii* 15, no. 9 (1961): 79–88. English summary (p. 185).

LOHMANN, HANS. "Die Technik und ihre Lehre" [Technology and its teaching], *Wissenschaftliche Zeitschrift der Technische Hochschule Dresden* 3 (1953–54): 601–29.

LÖSER, WOLFGANG. "Was ist Technik?" *Berufsbildung* 15 (1961): 63–66.

————. "Zur kybernetischen Darstellung von ökonomischen Systemen" [On the cybernetic description of the economic system], *Deutsche Zeitschrift für Philosophie* 14, no. 10 (1966): 1276–84.

————. "Bemerkungen zur Definition der Technik" [Remarks on the definition of technology], *Freiberger Forschungsheft*, Series D: *Kultur und Technik* (Leipzig) 53 (1967): 205–9.

LÜBBE, HERMANN. "Zur politischen Theorie der Technokratie," *Staat* 1, no. 1 (1962): 19–38.

MÁCHA, K. "The Problem of Isolated Man in an Industrial Civilization" (in Czech), *Filosofický časopis*, vol. 13, no. 5 (1965).

MARAKHOV, B. G., and Y. S. MELESHCHENKO. "The Present-Day Scientific and Technological Revolution and Its Consequences under Socialism" (in Russian), *Voprosy filosofii* 20, no. 3 (1966): 129–40. German translation in *Sowjetwissenschaft; Gesellschaftswissenschaftliche Beiträge* 10 (1966): 1065–76.

MASAL'SKII VALENTIN NIKOLAEVICH. *Protiv fal'sifikatsii posledstvii tekhnicheskogo progressa pri kapitalizme* [Against falsification of the consequences of technological progress under capitalism]. Moscow: "Mysl'," 1965. Pp. 157.

MECHANIK, G. "Die wissenschaftlich-technische Revolution und ihre Entwicklung auf die kapitalistische Wirtschaft" [The scientific-technological revolution and its development upon the capitalistic economy], *Sowjetwissenschaft; Gesellschaftswissenschaftliche Beiträge* 4 (1967): 342–56.

MEDUNIN, A. E. "The Influence of the Scientific and Technological Revolution on the Nature of Our Planet" (in Russian), *Voprosy filosofii* 23, no. 3 (1969): 23–33. English summary (p. 186). Recognizes the need to study man's interaction with the "bio-technological milieu," man's "secondary nature," but rejects the pessimistic attitude which concludes that man cannot adapt himself to his new artificial environment. All that is needed is strict scientific prognostication of the most probable trends of future development in order to introduce correctives into the main directions of technological progress.

MELESHCHENKO, YURII SERGEEVICH. *Chelovek, obshchestvo i tekhnika* [Man, society and technology]. Lenigrad: "Leninzdat," 1964. Pp. 342.

———. "Technology and the Regularities of Its Development" (in Russian), *Voprosy filosofii* 19, no. 10 (1965): 3–13. English summary (p. 183). German translation in *Sowjetwissenschaft; Gesellschaftswissenschaftliche Beiträge* 2 (1966): 121–32.

———. *Tekhnicheskii progress i ego zakonomernosti* [Technological progress and its regularities]. Lenigrad: "Leninzdat," 1967. Pp. 171.

———. "Character and Pecularities of Scientific and Technological Revolution" (in Russian), *Voprosy filosofii* 22, no. 7 (1968): 13–24.

———. V. I. Lenin and Problems of Technology" (in Russian), *Voprosy filosofii* 23, no. 6 (1969): 3–13.

MERKEL, G. "Erkenntnistheoretische Probleme in der Technik" [Epistemological problems of technology], *Technik* 15, no. 5 (1960): 325–30.

MEYER, G. "Ansatzpunkte einer Methodik der Lehre der Technik" [Toward a methodology of technological education], *Wissenschaftliche Zeitschrift der Technischen Hochschule Dresden* 8 (1958–59): 1111–18.

MEYER, HEINZ. "Zur näheren Bestimmung der Technik als gesellschaftliche Erscheinung" [Toward a clearer specification of technology as a social phenomenon], *Wissenschaftliche Zeitschrift der Technischen Hochschule Karl-Marx-Stadt* 10, no. 6 (1968): 703–11.

———. "Das Verhältnis Mensch und Technik—ein geistiges Grunproblem unserer Epoche" [The relation of man and technology—a fundamental spiritual problem of our epoch], *Wissenschaftliche Zeitschrift der Technischen Hochschule Karl-Marx-Stadt* 11, no. 2 (1969): 251–63.

MILLER, M. "Technik und Gesellschaft," *Die technische Gemeinschaft; Zeitschrift für Theorie und Praxis der freiwilligen Gemeinschaftsarbeit der deutschen technischen Intelligenz* (Berlin) 5 (1957): 531–35.

MILLIOSCHCHIKOV, M. *Socialism Speeds Scientific and Technical Progress.* Washington, D.C.: Joint Publications Research Service, Government Printing

Office, October 9, 1969. Pp. 7. From *Izvestiia* (September 4, 1969).

MINERVIN, G. B., and M. V. FËDOROV. "Concerning Technological Aesthetics" (in Russian), *Voprosy filosofii* 19, no. 7 (1965): 105–13.

MOISEEV, V. D. *Central Ideas and Philosophical Principles of Cybernetics.* Washington, D.C.: Joint Publications Research Service, Government Printing Office, April 21, 1966. Pp. 228.

MOVSEENKO, G. M. "The Scientific and Technological Intellectuals in Capitalist Society" (in Russian), *Voprosy filosofii* 25, no. 12 (1971): 43–54.

MÜLLER, HEINZ. "Grundfragen der freiwilligen Gemeinschaftsarbeit" [Basic questions of voluntary cooperative work], *Die technische Gemeinschaft; Zeitschrift für Theorie und Praxis der freiwilligen Gemeinschaftsarbeit der deutschen technischen Intelligenz* (Berlin) 4 (1956): 406.

MÜLLER, JOHANNES. "Über das Wesen konstruktiver Aufgaben in der Technik" [On the essence of constructive tasks in technology], *Deutsche Zeitschrift für Philosophie* 13, no. 9 (1965): 1094–1109.

———. "Zur Bedeutung des Strukturbegriffs in den technischen Wissenschaften" [The meaning of the notion of structure in the technological sciences], *Wissenschaftliche Zeitschrift der Humboldt-Universität; Mathematisch-naturwissenschaftliche Reihe* 16, no. 6 (1967): 933–38.

———. "Zur Bestimmung des Begriffe 'Technik' und 'technisches Gesetz' " [Concerning the definition of the terms "technology" and "technical law"], *Deutsche Zeitschrift für Philosophie* 15, no. 12 (1967): 1431–49.

———. "Operationen und Verfahren des problemlösenden Denkens in der konstruktiven technischen Entwicklungsarbeit — eine methodologische Studie" [Operations and procedures of problem-solving thinkers in the constructive technological development of work — a methodological study], *Wissenschaftliche Zeitschrift der Technischen Hochschule Karl-Marx-Stadt*, vol. 9, nos. 1–2 (1967).

———. "Zum Verhältnis von Naturwissenschaft und Technik" [On the relations of natural science and technology], *Freiberger Forschungsheft*, Series D: *Kultur und Technik* (Leipzig) 53 (1967): 163–70.

NOVIK, I. B. *Kibernetika, Filosofskie i sotsiologischeskie problemy* [Cybernetics, philosophical and sociological problems]. Moscow: "Gospolitizdat," 1963.

———. *Philosophical Questions of Cybernetic Modeling.* Washington, D.C.: Joint Publications Research Service, Government Printing Office, March 10, 1965. Pp. 44. Contains bibliography.

———. "Die Oktoberrevolution und Probleme der wissenschaftlich-technischen Revolution" [The October Revolution and problems of the scientific-technological revolution], *Deutsche Zeitschrift für Philosophie* 15, no. 10 (1967): 1149–67. Translated from Russian.

OLSHANSKII, V. B. "The Development of Technology and Social Progress" (in Russian), *Voprosy filosofii* 15, no. 9 (1961): 117–21.

OMAROV, ALIM MAGOMEDOVICH. *Teknika i chelovek; sotsial'no-èkonomicheskie problemy tekhnicheskogo progressa* [Technology and man; socioeconomic problems of technological progress]. Moscow: Izdatel'stvo sotsial'no-èkonomocheskoi literatury, 1965. Pp. 157. Devoid of gross ideological rhetoric, but nevertheless man is understood to be primarily economic man. Chapter titles: "Technology and the Relativity of Its Applications"; "Division, Unification and Changes in Labor"; "Automation and the Problem of Employment"; and "Conditional Labor in Production." — M. A. S.

OSIPOV, GENNADII VASIL'EVICH. *Tekhnika i obshchestvennyi progress; kriticheskii ocherk sovremennyz reformistskikh i revizionistskikh sotsiologicheskikh teorii* [Technology and social progress; a critical essay on contemporary reformist and revisionist socialist theories]. Moscow: Press of Academy of Sciences U.S.S.R., 1959. Pp. 261.

OSTROVITIANOV, IA. K. "The 'Post-Industrial Civilization' or Capitalism in the Year 2000?" (in Russian), *Voprosy filosofii* 23, no. 7 (1969): 30–41.

English summary (p. 183). English translation entitled " 'Post-Industrial Civilization' or Capitalism in the Year 2000?" *Soviet Studies in Philosophy* 8, no. 3 (Winter 1969-70): 252-72. Also in *Soviet Review* 9, no. 2 (Summer 1970): 99-119.

PAGEL, W., and ZIMMER, H. "Zur Bestimmung des Gegenstandes der Wissenschaft Technologie und ihrer allgemeinen Methodolgie" [How to determine the subject of scientific technology and its general methodology], *Deutsche Zeitschrift für Philosophie* 11, no. 3 (1963): 301-13.

PANOV, D. I. "The Interaction of Man and Machine" (in Russian), *Voprosy filosofii* 21, no. 1 (1967): 40-49. English summary (p. 184). English translation entitled "On the Interaction of Man and Machines," *Soviet Studies in Philosophy* 6, no. 3 (Winter 1967-68): 14-22. For other work by this author, see the Comey-Kerschner bibliography.

PAVLOV, TODOR. "Teorie odrazu a kybernetika" [Theory of reflection and cybernetics], *Filosofický časopis* 8, no. 2 (1960): 209-320. The author "regards Wiener's concept of information as the central concept of cybernetics; consequently, he considers the independence of information of matter and energy as inconsistent with a materialistic world view, and as a sympton of idealism." — Jiri Zeman, "Cybernetics and Philosophy in Eastern Europe," in R. Klibansky, ed., *Contemporary Philosophy; a Survey*, Vol. 2: *Philosophy of Science* (Florence: La Nuvoa Editrice, 1968), p. 409. For a critique of Pavlov, see E. Kolman, "Na obranu kybernetiky [In defense of cybernetics], *Filosofický časopis* 8 (1960): 515-29.

PERLO, VICTOR. "Social Consequences of the Technological Revolution" (in Russian), *Voprosy filosofii*, vol. 13, no. 11 (1959).

"Philosophical and Sociological Problems of the Scientific and Technological Revolution" (in Russian), *Voprosy filosofii* 23, no. 4 (1969): 3-15. An editorial.

Philosophy of Science and Technology.

Washington, D.C.: Joint Publications Research Services, Government Printing Office, January 6, 1966. Pp. 33. Articles from *Voprosy filosofii.*

PICK, V. *Some Philosophical Problems of Cybernetics.* Washington, D.C.: Joint Publications Research Service, Government Printing Office, August 9, 1965. Pp. 11. From *Lekarsky obzor*, vol. 14, no. 5 (May 1965).

POGODDA, HANS. "Technik und Natur, Technik und Kunst" [Technology and nature, technology and art], *Deutsche Zeitschrift für Philosophie* 18, no. 1 (1970): 57-76.

POVAROV, G. N. "Logic in the Service of Automation and Technological Progress" (in Russian), *Voprosy filosofii*, vol. 13, no. 10 (1959). For other works by this author, see the Comey-Kerschner bibliography.

Problems of Philosophy and Cybernetics. Washington, D.C.: Joint Publications Research Service, Government Printing Office, November 9, 1965. Pp. 70. Articles from *Voprosy filosofii.*

RAPOPORT, A. "Approaches to General Systems Theory" (in Polish), *Studie Filozoficzne*, vol. 1 (1963). English summary.

REIPRICH, KURT. "Zum Technikverständnis in Katholischer Sicht" [On the understanding of technology from a Catholic perspective], *Freiberger Forschungsheft*, Series D: *Kultur und Technik* (Leipzig) 53 (1967): 221-26.

RICHTER, MILOSLAV. *Plánování rozvoje vědy a techniky* [Planning the use of science and technology]. Prague: Statní nakladatelství technick literatury, 1963. Pp. 102.

ROUBINSKII, IA. I. "Theories of 'Technocracy' in France" (in Russian), *Voprosy filosofii*, vol. 14, no. 9 (1960).

RUBINSTEIN, MODEST ISOIFOVICH. *Burzhuazhaia nauka i tekhnika na sluzhbe amerikanskogo imperializma* [Bourgeois science and technology in the service of American imperialism]. Moscow: Press of Academy of Sciences U.S.S.R., 1951. Pp. 415.

RUHLE, DAGMAR, and WOLFGANG RUHLE. "Zum Problem det Mensch-Maschine-Symbiose," *Wissenschaftliche Zeitschrift der Humboldt-Universitat; Ges-*

sellschafts- und sprachwissenschaftliche Reihe 14, no. 2 (1965): 207-16.

RUTKEVICH, M. N. "Progress in Science and Technology in Relation to Art," *Soviet Studies in Philosophy* 11, no. 3 (Winter 1963-64): 44-50. Translation of a paper delivered to the XIIIth International Congress of Philosophy, Mexico City, 1963.

SĂHLEANU, V. "The Methodological Value of Cybernetics and the Theory of Information, in Light of Dialectical Materialism" (in Rumanian), *Cercetari filozofice*, vol. 8, no. 4 (1961).

SCHAUER, HEINZ. "Die technische Revolution und das Kollektiv im Blickwindel der Freiheit der Persönlichkeit" [The technological revolution and the collective in the perspective of the freedom of personality], *Wissenschaftliche Zeitschrift der Technischen Universität Dresden* 15, no. 1 (1966): 181-90.

———. "Zu den Problemen der Verantwortung in der technischen Revolution" [On the problems of responsibility in the technological revolution], *Wissenschaftliche Zeitschrift der Technischen Universität Dresden* 15, no. 4 (1966): 833-37.

SCHEININ, J. "Die wissenschaftlich-technische Revolution und einige Probleme der Gegenwart" [The scientific-technological revolution and some problems of the present], *Sowjetwissenschaft: Gesellschaftwissenschaftliche Beiträge* 9 (1968): 954-69.

SCHMIDT, H. "Die Rolle des Technologen und der Technologie in unseren sozialistischen Betrieben" [The role of the technologist and of technology on our socialist operations], *Feingeräte-Technik* 8 (1959): 7-10.

SCHULZ, ROBERT. "Der sozialistische Mensch meistert die Technik" [Socialistic man masters technology], *Deutsche Zeitschrift für Philosophie* 8, no. 8 (1960): 1009-12.

SCHWARZ, THEODOR. "Zur Technikphilosophie Ernst Jüngers" [Concerning Ernst Jünger's philosophy of technology], *Deutsche Zeitschrift für Philosophie* 15, no. 5 (1967): 528-35.

Scientific and Philosophical Applications of Cybernetics in Military Science. Washington, D.C.: Joint Publications Research Service, Government Printing Office, November 3, 1966. Pp. 17. From *Krasnaia Zvezda* (September 1, 1966); articles and bibliography.

Scientific and Technological Revolution and Its Social Consequences. Moscow: Progress Publishers, 1971. Pp. 192. A collection of articles by academicians M. Millionshchikov, N. Semyonov, N. Baibakov, G. Sorokin, and other Soviet economists, philosophers, historians, and sociologists.

SELBMANN, FRITZ. "Neue Probleme der Technik," *Bergakademie* 8 (1956): 312-18. Also in *Bergbautechnik*, new series 6 (1956): 411-15.

SEMËNOV, N. N. "The Humanism of Science," *Survey*, vol. 52 (July 1963). Translated from "Gumanizm nauki," *Nauka i Chelovechestvo*, vol. 2 (1963), as part of a special issue on "Prospects of Science."

SEVASTIANOV, V. I., and A. D. URSUL. "Space Age: New Relationship between Society and Nature" (in Russian), *Voprosy filosofii* 25, no. 3 (1971): 107-16. English summary (p. 186). English translation entitled "New Interrelations of Society and Nature in the Space Age," *Soviet Studies in Philosophy* 10, no. 2 (Fall 1971): 158-75.

SHALIUTIN, S. M. *Cybernetics and Religion.* Washington, D.C.: Joint Publications Research Service, Government Printing Office, October 13, 1964. Pp. 71.

SHKARATAN, O. I. "The Working Class of Socialist Society in the Era of the Scientific and Technological Revolution" (in Russian), *Voprosy filosofii* 22, no. 11 (1968): 14-25. English summary (p. 185).

SIMUSH, P. I. "The Impact of the Scientific and Technological Revolution on the Socialist Countryside" (in Russian), *Voprosy filosofii* 22, no. 11 (1968): 26-36. English summary (p. 185).

ŠINDELÁŘ, JAN. *Technika a humanismus v západoněmeckém myšlení; o německé kritice techniky se zvláštnim zřetelem k romantozujicim prodům* [Technology and humanism in West German thought; concerning the German criticism of technology with special attention to romantic notions]. Prague: Academia, 1967. Pp. 245.

SMIRNOV, A. D. "Socialism, the Scientific and Technological Revolution and Long-Term Forecasting" (in Russian), *Voprosy filosofii* 22, no. 9 (1968): 9-18. English summary (p. 186).

SMIRNOV, S. N. *Philosophical Basis for Subject Matter of Bionics.* Washington, D.C.: Joint Publications Research Service, Government Printing Office, September 29, 1969. Pp. 11. From *Filosofskie nauki,* vol. 4 (1969).

SMOLIAN, G. L. "Technology and Brain" (in Russian), *Voprosy filosofii* 19, no. 5 (1965): 83-94. English summary (p. 185). English translation entitled *Technology of Brain* (Washington, D.C.: Joint Publications Research Service, Government Printing Office, December 29, 1965).

SOBOLEV, S. L., A. I. KITOV, and A. A. LIAPUNOV. "Osnovnye čerty kibernitiki," *Voprosy filosofii* 9, no. 4 (1955): 136-48. English translation entitled *The Basic Features of Cybernetics* (Washington, D.C.: Joint Publications Research Service, Government Printing Office, February 25, 1958). For other articles by all three authors, see the Comey-Kerschner bibliography.

SOKOLOV, I. A. "Scientific and Technological Revolution and Revolutionary Progress" (in Russian), *Voprosy filosofii* 25, no. 4 (1971): 13-24. Part I of this article published in *Voprosy filosofii* 25, no. 3 (1971): 70-82. English summary of Parts I and II (p. 185).

SOKOLOV, V. V. "Francis Bacon—Philosopher-Innovator of His Age" (in Russian), *Voprosy filosofii* 15, no. 4 (1961): 86-97. English summary (p. 183).

Sotsial-no-èkonomicheskie problemy tekhnicheskogo progressa; materialy nauchnoi sessii [Social-economic problems of technological progress; materials from the science sessions]. Moscow: Press of Academy of Sciences U.S.S.R., 1961. Pp. 478.

SOUSEDIK, S. "Technology and the Philosophical Discussion of the 17th Century" (in Czech), *Filosofický časopis,* vol. 15, no. 1 (1967).

"Sovremennaia nauchno-tekhnicheskaia revoluitsiia sostoianie issledovanii (materialy soveshchaniia)" [The contemporary scientific-technical revolution; the state of studies (documents from a conference)], *Voprosy istorii estestvoznaniia i tekhniki,* no. 2 (1970), pp. 3-23. Summaries of papers presented at a June 1963 conference in Moscow.

STAUFENBIEL, FRED. "Zur Wechselwirkung von technischer Revolution und sozialistischer Kulturentwicklung" [On the correlation of technological revolution and development of socialist culture], *Deutsche Zeitschrift für Philosophie* 13, no. 12 (1965): 1474-86.

SZEFLER, STANISLAW. *Postęp techniczny a życie człowieka* [Technical progress and human life]. Warsaw: Książka i Wiedza, 1966. Pp. 165.

TARANSENKO, F. P. "Concerning the Definition of the Concept of Information in Cybernetics" (in Russian), *Voprosy filosofii* 17, no. 4 (1963): 76-84. English summary (p. 183). English translation entitled "Towards a Definition of 'Information' in Cybernetics," *Soviet Studies in Philosophy* 2, no. 4 (Spring 1964): 14-22.

TEICHMANN, DIETER. "Über das sozial-historische Wesen der Technik" [On the social-historical essence of technology], *Maschinenbau-Technik* 9 (1960): 273-76.

———. "Technik, Techniker und Klassenkampf" [Technology, Technician and the class struggle], *Maschinenbau-Technik* 10 (1961): 57-60.

———. "Die politische Verantwortung und politische Bewusstheit des Technikers" [The political responsibility and political consciousness of the technologist], *Maschinenbau-Technik* 10 (1961): 169-72.

———. "Zur Klassifikation der technischen Wissenschaften" [On the classification of technological science], *Freiberger Forschungsheft,* Series D: *Kultur und Technik* (Leipzig) 53 (1967): 199-203.

TESSMAN, KURT. "Vom Wesen des technischen Fortschritts in der Gegenwart" [On the essence of technological progress in the present], *Deutsche Zeitschrift für Philosophie* 7, no. 5 (1959): 743-58.

———. "Technische Revolution und Sozialismus," *Einheit* 20, no. 2 (1965): 15-22; and 20, no. 4 (1965): 36-42.

————. "Was ist technische Revolution?" *Die technische Gemeinschaft; Zeitschrift für Theorie und Praxis der freiwilligen Gemeinschaftsarbeit der deutschen technischen Intelligenz* (Berlin) 14, no. 2 (1966): 29–31.

————. "Die wissenschaftlich-technische Revolution und das System des Sozialismus" [The scientific-technical revolution and the system of socialism], *Deutsche Zeitschrift für Philosophie* 15, no. 3 (1967): 291–309.

————. "Zur Bestimmung der Technik als Gesellschaftliche Erscheinung" [Concerning the determination of technology as a social phenomenon], *Deutsche Zeitschrift für Philosophie* 15, no. 5 (1967): 509–27.

————. "Struktur, System und Prozess in der Technik," *Wissenschaftliche Zeitschrift der Humboldt-Universität; Mathematisch-naturwissenshaftliche Reihe* 16, no. 6 (1967): 925–28.

————. "Für die Überwindung mechanisch-materialistischer Auffassungen von der Technik" [Overcoming mechanistic-materialistic conceptions of technology], *Freiberger Forschungsheft, Series D: Kultur und Technik* (Leipzig) 53 (1967): 171–97.

————. "Wissenschaftlich-technische Revolution und philosophischer Revisionismus" [Scientific-technical revolution and philosophical revisionism], *Deutsche Zeitschrift für Philosophie* 17, no. 10 (1969): 1240–57.

TESSMAN, KURT, and HEINRICH VOGEL. "Die Struktur der Technik und ihre Stellung im sozialen Prozess" [The structure of technology and its place in the social process], *Deutsche Zeitschrift für Philosophie* 15, no. 12 (1967): 1493–1512.

THEIRETZ-BACHER, H. "Die Stellung der Technik zu Recht und Moral" [The position of technology regarding justice and morals], *Österreichische Ingenieur-Zeitschrift; Zeitschrift des Österreichischen Ingenieur- und Architekten-Vereines* (Vienna) 8, no. 11 (1965): 371–74.

"Theoretical Problems of the Scientific-Technical Revolution" (in German). *Wissenschaftliche Zeitschrift der Universität Rostock: Gesellschafts- und sprachwissenschaftliche Reihe* 14, nos. 5–6 (1965); 531–90. A symposium containing the following papers: H. Parthey, H. Tessmann, K. Tessmann, and H. Vogel's "Theses"; H. Tessmann's "The Contribution of the Socialist Unity Party to the Theory of the Scientifical-Technical Revolution"; H. Vogel's "On the Character of the Scientifical-Technical Revolution and on the Interaction of Its Material and Ideal Factors"; K. Tessmann's "On the Nature of Material and Ideal Factors of the Scientifical-Technical Revolution"; H. Parthey's "Science as a Form of Social Consequences and Its Function as Productive Force"; K. Tessmann's On the Social Content of the Scientifical-Technical Revolution"; H. Ley's "On Questions of the Nature of Material and Ideal Factors of the Scientifical-Technical Revolution"; and K. Tessmann's "Remarks on the Article of Hermann Ley." See also in this same issue two other symposia: (1) "Natural Science and Cybernetics," with the following papers: H. Vogel's "Philosophy and Cybernetics (On Philosophical Problems of Cybernetics)"; F.-H. Lange's "Cybernetics and Technics (Introduction into the Problems of Thinking Automatons)"; I.-O. Kerner's "The Efficiency of Counting Automatons"; I. Fenyo's "The Application of Electronic Counters in Medicine"; and K.-H. Otto's "On the Arrangement of a Literature Card Index by Means of Perforated Cards (Slot Punched Cards)." (2) "Structure and Function of the Experimental Method," with the following papers: K. Berka, H. Parthey, K. Tessmann, H. Vogel, W. Wächter, and D. Wahl's "Theses"; H. Parthey and W. Wächter's "Remarks on the Theory of the Experimental Method"; K. Berka's "Remarks on the Logical Basis of the Experimental Method"; H. Vogel's "On the Relation between Experiment and Theory (Under Special Consideration of the Views of Max Born)"; W. Wächter's "On the Problems of Methods in Natural Science in the Sight of Max Hartmann"; D. Wahl's "Application Problems of the Experimental Method in Social Sciences"; K. Tessmann's "The Application of the Experimental Method in the Technical Sciences"; J. Müller's "On the Problem of Isolation in Experimenting and on the Position of the Experiment in the Technical Sciences"; and W. Wächter's "Purposiveness and Conditionality." All papers are in German.

TIKHOMIROV, O. K. "Heuristics of Man and Machine" (in Russian), *Voprosy filosofii* 20, no. 4 (1966): 99–109. English summary (p. 185).

TIUKHTIN, V. S. "Reflection and Information" (in Russian), *Voprosy filosofii* 21, no. 3 (1967): 41–52. English summary (pp. 183–84). English translation entitled "Information Theory and the Cognitive Process," *Soviet Studies in Philosophy* 6, no. 2 (Fall 1967): 3–13.

_____. "System-and-Structural Approach and the Specificity of Philosophical Knowledge" (in Russian), *Voprosy filosofii* 22, no. 11 (1968): 47–58. English summary (p. 186). English translation entitled *Systematic-Structural Approach and Specific Nature of Philosophical Knowledge* (Washington, D.C.: Joint Publications Research Service, Government Printing Office, January 9, 1969).

TRAPEZNIKOV, S. P. "Leninism and the Present-Day Scientific and Technological Revolution" (in Russian), *Voprosy filosofii* 24, no. 4 (1970): 3–16.

UIZERMAN, T. "Technofobija—bolezn' socializma," *Kommunist* 42, no. 9 (1965): 103–11.

UKRAINTSEV, B. S. "The 'Activity' and 'Purpose' Categories in the Light of the Fundamental Concepts of Cybernetics" (in Russian), *Voprosy filosofii* 21, no. 5 (1967): 60–69. English summary (p. 184). For other work by this author, see the Comey-Kerschner bibliography.

ULBRICHT, G. "Zur Frage der Notwendigkeit der Planung geistiger Arbeit in der Technik" [On the question of the necessity of planning intellectual work in technology], *Technik* 14, no. 7 (1959): 449–52.

_____. "Über die Möglichkeit der Planung geistiger Arbeit in der Technik" [On the possibility of planning of in-

tellectual work in technology], *Technik* 15, no. 1 (1960): 1–5.

ULLE, DIETER. "Technik und Kultur in der westdeutschen bürgerlichen Kultursoziologie" [Technology and culture in West German bourgeois cultural sociology], *Deutsche Zeitschrift für Philosophie* 16, no. 1 (1968): 74–89.

_____. "Critical Notes to the Social Philosophy of Herbert Marcuse" (in Russian), *Voprosy filosofii* 23, no. 3 (1969): 77–86.

URSUL, A. D. "On the Problem of the Nature of Information" (in Russian), *Voprosy filosofii* 19, no. 3 (1965): 119–30. English summary (p. 185). English translation entitled "On the Nature of Information," *Soviet Studies in Philosophy* 5, no. 1 (Summer 1966): 37–46. See also the author's "The Non-statistical Approach in the Theory of Information" (in Russian), *Voprosy filosofii* 21, no. 2 (1967): 88–97.

VASIL'CHUK, IU. A. "The Present-Day Scientific and Technological Revolution and the Industrial Proletariat" (in Russian), *Voprosy filosofii* 23, no. 1 (1969): 11–23. English summary (p. 186).

VOGEL, HEINRICH. "Philosophical Aspects in the Work 'Automat und Mensch' by Steinbuch," *Wissenschaftliche Zeitschrift der Universität Rostock; Gesellschafts- und sprachwissenschaftliche Reihe* 18, no. 1 (1969): 67–71.

VOLKOV, GENRIKH NIKOLAEVICH. "Automation as a New Historical Stage in Development of Technology" (in Russian), *Voprosy filosofii* 18, no. 6 (1964): 15–26. English summary (p. 184).

_____. "Society in the Technical Age," *Diogenes* 55 (Fall 1966): 16–27.

_____. *Era of Man or Robot? The Sociological Problems of the Technical Revolution.* Translated by I. Sokolov. Moscow: Progress, 1967. Pp. 182. From *Èra rabotov ili èra chekoveka? Sotsiologicheskie problemy razvitiia tekniki.* Moscow: Uzdatel'stvo politicheskoi lit-ry, 1965. Chap. 1 contains a brief conceptual discussion of the definition of technology and the proper approach to its study. Conclusion: "Technology is the system of artificial organs of activity of social man, organs of his power over nature, formed in the historical process of the embodiment in natural material of labour functions, habits, experience and knowledge, and in the cognition and use in production of the forces and laws of nature" (pp. 33–34). Technique also defined as "potential technology." The rest of the book argues that "fully automated production is incompatible with capitalism, since ... it drives capitalism into a dead end of insoluble contradictions.... Automation is the technology of communism" (p. 18).

―――. "Interconnections of Science and Production" (in Russian), *Voprosy filosofii* 21, no. 2 (1967): 27–37.

―――. *Sotsiologiia nauki; Sotsiol ocherki nauchno-tekhnicheskoi deiatel'nosti* [Sociology of science; sociological essays on scientific-technological activities]. Moscow: "Politizdat," 1968. Pp. 328.

―――. "Changing the Social Orientation of Science" (in Russian), *Voprosy filosofii* 23, no. 1 (1969): 35–46. English summary (pp. 186–87).

Völz, H. von. "Einige Gedanken zum Begriff Information" [Some ideas concerning the term "information"], *Deutsche Zeitschrift für Philosophie* 16, no. 3 (1968): 336–48.

Wahl, Dietrich. "Zur Theorie der experimentellen Methode" [On the theory of the experimental method], *Deutsche Zeitschrift für Philosophie* 16, no. 1 (1968): 107–12.

Wambutt, Horst. "Technische Revolution und kompletter Anlagenbau" [Technological revolution and comprehensive planning], *Einheit* 21, no. 5 (1966): 650–59.

Wolkow, G. N. "Der Wandel in der sozialen orientierung der Wissenschaft" [The change in the social orientation of science], *Sowjetwissenschaft Gesellschaftswissenschaftliche Beiträge* 9 (1969): 709–20. Uses "science" in a sociological sense to mean the activity of producing new knowledge. In the development of science from ancient to modern times one can distinguish three main functions of science: (1) the ideological ori-

entation of science related to the people, (2) the orientation toward technology, and (3) the orientation toward the all-around developed people, toward the integration of science and sociology.

Wüstneck, Klaus Dieter. "Zur Definition der technischen Revolution" [On defining the technological revolution], *Wissenschaftliche Zeitschrift der Technischen Universität Dresden* 15, no. 4 (1966): 809–11.

―――. "Zur Bestimmung der technischen Revolution durch den historischen Materialismus" [On the definition of the technological revolution by historical materialism], *Deutsche Zeitschrift für Philosophie* 15, no. 10 (1967): 1229–34.

Zaleski, E., et al. *Science Policy in the U.S.S.R.* Paris: Organization for Economic Cooperation and Development, 1969. Pp. 615.

Zamoshkin, A., and N. V. Motroshilova. "Is Herbert Marcuse's 'Critical Theory of Society' Really Critical?" (in Russian), *Voprosy filosofii* 22, no. 10 (1968): 66–77. English translations entitled "Is Marcuse's 'Critical Theory of Society' Critical?" (in Russian), *Soviet Studies in Philosophy* 8, no. 1 (Summer 1969): 45–66. Also in *Soviet Review* 9, no. 1 (Spring 1970): 3–24.

Zeman, Jiří. "Philosophical Problems of Cybernetics" (in Czech), *Filosofický časopis,* vol. 3 (1961).

―――. *Poznání a informace* [Knowledge and information]. Prague, 1962. Pp. 218. "... deals with some philosophical questions of cybernetics, particularly with the application of the notion of an information channel to the problem of cognition and consciousness." — J. Zeman, "Cybernetics and Philosophy in Eastern Europe," in R. Klibansky, ed., *Contemporary Philosophy: a Survey,* Vol. 2: *Philosophy of Science* (Florence: La Nuova Italia Editrice, 1968), pp. 407–13.

―――. "Conception cybernétique de l'univers, de l'homme et de la pensée." In *Actes du 3e Congrès International de Cybernétique.* Namur, Belgium, 1965.

―――. "Technik und Menschheit," *Universum: Monatszeitschrift für Natur,*

Technik und Wirtschaft (Vienna) 20, no. 11 (1965): 443-48; 20, no. 12 (1965): 497-99.

ZEMAN, M. "Futurology and Philosophy" (in Czech), *Filosofický Časopis*, vol. 16, no. 4 (1969).

ZHUKOV, NIKOLAI. *Cybernetic Approach to Defining Essence of Life.* Washington, D.C.: Joint Publications Research Service, Government Printing Office, April 17, 1968. From *Nauka i Tekhnika* (January 1968). Other work cited in the Comey-Kerschner bibliography.

III. RELIGIOUS CRITIQUES

A. PRIMARY SOURCES

BANGERTER, OTTO. *Der "Geist der Technik" und das Evangelium; ein Beitrag zur Frage nach der religiösen Krisis des "modern" Menschen* [The "spirit of technology" and the Gospel; a contribution to the problem of the religious crisis of "modern" man]. Heidelberg: Evangelischer Verlag Jakob Comtesse, 1939. Pp. 177. "Is technology a hindrance to the Christian faith?" This means, first, does technology make the Christian revelation obsolete? No, the biblical "unveiling of the human essence" is still true — although it is difficult to demonstrate this to technological man because of the "spirit of technology." Second, then, does technology necessarily give birth to a *Weltanschauung* which is a hindrance to faith? No, because man is not determined externally but internally. "Technology cannot possibly be a hindrance to Christian faith. The hindrance to faith is an attempt by modern man to curse in God's presence." Technology is not autonomous but stands in need of faith, something which must be made apparent in order to end technological blaspheming. — W. C. Reviewed by L. Janssens, "De Geest van de Techniek," *Streven* (Bruges) 8 (1941): 293-98.

BARBOUR, IAN G. *Science and Secularity: the Ethics of Technology.* New York: Harper & Row, 1970. Pp. 151. An analysis of the challenges to religion posed by the scientific method, the autonomy of nature, and the technological mentality, followed by a discussion of three more specific issues: molecular biology, cybernetics, and science policy.

BARRY, ROBERT M. "Christian Metaphysics and a Technological World," *American Benedictine Review* 16, no. 4 (December 1965): 473-85. A contrast between the natural and technological worlds followed by the argument that technological advancement is both an outgrowth of and brings about renewed interest in the past. This thesis is illustrated by observing how the Hebraic notions of time, community, and the person have acquired a new vitality under the influence of technology. Emphasizes the reality of man's this-worldly action and concludes by calling upon metaphysicians "to explicate the created world as seriously as they have previously examined the natural world."

BERDYAEV, NICHOLAS. (Also spelled Berdiaeff, Berdiaev, Berdjajeff, Berdjajew, etc.) "Man and Machine." Pp. 31-64 in *The Bourgeois Mind and Other Essays.* London: Sheed & Ward, 1934. Short version published under the title "Man, the Machine, and the New Heroism," *Hibbert Journal* 33 (1934-35): 76-89. A historico-theological situating of the present technical age with an analysis of the influence of technology on religion. Technological problems call for a greater religiosity; and Christianity can be the foundation for the reassertion of human freedom in the face of technological tyranny. This essay included in C. Mitcham and R. Mackey, eds., *Philosophy and Technology* (New York: Free Press, 1972). Two works which provide a broader framework for Berdyaev's thought on technology are his *The Fate of Man in the Modern World* (Ann Arbor: Univ. Michigan Press, 1935) and *Slavery and Freedom* (New York: Scribner's, 1944). See also Berdyaev's "L'homme dans la civilisation technique," in *Progrès technique et progrès moral* (Neuchatel: Baconniere, 1947). For a number of other articles pre-1956, see Dessauer's bibliography.

BERGSON, HENRI. "Mechanics and Mysticism." In *The Two Sources of Morality and Religion.* Translated by R. A. Audra, C. Brereton, and W. H. Carter. London: Macmillan, 1935. Reprinted, Garden City, N.Y.: Doubleday Anchor, n.d. From *Les deux sources de la morale et de la religion* (Paris: Alcan, 1932). After suggesting that the distinction between closed and open societies is paralleled by that between static, myth-making religions and dynamic, mystical religions, Bergson analyzes the criticisms of industrial or mechanical civilization and argues that their solution lies with a revival of mysticism, which is at once ascetic (against luxuries) and charitable (for eliminating inequalities). Concludes with a brief description of some forces working for the revival of mysticism.

BRUNNER, HEINRICH EMIL. *Christianity and Civilization.* 2 vols. New York: Scribner's, 1948-49. Vol. 2, *Specific Problems,* contains a lengthy discussion of technology. "Modern technics is the product of the man who wants to redeem himself by rising above nature, who wants to gather life into his hand, who wants to owe his existence to nobody but himself, who wants to create a world after his own image, an artificial world which is entirely his creation.... Modern technics is, to put it crudely, the expression of the orld-voracity of modern man, and the tempo of its development is the expression of his inward unrest, the disquiet of the man who is destined for God's eternity, but has himself rejected this destiny" (pp. 4-5). See also the author's *Man in Revolt* (Philadelphia: Westminister, 1947).

BUCHANAN, R. A. "The Religious Implications of Industrialization and Social Change," *Technologist* 2, no. 3 (Summer 1965): 245-55. Science and technology have demythologized the world, and the churches must focus on the modern problems of individual identity and leisure instead of worrying about the loss of traditional values.

————. "The Churches in a Changing World." Pp. 1-20 in G. Walters, ed., *Religion in a Technological Society.* Bath, Somersetshire: Bath Univ. Press, 1968. Argues forcefully that the church must adapt both its theoretical doctrines and its practical institutional structures to the scientific and technological world. Included in C. Mitcham and R. Mackey, eds., *Philosophy and Technology* (New York: Free Press, 1972).

BYRNE, BARRY. "Art and the Machine," *Catholic Art Quarterly* 10, no. 3 (Pentecost 1947): 75-78.

CAREY, GRAHAM, and JOHN HOWARD BENSON. "The Wheel of Artifice: Part I," *Catholic Art Quarterly* 15, no. 2 (Easter 1952): 43-50. Part II, ibid., 15, no. 3 (Pentecost 1952): 93-108; Part III, ibid., 16, no. 3 (Pentecost 1953): 101-9. Contains an analysis of the relation between art and technique.

CASTELLI, ENRICO, ed. *Tecnica e casistica; technica, escatologia e casistica* [Technology and casuistry; technology, escatology, and casuistry]. Padua: CEDAM, 1964. Pp. 372. Essays include J. Hollak's "Considerazioni sulla natura della tecnica odierna; l'uomo e la cibernetica nel quadro della filosofia sociologica" (pp. 121-46, with discussion pp. 147-52), W. Biemel's "L'ambiguïté de la technique" (pp. 319-28), R. Lazzarini's "Tecnica, costume e iniziazione escatologica" (pp. 329-36).

CHENU, M.-D. "Vers une théologie de la technique," *Recherches et débats* 31 (June 1960): 157-66. A theological vision of technology as a participation in God's creation.

CLARKE, W. NORRIS, S. J. "Technology and Man: a Christian Vision." Pp. 38-58 in Carl F. Stover, ed., *The Technological Order.* Detroit: Wayne State Univ. Press, 1963. An analysis of the place of technology in human affairs leading up to the argument that "only men with something like the Christian virtue of self-denial ... would really be safe enough to entrust with the responsibility of ... technology." Slightly revised version included in C. Mitcham and

R. Mackey, eds., *Philosophy and Technology* (New York: Free Press, 1972). See also Fr. Clarke's "Christian Humanism for Today," *Social Order* 3 (May–June 1953): 269–88.

COULSON, C. A. *Science, Technology and the Christian.* Nashville, Tenn.: Abingdon, 1960. Pp. 111. A noted scientist and committed Christian argues that Christianity must not abdicate its this-worldly responsibilities in the second industrial revolution as it did in the first. A moderate statement that seeks to steer a course between fundamentalist evangelism and secular theology. See also the same Coulson's two pamphlets, *Nuclear Knowledge and Christian Responsibility* (London: Epworth, 1958) and *Some Problems of the Atomic Age* (London: Epworth, 1957), as well as his *Faith and Technology* (London: Chester House, 1969), the inaugural lecture of the Luton Industrial College, Sept. 14, 1968.

DONDEYNE, CANON ALBERT. "Technology and Religion," *Albertus magnus* 6, no. 5 (November 1959): 5–6. Brief English summary of a lecture, "Technique et religion," which can be found in full in *Recherches et débats* 31 (1960): 124–35. Uses the Prometheus myth to reflect on the paradoxical character of technology as gift and threat to religion. The threat can be overcome by Christian hope which is "a synthesis of hope founded on technology and the theological virtue of hope which is a gift of the Holy Spirit."

FURFEY, PAUL HANLY. "Art and Machines," *Catholic Art Quarterly* 7, no. 3 (Pentecost 1944): 1–6.

GILL, ERIC. *Christianity and the Machine Age.* London: Sheldon, 1940. Pp. 72. The essence of Christianity is poverty, and the Christian lives in the world by means of art. He is thus opposed to technology on two counts: (1) because, as an artist, he is concerned with the good of the work to be made, not making money; and (2) because, as a man of poverty, he is opposed to the corrupt leisure of the welfare state. Included in C. Mitcham and R. Mackey, eds., *Philosophy and*

Technology (New York: Free Press, 1972). See also "The Factory System and Christianity," "Sculpture on Machine-made Buildings," and "Secular and Sacred in Modern Industry" in *It All Goes Together; Selected Essays, 1918–1940* (New York: Devin Adair, 1944).

GREINACHER, NORBERT, and THEODOR SEEGER, eds. *Die Frohbotschaft; Christi im Reiche der Arbeit; Wege zur katholischen Aktion der Arbeiter.* Colmar, 1959. Includes H. Daniel-Rop's "Der Mensch und die Technik" (pp. 107–24), and M.-D. Chenu's "Die christliche Bewältigung der technischen Welt" (pp. 125–49).

GUARDINI, ROMANO. *The End of the Modern World.* Translated by J. Thomas and H. Burke. New York: Sheed & Ward, 1956. Pp. 133. From *Das Ende der Neuzeit; ein Versuch zur Orientierung* (Basel: Hess, 1950). A theologian's argument that the technological world destroys all traditional values. Cf. also the review by W. Norris Clarke, "The End of the Modern World?" *America* 49, no. 3 (April 19, 1958): 106–8; and the exchange between Clarke and Frederick D. Wilhelmsen, "Modern World: End or Beginning?" *America* 49, no. 10 (June 7, 1958): 310–12.

———. *Power and Responsibility: a Course of Action for the New Age.* Translated by E. C. Briefs. Chicago: Regnery, 1961. Pp. 104. From *Die Macht; Versuch einer Wegweisung* (Wüzburg: Werkbund-Verlag, 1952). A companion volume to *The End of the Modern World.* Contrasts the theological and the technological concepts of power.

HATT, HAROLD E. *Cybernetics and the Image of Man; a Study of Freedom and Responsibility in Man and Machine.* Nashville, Tenn.: Abingdon, 1968. Pp. 304. A high-quality theological study of the relationship between two images of man—the *imago machinae* and *imago dei*—which, while it relies heavily on E. Brunner's interpretation of the *imago dei*, is critical of Brunner's own pessimistic conclusions. Particularly concerned with the problem of freedom and determinism.

HILDEBRAND, DIETRICH VON. "Efficiency

and Holiness." Pp. 205-43 in *The New Tower of Babel*. New York: P. J. Kenedy, 1953. Other essays in this volume are also important. See too Hildebrand's "Technology and Its Dangers" in R. P. Mohan, ed., *Technology and Christian Culture* (Washington, D.C.: Catholic Univ. America Press, 1960).

HOLLOWAY, JAMES Y., ed. *Introducing Jacques Ellul*. Grand Rapids, Mich.: Eerdmans, 1970. Pp. 183. First published as a special issue of *Katallagete: Be Reconciled* (Winter-Spring 1970). A collection of articles ostensibly devoted to Ellul and the problems which the technological society poses for Christianity, but actually somewhat wider in scope. Contains a brief autobiographical note by Ellul plus the following: Holloway's "West of Eden," G. Vahanian's "Theology, Politics and the Christian Faith," C. Lasch's "The Social Thought of Jacques Ellul," J. Lester's "The Revolution: Revisited," S. Rose's "Whither Ethics, Jacques Ellul?" W. Stringfellow's "The American Importance of Jacques Ellul," J. W. Douglas's "On Transcending Technique," J. Branscome's "The Educational Illusion," and J. Wilkinson's "The Divine Persuasion: an Interview on Jacques Ellul."

JOSEPH, ELLIS A. "Jacques Maritain on Reason, Technology, and Transcendence," *Spiritual Life* 10 (Fall 1964): 170-75. A good, scholarly summary of Maritain's position.

LALOUP, JEAN, and JEAN NELIS. *Hommes et machines: initiation a l'humanisme technique* [Men and machines: introduction to a technical humanism]. Tournai: Castermann, 1953. Pp. 317. "Starting from a summary description and a history of the *reality* of modern technology, the authors study its affects or *incidental* effects upon diverse aspects of human life: economics, biology, psychology, aesthetics, and sociology. On the basis of the conclusions which little by little emerge from this lengthy pheonomenological examination, they push their reflections further and investigate the *norms* which ought to

preside over human reactions to technology, first in respect to philosophy and humanism, then relative to Christian revelation and present tendencies in Catholic theology" (preface).

LEEUWEN, AREND THEODORE VAN. *Prophecy in a Technocratic Era*. New York: Scribner's, 1968. Pp. 130. A rather oracular and cryptic work by a noted secular theologian. Requires for background the author's more substantial *Christianity in World History: the Meeting of the Faiths of East and West*, translated by H. H. Hoskins (New York: Scribner's, 1966), with its vision of the Christian as working with God to desacralize the world and its religious institutions.

LEROI-GOURHAN, ANDRÉ. "L'illusion technologique," *Recherches et débats* 31 (June 1960): 65-74. It is an illusion to think that because technology discovers all the secrets of the natural world it therefore alters the fundamental situation of man before the supernatural mystery.

MARCEL, GABRIEL. "Some Remarks on the Irreligion of Today." Pp. 176-98 in *Being and Having: an Existentialist Diary*. Translated by K. Farrer. Westminster: Dacre, 1949. Reprinted, New York: Harper & Row, 1965. From *Être et avoir* (Paris: Aubier, 1935). "Pure religion, religion as distinct from magic and opposed to it, is the exact contrary of an applied science; for it constitutes a realm where the subject is confronted with something over which he can obtain no hold at all" (p. 187).

──────. "The Limitations of Industrial Civilisation." Pp. 1-20 in *The Decline of Wisdom*. Translated by M. Harari. London: Harvill, 1954. Reprinted, New York: Philosophical Library, 1955. From *Le déclin de la sagesse* (Paris: Plon, 1954). While remaining critical of technological civilization, Marcel seeks to credit its obvious achievements. The "exorcism" of its evils is Christian love, which can only be achieved through grace. See also *Man against Mass Society*, translated by G. S. Fraser (Chicago: Regnery, 1952).

MARSCH, WOLF-DIETER. "Tragik technischen Daseins?" [Tragedy of technological existence?], *Radius* 1 (1964): 28-36. "Do we find or lose our existence in complete achievement or in the destruction of relations, so that the question of God is simply outdated? Is the technological present not necessarily Godless? Or does it contain its own question of God, which we have not yet learned to ask? These questions shall be investigated here."—P. N.

————. *Christliche Ethik in der Technischen Welt.* Berlin: Wichern-Verlag, 1966. Pp. 24. A theologian's reflections on what appear to be the two contemporary secular moral options in the technological world—namely, that represented by technologists and by revolutionaries.

Mensch und Technik. Vorträge und Diskussionen, gehalten anlässlich der 9. Arbeitstagung des Instituts der Görres-Gesellschaft für die Begegnung von Naturwissenschaften und Theologie. Freiburg: Alber, 1967. Pp. 158. Includes: P. Koessler's "Technik aus der Sicht des Ingenieurs" (pp. 5-34), I. Moeller's "Mass und Zahl in der Technik" (pp. 35-66), D. Dubarle's "Technique et création" (pp. 67-87), J. H. Walgrave's "Die Technik in der Perspektive des Theologen" (pp. 110-39), and N. A. Luyten's "Technik und Selbstverständnis des Menschen" (pp. 140-58).

MESTHENE, EMMANUEL G. "Technology and Religion," *Theology Today* 23, no. 4 (January 1967): 481-95. Technology gives man the power to liberate himself from servitude to nature. After a critique of those who would reject technology, the author concludes that "man's newfound power and confidence enable him to pick up once more his partnership with God in doing the work of the world." At the same time, "the churches are freed of the burden of doing man's work, and may find their new vocation in doing God's." Article is basically an extension of the author's earlier "Technology and Wisdom," in *Technology and Social Change* (Indianapolis: Bobs-Merrill, 1967). Reprinted under the title "Religious

Values in the Age of Technology," in J. Metz, ed., *The Evolving World and Theology* (New York: Paulist Press, 1967).

————. "Technological Change and Religious Unification," *Harvard Theological Review* 65, no. 1 (January 1972): 29-51. Also in Harvard University Program on Technology and Society, Reprint no. 11.

MOHAN, ROBERT PAUL, ed. *Technology and Christian Culture.* Washington, D.C.: Catholic Univ. America Press, 1960. Pp. 144. A collection of five lectures of unequal value: R. Allers's "Technology and the Human Person," T. P. Neill's "Automation and Christian Culture," D. von Hildebrand's "Technology and Its Dangers," J. C. H. Wu's "Technology and Christian Culture: an Oriental View," and F. J. Connell's "Technology and the Mystical Body of Christ." Allers and Hildebrand raise the most important issues. Reviewed in *Dominicana,* vol. 45, no. 4 (Winter 1960) and in *Technology and Culture,* vol. 3, no. 3 (Summer 1962).

MÜLLER-SCHWEFE, HANS-RUDOLF. "Neue Welt durch Technik," Pp. 127-32 in Heinrich Griese, ed., *Der mündige Christ.* Stuttgart: Kreuz-Verlag, 1956.

————. "Geist der Technik und Religion der Technik," *Zeitschrift für evangelische Ethik* 7 (1963): 305-15.

————. *Technik als Bestimmung und Versuchung* [Technology as determination and temptation]. Göttingen: Vandenhoeck, 1965. Pp. 64.

ONG, WALTER J., S.J. "Technology and New Humanist Frontiers." In *Frontiers in American Catholicism: Essays on Ideology and Culture.* New York: Macmillan, 1957. An optimistic appraisal with affinities to T. de Chardin and M. McLuhan.

————. *In the Human Grain; Further Exploration of Contemporary Culture.* New York: Macmillan, 1967. Pp. 207. "A marvelous, perhaps major, testament of Christian optimism" with regard to the effects of technological media." —Edward R. F. Sheehan, *New York Times Book Review* (February 19, 1967), p. 41.

————. "The Spiritual Meaning of Technology." Pp. 29-31 in *Technology and Culture in Perspective*. Cambridge, Mass.: Church Society for College Work, February 1967. See also "The Challenge of Technology," *Sign* (February 1968), pp. 21-24.

ORNA, MARY VIRGINIA. *Cybernetics, Society and the Church*. Dayton, Ohio: Pflaum, 1969. Pp. 177. An optimistic Catholic analysis. See also Orna's "Cybernetics and Society," *Downside Review* (July 1968).

PIEPER, JOSEF. "Leisure, the Basis of Culture." Pp. 19-68 in *Leisure, the Basis of Culture*. Translated by A. Dru. New York: Pantheon, 1952. From *Musse und Kult* (Munich: Kosel-Verlag, 1948). An extended critique, from the Catholic perspective, of E. Jünger's apotheosis of technological work. Argues that leisure rather than technological activity is the necessary basis of any culture—leisure being identified with the freedom and detachment that are part of play, contemplation, feast and worship.

QUEFFÉLEC, HENRI. *Technology and Religion*. Translated by S. J. Tester. New York: Hawthorn, 1964. Pp. 110. From *La Technique contre la foi?* (Paris: Librairie Arthème Fayard, 1962). A Catholic appraisal of the relationship between technology and faith which seeks to reject all false reconciliations which either deny the antagonism or make technology supreme. "Technology can help charity greatly, but it must be shown how, since alone it will go on falling into the sin of pride" (p. 104).

RUSSO, FRANÇOIS. "Concezione cristiana e umanistica della technica" [Christian and humanistic conceptions of technology], *Civiltà cattolica* 118, vol. 1, no. 4 (February 18, 1967): 339-52.

SCHAPTIZ, EBERHARD. *Schwärmertum am Rande der Technik* [Fanaticism on the edge of technology]. Munich: Claudius-Verlag, 1954. Pp. 87. Christianity is more important than technology. Technology is man's work, not (contra Dessauer) a continuation of divine creation but a means which must be employed within the frame-work of God's order. A reasoned critique of mythic optimism.—R. J. R.

———. *Ingenieur und Theologe; Grundlagen für einen Dialog über geistige Existenz in der technischen Welt* [Engineer and theologian; foundations for a dialogue on spiritual existence in the technological world]. Würzburg: Werkbund-Verlag, 1968. Pp. 219.

SCHUMACHER, E. F. "Industrial Society," *Good Work* 30, no. 2 (Spring 1967): 42-49. A Catholic critique.

———. "Intermediate Technology," *Good Work* 30, no. 4 (Autumn 1967): 106-12. Argues for a decentralized or intermediate technology especially for developing countries. Similar arguments contained in "A Saner Technology," *Liberation* (August 1967), pp. 15-19. See also A. Lathan-Koenig's "The Church and Intermediate Technology," *Clergy Review* 54 (July 1969): 519-26.

———. "Small Is Beautiful: Toward a Theology of 'Enough'," *Christian Century* 88, no. 30 (July 28, 1971): 900-902.

"Science-Technology and the Church." Paper of the Ad Hoc Committee of the Division of Christian Life and Mission of the National Council of Churches. New York: Council Press, 1968. Pp. 31.

SHAULL, RICHARD. "The Christian World Mission in a Technological Era," *Cross Currents* 15, no. 4 (Fall 1965): 461-72. After a review of van Leeuwen's *Christianity in World History* (London: Edinburgh House, 1964), Shaull considers some concrete ways in which "the Christian is called today to give shape to community in the midst of the secular"—first in the churches in developing countries, second in the Western churches. First published in the *Ecumenical Review* (July 1965); a shorter version published under the title "Technology and Theology," *Theology Today* 23, no. 2 (1966-67): 271-75.

———. "Revolutionary Change in Theological Perspective." In J. C. Bennett, ed., *Christian Social Ethics in a Changing World*. New York: Association Press, 1966. The Christian must become a revolutionary.

_____. "Revolution: Heritage and Contemporary Options." In C. Oglesby and R. Shaull, *Containment and Change.* New York: Macmillan, 1967.

_____. *Christian Faith as Scandal in a Technocratic World.* Saint Louis: United Ministries in Higher Education Publications Office, 1969. A short pamplet. Included in M. E. Marty and D. G. Peerman, eds., *New Theology No. 6* (New York: Macmillan, 1969).

SIU, R. G. H. *The Tao of Science.* Cambridge, Mass.: M.I.T. Press, 1957. Pp. 108. "An essay on Western knowledge and Eastern wisdom" by an Oriental who became first a scientist and then an administrator of Western science.

SLUSSER, DOROTHY M., and GERLAD H. SLUSSER. *Technology—the God That Failed.* Philadelphia: Westminster, 1971. Pp. 169.

TAYLOR, ALVA W. *Christianity and Industry in America.* New York: Friendship Press, 1933. Pp. 212. Represents the social gospel tradition. "If industrial relationships are not Christianized the machine may prove a Frankenstein turning to destroy what it has helped to create." Includes bibliography.

TILLICH, PAUL. "The Person in a Technical Society." In John A. Hutchinson, ed., *Christian Faith and Social Action.* New York: Scribner's, 1953.

TYE, IRVIN, O.F.M. "The Two Roads of Technology," *Duns Scotus Philosophical Association Convention Report* 23 (1959): 22-45. Technology can lead to dehumanization or true humanization. What happens depends on the philosophy of those who guide it. The true philosophy is that taught by the Catholic church.

VAUX, KENNETH. *Subduing the Cosmos; Cybernetics and Man's Future.* Richmond, Va.: Knox, 1970. Pp. 197. Analyzes the humanizing and dehumanizing potential of cybernetics and the implications of man's role as co-creator with God.

WALTERS, GERALD, ed. *Religion in a Technological Society.* Bath, Somersetshire: Bath Univ. Press, 1968. Pp. 75. Four papers delivered at Bath University in May 1967, along with an appended sermon and reports on seminar groups. The papers: R. A. Buchanan's "The Churches in a Changing World," G. Walter's "The Secular Premise," D. Martin's "The Secularization Pattern in England," and N. Smart's "The Church and Industrial Society." The tone of the book is set by Buchanan when he argues that the church must accommodate itself to the technological society if it is to be of any significance in the future. Smart is the most straightforward critic of this position.

WHITE, HUGH C., JR., ed. *Christians in a Technological Era.* New York: Seabury, 1964. Pp. 143. Contains seven essays: M. Mead's "Introduction," M. Polanyi's "The Scientific Revolution," J. Ladrière's "Faith and the Technician Mentality," B. Morel's "Science and Technology in God's Design," F. Russo's "Modern Science and the Christian Faith," J. de la Croix Kaelin's "Faith and Technology," and S. I. Paradise's "Christian Mission and the Technician Mentality in America." Primarily the result of a symposium held at Louvain in April 1961; the papers by Ladrière, Morel, Russo, and Kaelin give the best idea of European theological reflection on this subject available in English in a single volume.

WHITE, LYNN, JR. "The Historical Roots of Our Ecologic Crisis." In *Machina ex Deo.* Cambridge, Mass.: M.I.T. Press, 1969. First published in *Science* 155 (March 10, 1967): 1203-7, and as "St. Francis and the Ecologic Backlash," *Horizon* 9 (Summer 1967): 42-47. Argues that the Christian theology of the Latin West is largely responsible for modern technology. Included in P. Shepard and D. McKinley, eds., *The Subversive Science, Essays toward an Ecology of Man* (Boston: Houghton Mifflin, 1969); and in C. Mitcham and R. Mackey, eds., *Philosophy and Technology* (New York: Free Press, 1972). Cf. also Yi-Fu Tuan, "Our Treatment of the Environment in Ideal and Actuality," *American Scientist* 58 (May-June 1970): 244, 247-49, which points out the similarities between the exploitation of resources in both East and West. A number of other works on the theology of ecology are cited under III, B.

YORK, JEREMY FRANCIS. "The Bent World of the Mental Technician," *Good Work* 29, no. 4 (Autumn 1966): 108-14.

B. SECONDARY SOURCES

ALFARO, JUAN. "Tecnopolis e christianesimo," *Civiltà cattolica* 120, vol. 2, no. 6 (June 21, 1969): 533-48.

ALPERS, KENNETH A. "Starting Points for an Eco-Theology: a Bibliographical Survey." Pp. 292-312 in Martin E. Marty and Dean G. Peerman, eds., *New Theology No. 8.* New York: Macmillan, 1971. First published in a special issue of *Dialog* (Summer 1970) devoted to the environmental crisis. A survey, first, of the main works relevant to an appreciation of the ecological crisis itself, followed by an outline of theological and ethical assessments. Both are very helpful. Some other articles relevant to a theology of ecology: Rudy Abramson, "The Reckless Ruination of Creation," *Mennonite* 86, no. 48 (December 28, 1971): 779-81; Richard A. Baer, Jr., "Ecology, Religion and the American Dream," *American Ecclesiastical Review* 165, no. 1 (September 1971): 43-59; Richard A. Baer, Jr., "Environmental Turnabout," *Stewardship 72* (Council Press for Section on Stewardship and Benevolence, National Council of Churches), pp. 53-56; Olov Hartman, "Christ and the Garden," *Lutheran Forum* 6, no. 1 (February 1972): 26-28; Philip N. Joranson, "What Is Next on Earth?" *Stewardship 72* (Council Press for Section on Stewardship and Benevolence, National Council of Churches), pp. 38-41; Charles W. Martin, "Ecology and Theology," *Living Church* (November 1, 1970); Joseph Sittler, "A Theology for Earth," *Christian Scholar*, vol. 37, no. 3 (September 1954); Joseph Sittler, "Creation and Redemption," *Dialog* (Autumn 1964); and Richard A. Underwood, "Ecological and Psychedelic Approaches to Theology," in Martin E. Marty and Dean G. Peer-

man, eds., *New Theology No. 8* (New York: Macmillan, 1971), pp. 137-72, first published in *Soundings* (Winter 1969).

ALVAREZ TURIENZO, SATURNINO. "Inquieted ética y mundo técnico," *Religion y cultura* 9 (1964): 9-40.

AYEL, VINCENT. "Technical Studies and Religious Teaching," *Lasallian Digest* 1 (September 1959): 30-44.

BARBOUR, IAN G., ed. *Earth Might Be Fair.* Englewood Cliffs, N.J.: Prentice-Hall, 1972. Pp. 168. Essays on the theology of ecology.

BAUHOFER, OSCAR. "Technik und Kultur," *Gloria Dei* 4 (1949-50): 232-38.

BAYART, PIERRE. *L'homme et la machine dans notre civilisation industrielle* [Man and machine in our industrial civilization]. Lille: Ecole d'administration des affaires des facultés catholiques, 1958. Pp. 112. Divided into three parts: "A New Type of Businessman," "A New Type of Worker," and "The Problem That Is Posed to the Church." Rather conventional sociological analysis followed by an attempt to assess developments in terms of the social teachings of Leo XIII and Pius XII. — W. C.

BECK, HUBERT F. *The Age of Technology.* Saint Louis: Concordia, 1970. Pp. 133.

BELLAH, ROBERT N. *Beyond Belief: Essays on Religion in a Post-traditional World.* New York: Harper & Row, 1970. Pp. 298. An autobiographical sketch followed by chapters on the development of religion: first generalized, then in specific countries, and finally in our own situation. Underlying the text is the search for a personal religion which, in the traditional sense, seems to have been lost among the complexities and problems of modern society.

BENNE, ROBERT. "The Technological Backlash," *Lutheran Quarterly* 19, no. 4 (November 1967): 341-56. "If the church could develop a living illustration of what it means to live with real 'sensibility,' it would perform a much needed service for our flat and banal mind-set. In order for it to move in this direction, though, the church itself must put away its own infatuation with the technology of evangelism . . . and work toward the creation of its own characteristic ethos. . . ."

BENNETT, JOHN COLEMAN, ed. *Nuclear Weapons and the Conflict of Conscience.* New York: Scribner's, 1962. Pp. 191.

―――. *Christian Social Ethics in a Changing World.* New York: Association Press, 1966. Pp. 381. Preparatory volume for the World Council of Churches Geneva Conference (1966) on "Christians in the Technological and Social Revolutions of Our Time."

BERNANOS, GEORGES. *La France contre les robots.* Paris: Laffont, 1947. Pp. 222. A classic indictment of technology by a well-known French Catholic novelist.

BERRIGAN, DANIEL, S. J. "Man's Spirit and Technology." Pp. 163–81 in *They Call Us Dead Men: Reflections on Life and Conscience.* New York: Macmillan, 1968. First published in *Fellowship,* vol. 31, no. 5 (May 1965). A meditation on the problem of integrating technology into human life by means of conscience and judgment. With regard to technological war and the specter of total annihilation: "It is this appalling limitation of human possibility, brought about by techniques that had promised men all heaven and earth, that must give pause to thinking men" (p. 172).

BIEMEL, WALTER. "Fortschritt der Technik―Fortschritt der Menschheit?" [Progress of technology―progress of humanity?]. Pp. 53–68 in M. V. Schöndorfer, ed., *Der Fortschrittsglaube: Sinn und Gefahren.* Studies of the Vienna Katholischen Akademie, vol. 5. Graz: Verlag Styria, 1965.

BISER, E. "Glaube im Zeitalter der Technik; der heutige Mensch auf der Suche nach sich selbst" [Faith in the epoch of technology; modern man in search of himself], *Wort und Wahrheit* 19 (Fall 1964): 89–102.

BLACK, JOHN. *The Dominion of Man: the Search for Ecological Responsibility.* Edinburgh: Edinburgh Univ. Press; Chicago: Aldine, 1970. Pp. 169. "This book traces the power of the biblical notion of man's dominion over nature into (*a*) the various rationalizations of the original myth, (*b*) philosophical-political thought, and

(*c*) practical consequences. I know of no treatment of the intellectual and cultural career of this powerful and penetrating notion to match Professor Black's chapter on the matter."―Joseph Sittler, *Zygon,* vol. 5, no. 4 (December 1970). Good background on the ecology and religion debate.

BLÖCHLINGER, ALEX. "Kann ein Techniken Christ sein?" [Can a technologist be a Christian?], *Technische Rundschau* (Bern) 54, no. 39 (1962): 1–2.

BLOY, MYRON B., JR. *The Crisis of Cultural Change: a Christian Viewpoint.* New York: Seabury, 1965. Pp. 139. A collection of essays.

―――. "The Christian Function in a Technological Culture," *Christian Century* 83, no. 8 (February 23, 1966): 231–34. "Guided by the Christian perspective, men can use their new freedom to cope creatively with a world of constant change." See also "The Christian Norm," in J. Wilkinson, ed., *Technology and Human Values* (Santa Barbara, Calif.: Center for the Study of Democratic Institutions, 1966); and "Technology and Theology," in R. Theobald, ed., *Dialogue on Technology* (Indianapolis: Bobbs-Merrill, 1967).

―――. "Theology for the Space Age." In Malcom Boyd, ed., *On the Battle Lines.* New York: Morehouse-Barlow, 1964.

BONIFAZI, CONRAD. *A Theology of Things; a Study of Man in His Physical Environment.* Philadelphia: Lippincott, 1967. Pp. 237. A "prolegomena to any future theology of matter" which raises important issues relevant to the nature of artifacts or technological objects.

BONNER, HUBERT. "Spiritual Man in a Technological Age," *Humanities* 6, no. 3 (Winter 1971): 277–93.

BORGMANN, A. "The Place of Theology in a Technological World," *National Catholic Educational Bulletin* 64 (May 1968): 28–33.

BOURBECK, CHRISTINE. *Kommunismus, Frage an die Christen; der angefochtene Mensch des technischen Zeitalters in Ost und West* [Communism, the question as addressed to Christians; the tempted man of the technical era in East and West]. Nürenberg: Laetare-Verlag, 1957. Pp. 142. Contains chapters on "The Technological Age," "The East and the Technological Age," "The West and the Technological Age," "The Gospel Today." Very theological. — R. J. R.

BOVET, THEODOR. "Technik als Geschenk und Versuchung" [Technology as gift and temptation], *Schweizerische Bauzeitung* 76, no. 18 (1958): 271-74.

BROCKMÖLLER, KLEMENS. *Industriekultur und Religion*. Frankfurt: Knecht, 1964. Pp. 288.

BRUNSTÄD, FRIEDRICH. "Vom Geist und Sinn des technischen Zeitalters im Lichte des Christentums." Pp. 364-76 in E. Gerstenmaier and C. G. Schweitzer, eds., *Gesammelte Aufsätze und kleinere Schriften*. Berlin: Lutherisch Verlagshaus, 1957.

BURKE, P. "The Technologist and the Christian Life," *Doctrine and Life* 12 (June 1962): 295-300.

BURHOE, RALPH WENDELL, ed. *Science and Human Values in the 21st Century*. Philadelphia: Westminster, 1971. Essays by five scientists and theologians.

CALON, J. A. "Seelische Krisen im technischen Zeitalter" [Spiritual crises in the technological age], *Arzt und Christ* (Salzburg), special issue (1961), pp. 178-94.

CAMARA, DOM HELDER. *Revolution through Peace*. New York: Harper & Row, 1971. Pp. 359. A Brazilian archbishop argues that faith and technology need to unite to overcome the crises of the Third World.

"Catechesis in a Technical World." *Lumen Vitae* (English ed.), vol. 13, no. 4 (October-December 1958), An introduction, A. Dondeyne's "The Christian in Face of the World To-day," followed by three symposia: I, on "Technical Mind and Method"; II, on "How to Teach Religion in Vocational Schools"; and III, specific reports on "Religion in Vocational Schools of Some Countries." Part I contains A. M. Besnard's "Is Our Technical Civilisation Open to the Gospel?" B. Haering's "Technical Attitude and Approach to the Liturgical

World," A. Brien's "Technical Mentality and the the Teaching of Religion," and A. Nachtergaele's "Scientific Method and Spiritual Experience." See also E. Rideau's "Technique and the Eucharist" in Part II.

CAUTHEN, KENNETH. *Science, Secularization and God; toward a Theology of the Future*. Nashville, Tenn.: Abingdon, 1969. Pp. 237.

Christians and the Good Earth. Addresses and discussions at the Third National Conference (November 1967) of the Faith-Man-Nature Group. 800 South Royal St., Alexandria, Va.: Faith-Man-Nature Group, n.d. Pp. 190. Good introduction to the developing theology of ecology. Includes D. A. Williams's "Christian Stewardship of the Soil," L. H. DeWolf's "Elements in a Program," E. W. Mueller's "Man and Divine Confrontation," P. Knight's "The Politics of Conservation," T. L. Kimball's "The Place of Preservationists," M. Bush's "Progress in the Present," C. Bonifazi's "The Inwardness of Things," D. D. Williams's "The Good of All," R. F. Faricy's "Christ and Nature," R. A. Baer's "The Church and Conservation: Talk and Action," J. L. Fisher's "For a Systematic View," A. A. Lindsey's "Preservation and Conservation," J. C. Logan's "The Secularization of Nature," H. P. Santmire's "The Integrity of Nature," H. R. Glascock's "In Harmony with Nature," R. Anderson's "An Ecological Conscience for America," G. H. Hansen's "The Role of the Church," and P. N. Joranson's "Coming to Grips." The Faith-Man-Nature Group also publishes a bulletin with continuing bibliographical information on the theology of ecology.

"Christians Confront Technology," *America* 101, no. 26 (September 26, 1959): 761-65. Short essays by W. N. Clarke, G. Weigel, and W. J. Ong. A generally optimistic assessment by three Catholic scholars writing in a more popular vein. Cf. also "The Problem of Our Age," *America* 106, no. 20 (February 24, 1962): 676-77, an editorial on the technological revolution and the role of natural law.

CLARKE, ARTHUR C., and ALAN

WATTS. "At the Interface: Technology and Mysticism," *Playboy* (January 1971), pp. 95 ff. A superficial dialogue: two famous writers alone with a tape recorder talking just for the money.

COLONNETTI, G. "La technica e la sua influenza nel regno dello spirito," *Studium* (Federazione universitaria cattolica italiana, Rome) 31 (1935): 517–34.

_____. "La technica, strumento di elevazione dell'uomo," *Studium* (Federazione universitaria cattolica italiana, Rome) 47 (1950): 604–17.

COOPER, BRIAN G. "Religion and Technology — toward Dialogue," *Main Currents in Modern Thought* 26, no. 1 (September–October 1969): 10–13. "The Christian religion must question the all-sufficiency of the technologist's empiricism; the intellectual posture of technology must question the adequacy of traditional forms of religious statement."

CORMAN, GILBERT. "Das katholische Bildungsideal in der technischen modernen Welt" [The Catholic ideal of education in the modern technological world]. Pp. 45–50 in *Schule und moderne Arbeitswelt; ein Tagungsbericht*. Düsseldorf, 1962.

COX, HARVEY G. *The Secular City*. New York: Macmillan, 1965. Pp. 276. The secular (or technological) city exhibits some positive religious values. See also D. Callahan, ed., *The Secular City Debate* (New York: Macmillan, 1967). Cox's later *On Not Leaving It to the Snake* (New York: Macmillan, 1968) is also relevant.

_____. "Tradition and the Future: I and II," *Christianity and Crisis* 27 (October 2, 1967): 218–20, and 27 (October 16, 1967): 227–31. On the theology of futurology. Of the three basic attitudes toward the future — the apocalyptic, teleological, and prophetic — "only a recovery of the prophetic perspective will supply the ethos required for the political ethic required today." The prophetic, as opposed to other attitudes, views the future as both "open and responsible."

_____. "The Future of Christianity and the Church," *Futurist* 4, no. 4 (August 1970): 122–29. Note also in this issue

T. J. Gordon's "Some Possible Futures of American Religion" (pp. 131–33), and D. M. Shore's "Ministry: Year 2000" (pp. 134–38).

CUNLIFFE-JONES, HUBERT. *Technology, Community and Church*. London: Independent Press, 1961. Pp. 150.

DANIELOU, JEAN. "Technical Civilization and Atheistic Humanism," *Christ to the World* 4 (1959): 221–29.

DANIEL-ROPS, H. *Jenseits unserer Nacht; ein Christ vor der Technik* [Beyond our night; Christ before technology]. Mainz: Matthias Grünewald, 1948. Pp. 142.

DANUSSO, A. "La tecnica e la morale," *Studium* (Federazione universitaria cattolica italiana, Rome) 30 (1934): 460–74.

DAWSON, CHRISTOPHER. *Progress and Religion*. New York: Sheed & Ward, 1931. Reprinted, New York: Doubleday Image, 1960. Pp. 200. See especially chap. 9, "The Age of Science and Industrialism: the Decline of the Religion of Progress."

DERISI, OCTAVIO N. "Técnica y espíritu," *Sapientia* 20 (1965): 3–7.

DERR, THOMAS SIEGER. "Man against Nature: Hidden Assumptions in the Argument over Environmental Control," *Cross Currents* 20, no. 3 (Summer 1970): 263–75. A critique of the philosophical assumptions of the romantic ecologists who reject environmental control. Argues instead for a moderate humanism which, Derr holds, "is not very different from . . . the Biblical idea that the whole of creation awaits its transformation . . . [and that] man is to be the priest of the transformation."

DIRKS, WALTER. "Plädoyer für die Technik; das Heil und der Weltauftrag der Christen" [Plea for technology; the salvation and mission of the Christian], *Kontexte* (Stuttgart) 1 (1965): 33–40.

D'ORSI, D. "L'origine cristiana della scienza moderna," *Sophia*, vol. 34, nos. 3–4 (July–December 1966).

DOUGLAS, TRUMAN B. "Christ and Technology," *Christian Century* 78, no. 4 (January 25, 1961): 104–7. "The problem of our so-called 'materialistic' age is not our inventions but man himself and the uses to which he puts his technological achievements." W. Temple argued that Christianity is the most materialistic of all religions (since it begins with an act of materialization, the incarnation). Using this as a starting point, Douglas argues for Christ's lordship in a technological age.

DUBOS, RENE. *A Theology of the Earth.* Washington, D.C.: Government Printing Office, 1969. A lecture given at the Smithsonian Institution.

DUVEAU, G. "La technocratie menace-t-elle la liberté humaine?" *Foi et vie* 52 (1954): 97–107.

ELDER, FREDERICK. "A Different 2001," *Journal* 8, no. 4 (January–February 1970): 8–14. A theological examination of desirable futures in the light of current ecological problems. Based on a paper given at the Faith-Man-Nature Group Conference, November 1969, and included in Glenn C. Stone, ed., *A New Ethic for a New Earth* (New York: Friendship Press, 1971).

———. *Crisis in Eden: a Religious Study of Man and Environment.* Nashville, Tenn.: Abingdon, 1970. Distinguishes two theories of the man-nature relationship: inclusivists (represented by Loren Eiseley) and exclusivists (represented by Teilhard de Chardin, Harvey Cox, and Herbert Richardson).

ELL, ERNST. "'Mensch und Technik'—Freund oder Feind" [Man and technology—friend or enemy], *Jugendwohl; Katholische Zeitschrift für Kinder- und Jugendfürsorge* (Freiburg) 37 (1956): 162–72.

Espoir humain et espérance chrétienne. Paris: Flore, 1951. Contains: A. George's "Science et technique, espoir de l'humanité?" (pp. 135–48), V. Ducatillon's "Le progrès scientifique et technique, phénomene de maturation" (pp. 148–66), L. Leprince-Ringuet's "Sciences et techniques, espoir de l'humanité?" (pp. 166–70), and M. Javillier's "Science et technique, espoir de l'humanité?" (pp. 183–87).

FAHRBACH, GEORG, ed. *Der Mensch zwischen Natur und Technik.* Stuttgart:

Fink, 1967. Pp. 100. Papers delivered at a conference sponsored by the Schwäbischer Albverein and the Evangelishe Akademie concerned with the conservation of natural resources.

FARAMELLI, N. J. *Technethics.* New York: Friendship Press, 1971.

FLAM, LEOPOLD. "The Sacred and Desacralization in Contemporary Thought," *Philosophy Today* 7, no. 3 (Fall 1963): 209–15.

FLEMING, REV. LAUNCELOT. "Living with Progress," *New Scientist* 36 (October 19, 1967): 166–68. An outline, by the Bishop of Norwich, of possible dangers of technological and scientific progress. Suggests a moral basis from which scientists should view their future responsibilities. Three examples—the sea, genetics, wild life—are used to illustrate areas which science is likely to affect profoundly.

Foi et technique. Paris: Plon, 1960. Pp. 177. Contains essays by R. Sugranyes de Franch, C. A. Dondeyne, J. Kaelin, and O. Costa de Beauregard.

GIBSON, ARTHUR. "New Heaven and New Earth," *Commonweal* (October 31, 1969). Optimistic Christian futurology. Reprinted in Martin E. Marty and Dean G. Peerman, eds. *New Theology No. 8* (New York: Macmillan, 1971), pp. 104–18.

HAIGERTY, L. J., ed. *Pius XII and Technology.* Milwaukee: Bruce, 1962. Pp. 244. A collection of addresses by the Pope to various groups on various problems of technology. The general thesis is that technology itself is morally neutral and that man must refer to extratechnological values in order to determine its proper use. Reviewed in *Technology and Culture* 4, no. 2 (Spring 1963): 241–43.

HALL, CAMERON P. *Community Leaders Confront Community Problems in a Technology-Human Values Perspective.* New York: National Council of Churches, 1967.

———, ed. *Human Values and Advancing Technology; a New Agenda for the Church in Mission.* New York: Friendship Press, 1967. Pp. 175. The result of a National Council of Churches conference in Chicago (May 1967). Six papers plus working group reports. The papers: J. E. Carothers's "Man the Manipulator," H. Smith's "Technology and Human Values: This American Moment," R. Theobald's "Compassion or Destruction: Our Immediate Choice," Th. Dobzhansky's "Human Values in an Evolving World," J. M. R. Delgado's "Brain Technology Psychocivilization," and D. N. Michael's "Twenty-first Century Institutions: Prerequisites for a Creative and Responsible Society." The book as a whole reflects the American liberal Protestant's determination to take a positive stand toward the moral problems raised by technology.

HALL, D. J. "Theology of Hope in an Officially Optimistic Society," *Religion in Life* 40 (Autumn 1971): 376-90.

HAMILL, ROBERT H. *Plenty and Trouble: the Impact of Technology on People.* Nashville, Tenn.: Abingdon, 1971. Pp. 192. A Christian analysis which discovers a "new lifestyle in the making."

HAMMER, VICTOR. "Industrial Methods of Work and Socialism," *Catholic Art Quarterly* 13, no. 1 (Christmas 1949): 9-10.

HASENFUSS, J. "Religion und Technik," *Die Kirche in der Welt* 8, no. 2 (1956): 137-42.

HECKMANN, JUSTUS. "Neue Welt durch Technik," *Zeichen der Zeit; Evangelische Monatsschrift* (Berlin) 8, no. 9 (1954): 346-48.

HEFNER, PHILIP. "Toward a New Doctrine of Man: the Relationship of Man and Nature," *Zygon* 2 (1967): 127-51.

HESSELINK, I. J. "The Future of the Church in a Technological Society," *Japan Missionary Bulletin* (1971), pp. 562-68.

HÖFFNER, JOSEPH. *Der technische Fortschritt und das Heil der Menschen* [Technological progress and the welfare of hunanity]. Schriftenreihe der Arbeitsgemeinschaft katholisch-sozialer Bildungswerke, vol. 3, no. 4. Paderborn: Bonifacius, 195?. Pp. 31.

———. "Die moderne Technik in theologischer Sicht," *Offene Welt* (1956), pp. 126-33.

HÖRRMANN, MARTIN. "Siegeszug der Technik—Siegeszug des Unglaubens?" [Triumphal procession of technology—triumphal procession of faith?], *Kirche im Dorf; Zweimonatsschrift zur evangelischen Besinnung und Arbeit in den Dorfgemeinden* (Himmelpforten-Niederelbe) 12 (1961): 249-56.

HORTON, WALTER MARSHALL. "Conditions and Limits of Man's Mastery over Nature." Pp. 89-103 in *Das Menschenbild im Lichte des Evangeliums.* Zurich: Zwingli-Verlag, 1950. Suggests that the reason behind the opposition between Christianity and technology is "the fact that the Bible lays down moral conditions and acknowledges metaphysical limits to man's mastery over nature, while modern culture increasingly has ignored these conditons and limits, with disastrous consequences." But if man will recognize his dependence on God, technology may be used to construct "a new heaven and a new earth."

HUMPHREYS, W. LEE. "Pitfalls and Promises of Biblical Texts as a Basis for a Theology of Nature." Pp. 99-118 in Glenn C. Stone, ed., *A New Ethic for a New Earth.* New York: Friendship Press, 1971. A textual study arguing that the Old Testament view of nature is a dialectical one. Nature is seen both as mechanical and as living; God is found both above and in nature; and man is viewed as both separate from and united with nature. On this same subject see also the following articles: David T. Asselin, S.J., "The Notion of Domination in Genesis 1-3," *Catholic Biblical Quarterly* 16 (1954): 277-94; Gertrude Bayless, "Is the Bible to Blame for Pollution?" *Christian Science Sentinel* 73, no. 48 (November 27, 1971): 2076-81; and Andre Lacocque, "The World Dies at Dawn: the Old Testament and the Environmental Crisis," *President's Newsletter* (Chicago Theological Seminary) 5, no. 3 (January 1971): 2-3.

HUNTEMANN, GEORG HERMANN. *Provozierte Theologie in technischer Welt* [Provocation theology in a technological world]. Wuppertal: R. Brockhaus, 1968. Pp. 295.

ILLICH, IVAN. *The Church, Change and Development.* Chicago: Urban Training Center Press, 1970. Pp. 125. In a series of collected essays the author argues that the "specific function of the Church must be a contribution to development which could not be made by any other institution." This is faith in Christ. And "applied to development, faith in Christ means the revelation that the development of humanity tends toward the realization of the kingdom, which is Christ already present in the Church." See also Illich's *Celebration of Awareness: a Call for Institutional Revolution* (Garden City, N.Y.: Doubleday, 1970). And, finally, one might compare Illich's ideas with those of W. D. Campbell and J. Y. Holloway in *Up to Our Steeples in Politics* (New York: Paulist Press, 1970).

JACCARD, PIERRE. "Progrès technique et destin de l'homme," *Cahiers protestants* (Lausanne) 42 (1958): 196-99.

JACOB, OSKAR. "Der Christ im Zeitalter der Technik," *Ordo Socialis; Carl Sonnenschein-Blätter; Zeitschrift für christliche Soziallehre* (Münster) 5 (1958): 207-9.

JOHNSON, ARTHUR L. "Technologie und Arbeit," *Lutherische Rundschau; Zeitschrift des Lutherischen Weltbundes* (Stuttgart) 18, no. 2 (1968): 115-39.

KAULHAUSEN, D. "Zu einer christlichen Besinnung auf die Technik" [A Christian deliberation upon technology], *Neue deutsche Schule; Organ der Gewerkschaft Erziehung und Wissenschaft* (Essen) 12, nos. 14-15 (1960): 10-12, addition 3.

KEYS, DONALD, ed. *God and the H-Bomb.* New York: Macfadden-Bartell, 1962. Pp. 176. First published 1961.

KLAUSLER, ALFRED P. "Radiation and Social Ethics," *Christian Century* 80 (February 13, 1963): 199-200. Report on a conference of scientists and theologians held at the University of Chicago (January 16-19, 1963) to discuss

religious ethics in the context of problems created by nuclear radiation. Cf. also the second entry under William A. Wallace (this Section).

KLOSEK, IGNATIUS S., O.S.B. "II Vatican Council and the Age: Contemporary Technological Revolution," *American Benedictine Review* 12, no. 2 (June 1961): 147-63.

KOSER, KONSTANTIN. "Teologia da técnica," *Rivista ecclesiástica brasileira* (Petropolis, Brazil) 15 (1955): 573-94.

_____. "Dämonie oder Gottgerichtetheit der Technik" [Demony or divine legality of technology], *Sanctificatio nostra* 22 (1957): 65-75, 109-14, 175-79, 215-21.

LANGEN, KURT. "Vom Sinn der Technik; 'Gott regiert!'" [On the meaning of technology; "God rules!"]. Pp. 24-25 in *Kurt Langen: Lebenserfahrungen; 3 Vorträge.* Dortmund, 1954.

LEE, R. *Religion and Leisure in America: a Study in Four Dimensions.* Nashville, Tenn.: Abingdon, 1966. Pp. 271.

LEFORT, T. "Plea for Wisdom," *Dominicana* 46 (Spring 1961): 43-50.

LEFRINGHAUSEN, KLAUS. *Christlicher Glaube und industrielle Wirklichkeit* [Christian faith and industrial reality]. Wuppertal: Brockhaus, 1967. Pp. 79. "The Life of Faith in a Technico-Scientific World," *Albertus magnus* 6 (April 1959): 1-2. Results of a questionnaire survey of Catholic intellectuals.

LOTTMANN, WERNER. "Leben mit der Technik" [Living with technology], *Zeitschrift für evangelische Ethik* 5 (September 1962): 257-76.

LUTZ, HANS. "Die technische Entwicklung und die Sicherheit des Menschen" [Technological development and the security of man], *Zeitschrift für evangelische Ethik* 4, no. 6 (November 1960): 341-55.

LYNCH, JAMES J., S.J. "Christian Technology," *Messenger of the Sacred Heart* 96 (November 1961): 54-55 and 73. In the past, education was both religious and technical; "instruction in hunting had to do not merely with the fashioning of arrowheads, . . . but also with the gods of the hunt." This unity has been lost. But "technology, divorced from religion, and hostile to it, cannot improve the essential condition of man."

MAHR, FRANZ. *Der Christ in der Welt der Apparate.* Wurzburg: Echter, 1958. Pp. 178.

MAIRINGER, HANS. "Der Christ und die Technik." Pp. 291–329 in Josef Stradlmann and Ludwig Hänsel, eds., *Christentum und moderne Geisteshaltung.* Vienna and Munich, 1954.

MANN, KENNETH W. *Deadline for Survival: a Survey of Moral Issues in Science and Medicine.* New York: Seabury, 1970. Pp. 147. An overview of selected technologies in the scientific and medical fields. Concerned primarily with the social comprehensiveness of technical applications and the latter's effect upon the nature of man, as well as the responsibility of man to himself and to his environment. Also discusses the development of new, social oriented policies within the scientific community and man's conceptions of himself, his religious and moral perspectives in a technological age, and his responsibility to implement relevant ethical decisions.

MARTI, CASIMIRO, and JOSE BIGORDA. *Los hombres en la era industrial; iniciacion al problema social a la luz de la doctrina pontificia* [Men in the industrial era; initiation to the social problems in light of pontifical doctrine]. Barcelona: Editorial Teide, 1963. Pp. 221.

METZ, JOHANNES B. "Future of Faith in a Hominized World," *Philosophy Today* 10, no. 4 (Winter 1966): 236–45.

———, ed. *The Evolving World and Theology.* New York: Paulist Press, 1967. Pp. 184. Contents include: W. Broker's "Aspects of Evolution," H. Cox's "Evolutionary Progress and Christian Promise," A. van Melsen's "Natural Law and Evolution," J. Ellul's "The Technological Revolution and Its Moral and Political Consequences," E. G. Mesthene's "Religious Values in the Age of Technology," and E. Mascall's "The Scientific Outlook and the Christian Message."

MICHAEL, T., and D. NG. *How to Cope in a Computer Age without Pulling the Plug.* New York: Friendship Press, 1971.

MONTANA, ILENE, ed. *Technology and Culture in Perspective.* Cambridge, Mass.: Church Society for College Work, February 1967. Pp. 44. Comments by H. Cox, H. Hoaglund, W. Ong, and G. Kepes at the Seminar on Technology and Culture at M.I.T. Another occasional collection of dubious value to outsiders.

MORTON, A. Q., and JAMES MCLEMAN. *Christianity in the Computer Age.* New York: Harper & Row, 1965. Pp. 95. Reviewed by John W. Ellison, *Journal of Biblical Literature,* vol. 85 (1965).

MOSLEY, JOHN BROOKE. *Christians in the Technical and Social Revolutions; Suggestions for Study and Action.* Cincinnati: Forward Movement Publications, 1966. Pp. 141.

MOULE, CHARLES FRANCIS DIGBY. *Man and Nature in the New Testament: Some Reflections on Biblical Ecology.* Philadelphia: Fortress Facet Books, 1967. Pp. 27. First published, London: Univ. London Athlane Press, 1964. Pp. 22.

O'REILLY, J. "The Christian in a Technological World," *Homiletic and Pastoral Review* 67 (August 1967): 951–55.

OTTAVIANO, C. "Il Christianesimo el l'origine della scienza moderna," *Sophia,* vol. 34, nos. 3–4 (July–December 1966). Note also M. Rocca's "Ultime novità della scienza della tecnica" in this same issue.

"Papers of the 1970 Conference on Ethics and Ecology of the Institute on Religion in an Age of Science," *Zygon,* vol. 5, no. 4 (December 1970). Six essays: J. L. Fisher's "Dimensions of the Environmental Crisis," J. B. Bresler's "Three Man-made Ecological Factors and Their Implications on Human Heredity and Health," K. H. Hertz's "Ecological Planning for Metropolitan Regions," W. E. Martin's "Simple Concepts of Complex Ecological Problems," W. W. Robbins's "The Theological Values of Life and Nonbeing," and D. F. Martensen's "Concerning the Ecological Matrix of Theology."

PASTORE, ANNIBALE. "La spiritualità della technica," *Rivista di filosofia neo-scolastica* 43, no. 3 (1952): 289–300.

PAUL VI, POPE. "Kirche und Errungenschaften der Technik" [The church and the gains of technology], *Seelsorger; Monatsschrift für alle Bereiche priesterlicher Reich-Gottes-Arbeit* (Vienna) 33 (1963): 532.

PHILIP, ANDRÉ. "Techniques modernes, mutations de structures et valeurs de civilisation" [Modern techniques, mutations of structures and values of civilization], *Christianisme social* 74, nos. 1-2 (1966): 3-23.

PLATZECK, ERHARD, WOLFRAM. *Técnica y espíritu.* Madrid, 1942. Pp. 100.

POLLARD, WILLIAM G. "The Key to the Twentieth Century." Pp. 45-57 in *Proceedings of the American Catholic Philosophical Association,* vol. 62. Washington, D.C.: Catholic Univ. America Press, 1968. The key is the biblical statement of man's destiny to "be fruitful and multiply and fill the earth and subdue it." "Only in the 20th century has this been realized."

POPPI, A. "Il disagio della civiltà tecnologica: Prospettive morali," *Studia patavina* (1971), pp. 484-504.

POWER, L. R. "Il Poverello and Technology," *Cord* 11 (October 1961): 297-303.

———. "Technology as a Means of Sharing in the Act of Creation," *Interest* 1 (Summer 1961): 2-3.

POZZO, GIANNI M. "Tecnicismo e morale: crisi del mondo moderno," *Humanitas* 6 (1961): 861-69.

PRESTON, RONALD H., ed. *Technology and Social Justice.* Valley Forge, Pa.: Judson, 1971. Pp. 472. Papers from an international symposium on the social and economic teaching of the World Council of Churches from Geneva 1966 to Uppsala 1968. Part II, "Towards a Theology of Development," contains E. Hoffmann's "The Challenge of Economic and Social Development," S. L. Parmar's "Reflections of a Lay Economist from a Developing Country," and N. A. Nissiotis's "Introduction to a Christological Phenomenology of Development." Some other essays in Parts I and III-VII are also relevant.

RAHNER, KARL, S.J. "The Mass and Television." Pp. 205-18 in *The Christian Commitment.* New York: Sheed & Ward, 1963. Argues against televising the mass by considering the nature of television per se in a way that raises larger questions for technology as a whole.

RAMSEY, PAUL. *Fabricated Man: the Ethics of Genetic Control.* New Haven, Conn.: Yale Univ. Press, 1970. Pp. 174. Inquiry by a Protestant theologian. Good introduction to this area of theological concern. See also Ramsey's "Moral and Relgious Implications of Genetic Control," in John D. Roslansky, ed., *Genetics and the Future of Man* (New York: Appleton-Century-Crofts, 1966), pp. 109-69; and Ramsey's *The Patient as Person* (New Haven, Conn.: Yale Univ. Press, 1972).

RAUNSCHENBUSCH, WALTER. *Christianity and the Social Crisis.* New York: Macmillan, 1907. Reprinted, New York: Harper Torchbook, 1964. Pp. 429. A Christian attempt to come to terms with industrialization by one of the important thinkers in the Christian Socialist tradition. An early advocate of the social gospel, the theory that Christianity finds its true meaning in social action. See also *Christianizing the Social Order* (New York: Macmillan, 1912).

REDEKOP, CALVIN. *The Free Church and Seductive Culture.* Scottsdale, Pa.: Herald Press, 1970. Pp. 189. Analyzes the relation between the free church, i.e., Anabaptist tradition, and technologically affluent society. Although it does not deal directly with technology, it is a good background volume containing some vital observations.

"Religion in a Technological Society," *Technologist,* vol. 2, no. 3 (1965). A symposium.

RIAZA, FERNANDO. "Aportaciones teilhardianas a una filosofía de la técnica," *Pensamiento* 24 (April-June 1968): 109-24.

RICHARDSON, HERBERT W. *Toward an American Theology.* New York: Harper & Row, 1967. Pp. 170. Attempts to develop a "systems theology," or a theology based on systems theory.

RIQUET, MICHÈLLE. "Progrès des techniques et royaume de Dieu" [Technological progress and the realm of God], *Arzt und Christ* (Salzburg), special issue (1961), pp. 310-16.

ROHRER, W. "Gedanken eines Christen zum Phänomen der Technik" [Thoughts of a Christian on the phenomenon of technology], *Civitas: Monatsschrift des Schweizerischen Studentenvereines* 19, no. 11 (1964): 472-80.

ROTUREAU, GASTON. *Conscience religieuse et mentalité technique* [Religious consciousness and technical mentality]. Tournai: Desclée, 1962. Pp. 143.

ROVASENDA, E. D. "I valori della tecnica," *Sapienza* 13, nos. 3-4 (May-August 1960): 161-80.

RUST, ERIC C. *Nature, Garden or Desert; an Essay in Environmental Theology.* Waco, Tex.: Word Books, 1971. Pp. 150. See also Rust's *Nature and Man in Biblical Thought* (London: Lutterworth, 1953).

SANTMIRE, PAUL H. *Brother Earth; Nature, God and Ecology in a Time of Crisis.* New York: Nelson, 1970. Pp. 236.

SCHEPERS, M. B. "Protestant Theology Contemporary Trends In." Pp. 886-91 in *New Catholic Encyclopedia.* Vol. 11. New York: McGraw-Hill, 1967. Surveys attitudes toward technology in the work of K. Barth, E. Brunner, R. Niebuhr, H. R. Niebuhr, R. Bultmann, P. Tillich, etc.

SCHLEMMER, HANS. *Die Technik und das Evangelium.* Leipzig: Hinrichs, 1941. Pp. 30. Reviews: Helmuth Thielicke, *Theologisches Literaturblatt* (Leipzig) 66 (1941): 367-70; and Else Lüders, *Soziale Praxis* (Leipzig), vol. 50 (1941).

SHAUGHNESSY, MOTHER C. "Future of Technology," *Perspectives* 6 (December 1961): 25-27.

SIEGFRIED, THEODOR. "Schöpfung und Technik" [Creation and technology]. In R. Paulus, ed., *Glaube und Ethos.* Stuttgart: Kohlhammer, 1940.

SILVA, LÚCIO CRAVEIRO DA. "Filosofia e teologia da técnica," *Rivista portuguesa de filosofia* 15 (1959): 149-62.

STERN, KARL. "Christian Humanism in an Age of Technology," *Critic* 17, no. 5 (April-May 1959): 18 ff.

STÖCKLEIN, ANSGAR. *Leitbilder der Technik; Biblische Tradition und technischer Fortschritt* [Guiding images of technology; biblical tradition and tech-

nological progress]. Munich: Heinz Moos Verlag, 1969. Pp. 198. A historical survey originally entitled "Biblical Images and Their Counter-Images in Machine Books from about 1550 to 1750." Documents the way in which both those who opposed and those who favored the modern adoption of machines used scripture to justify their positions.

STONE, GLENN C., ed. *A New Ethic for a New Earth.* New York: Friendship Press (for Faith-Man-Nature Group), 1971. Pp. 172. Faith-Man-Nature Group papers no. 2. Includes D. D. William's "Philosophical and Theological Concepts of Nature," F. J. Ayala's "A Biologist's View of Nature," T. F. Driver's "The Artist Looks at Nature," R. J. Seeger's "The Concept of Nature in Physical Science," J. N. Hartt's "Faith and the Informed Use of Natural Resources," P. A. Jordan's "An Ecologist Responds," W. L. Humphrey's "Pitfalls and Promises of Biblical Texts as a Basis for a Theology of Nature," F. Elder's "A Different 2001," S. I. Paradise's "Rehabilitation for Cosmic Outlaws," and R. Theobald's "The Changing Environment: Does the Church Have a Major Responsibility?"

STORCK, HANS. "Technik als Aufgabe" [Technology as task], *Die Mitarbeit; Monatshefte der Aktion evangelischer Arbeitnehmer* (Berlin) 10 (1961): 191-96.

THALHAMMER, DOMINIKUS. "Theologie der Technik," *Der grosse Entschluss* 16 (December 1960): 117-19; 16 (January 1961): 166-70; 16 (February 1961): 208-12. Also in *Aus Hertens Vergangenheit* 6 (1960): 152-62.

THIELECKE, HELMUT. "Theologische Probleme des technischen Zeitalters," *Universitas; Zeitschrift für Wissenschaft, Kunst und Literatur* (Stuttgart) 1 (1946): 567-72, 817-28.

_____. *Fragen des Christentums an die moderne Welt.* Tübingen: Mohr Siebeck, 1947. Pp. 274.

_____. "Dämon der Technik, eine Modephase?" [Demon of technology, a fashionable phase?], *Wirtschaft und Wissenschaft* 23 (1958): 1-2.

_____. "Freiheit und Schicksal in der

technischen Welt" [Freedom and fate in the technological world], *Glückauf; Berg- und hüttenmännische Zeitschrift* 95 (1959): 1561-66.

———. *Der Einzelne und der Apparat; von der Freiheit des Menschen im technischen Zeitalter* [The individual and the apparatus; on the freedom of man in the technological age]. Studenbücher no. 34. 2d ed. Hamburg: Furche-Verlag, 1966. Pp. 125.

TILMAN-TIMON, ALEXANDRE. "La technique et le spirtuel," *Etudes philosophiques* 22 (April-June 1967): 310-16.

TOURNIER, PAUL. *Technik und Glaube* [Technology and faith]. Tubingen: Furche-Verlag, 1947. Pp. 256.

"Towards a Theology of Environment," *Saint Louis University Magazine*, special suppl., vol. 34, no. 1 (Spring 1970). Contents: K. P. Shea's "Experiments in Global Ecology," R. R. Maginn's "Prospects for Human Development through Surgical Techniques," J. A. Mulligan's "Prospects for Bioengineering," B. Korol's "Mind Forming Drugs," N. J. Colarelli's "Psychology: the Image Maker," E. J. F. Arndt's "The Vision of Man," J. C. Futrell's "Spiritual Continuity and Cultural Shock," and A. von R. Sauer's "Man as Steward of Creation."

TRAUPEL, WALTER. "Technique contre esprit," *Cahiers protestants* (Lausanne) 46 (1962): 29-36.

TROELTSCH, ERNST. *Protestantism and Progress: a Historical Study of the Relation of Protestantism to the Modern World.* Translated by W. Montgomery. Boston: Beacon, 1958. Pp. 210. First published 1912. Material on the general problem of the relation between Christianity and the world which can serve as a background for dealing with the more specific issue of Christianity in the technological world. See also Troeltsch's *The Social Teachings of the Christian Churches* (London: Allen & Unwin, 1931; reprinted, New York: Harper Torchbook, 1960); H. Richard Neibuhr's *Christ and Culture* (New York: Harper & Row, 1951); and Harry Emerson Fosdick's *Christianity and Progress*

(New York: Fleming H. Revell Co., 1922). Max Weber's *The Protestant Ethic and the Spirit of Capitalism* (London: Allen & Unwin, 1930) is also relevant. But cf. also G. Grant's "In Defense of North America," *Technology and Empire* (Toronto: House of Anansi, 1970), where Weber's thesis is extended in important respects.

WALLACE, WILLIAM A. "Science, Technology and God," *Albertus magnus* 9 (November 1961): 1 ff.

———. "Some Moral and Religious Implications of Nuclear Technology," *Journal of the Washington Academy of Sciences* 55 (1965): 85-91. Discussion of an interdisciplinary conference held at the University of Chicago (January 16-19, 1963) to consider social ethics in the context of radiation effects.

WEAVER, WARREN. "Some Moral Problems Posed by Modern Science," *Zygon* 1, no. 3 (September 1966): 286-300. After defining the words "morals," "science," and "religion," the author considers several moral problems posed by medicine, physics, biology, and science in general. Modern man "is involved in decisions that affect millions . . . of persons." Yet "the extent of the individual involvement is often obscure" even while the "overall effect on mankind may be catastrophic." Reprinted in Weaver's *Science and Imagination* (New York: Basic, 1967).

WESSEL, R. "Apologie des technischen Menschen," *Unitas; Monatsschrift des Verbandes der wissenschaftlichen katholischen Studentenvereine* (Cologne) 98 (1958): 11-18, 50-51.

WILKES, KEITH. "Religion and Technology: Aspects of a Dynamic Relationship," *Technology and Society* 6, nos. 3-4 (March 1971): 112-16. Historical glimpses into the influence of religious beliefs on the encouragement and spread of technological innovations. As in the past, technological man needs the guidance of religion.

WILLIAMS, GEORGE H. *Wilderness and Paradise in Christian Thought; the Biblical Experience of the Desert in the History of Christianity and the Paradise Theme in the Theological Idea of the University.* New York: Harper, 1962. Pp. 245.

WREN-LEWIS, JOHN. "Faith in the Technological Future," *Futures* 2, no. 3 (September 1970): 258-62. Technology liberates man; "the grounds for hope in the future is the notion that the patterns of nature are made for man, not man for them."

ZIETEMANN, WINFRIED. "Mensch und Technik," *Die berufsbildende Schule* (Wolfenbüttel) 9 (1957): 430-35.

"Zum theologischen Problem der Technik," *Lutherische Monatshefte* (Stuttgart) 1 (1962): 422-29.

[GENERAL NOTE: For a number of other German works pre-1956 on the relation between religion and technology, see Dessauer's bibliography.]

IV. METAPHYSICAL AND EPISTEMOLOGICAL STUDIES

ADLER, MORTIMER J. *The Difference of Man and the Difference It Makes.* New York: Holt, Rinehart & Winston, 1967. Pp. 395.

AGASSI, JOSEPH. "The Confusion between Science and Technology in the Standard Philosophies of Science," *Technology and Culture* 7, no. 3 (Summer 1966): 348-66. A critique of inductivist and instrumental philosophies of science which equate applied science and technology. "Contrary to most, if not all, writers in the field, I hold that confirmation plays no significant role in science, pure or applied. The contrary impression seems to me to stem from the fact that both invention and the implementation of novelties—from applied science or from invention—require confirmation. The standards of confirmation are legal, and they are set by patent offices in the case of invention and by institutions in charge of public safety and of commercial practices in the case of implementation." See also J. O. Wisdom's "The Need for Corroboration; Comments on Agassi's Paper," *Technology and Culture* 7, no. 3 (Summer 1966): 367-70; and Agassi's

"Planning for Success: a Reply to Professor Wisdom," *Technology and Culture* 8, no. 1 (January 1967): 78-81.

———. "The Logic of Technological Development." Pp. 483-88 in *Proceedings of the XIVth International Congress of Philosophy; Vienna, Sept. 2-9, 1968.* Vol. 2. Vienna: Herder, 1968. A good statement of Agassi's views on the irrelevance of positive evidence (or verification, or confirmation, or corroboration) to scientific progress. "The distinct role which positive evidence plays, then is in the field of social implementation of technological innovations and alterations (physical, biological, and social)."

———. "Positive Evidence in Science and Technology," *Philosophy of Science* 37, no. 2 (June 1970): 261-70. See also *Towards an Historiography of Science* (The Hague: Mouton, 1963) (*History and Theory*, Studies in the Philosophy of History, Suppl. 2).

———. "Positive Evidence as Social Institution," *Philosophia*, vol. 1 (1971).

AGAZZI, EVANDRO. "Alcune osservazioni sul problema dell'intelligenza artificiale," *Rivista di filosofia neo-scolastica* 59, no. 1 (January-February 1967): 1-34.

AHLERS, ROLF. "Is Technology Intrinsically Repressive?" *Continuum* 8, no. 1 (Spring-Summer 1970): 111-22. Marcuse's and Heidegger's critiques of technology are both romantic flights from the kind of Hegelian acceptance and understanding that could really overcome technology's present repressiveness. Later published in *Tijdschrift voor filosofie* 32 (December 1970): 651-700. See also "Technologie und Wissenschaft bei Heidegger und Marcuse," *Zeitschrift für Philosophische Forschung* 25, no. 4 (October-December 1971): 575-90.

ALBRITTON, ROGERS. "Mere Robots and Others," *Journal of Philosophy* 61 (November 1964): 691-94. Reprinted in F. J. Crosson, ed., *Human and Artificial Intelligence* (New York: Appleton-Century-Crofts, 1970).

ALDERMAN, HAROLD. "Heidegger: Technology as Phenomenon," *Personalist* 51 (Fall 1970): 535-45. According to

Heidegger, technology is the perfect expression of modern man's aggressive and calculative rationalism. And contemporary thinkers must recognize that in spite of themselves history has made them all technicians. Only from within such a recognition is a responsible critique of technology possible.

ALT, FRANZ L., ed. *Advances in Computers.* Vols. 1-11. New York: Academic, 1960-71.

ANDERSON, ALAN ROSS, ed. *Minds and Machines.* Englewood Cliffs, N.J.: Prentice-Hall, 1964. Pp. 114. Contains A. M. Turing's "Computing Machinery and Intelligence," M. Scriven's "The Mechanical Concept of Mind," J. R. Lucas's "Minds, Machines and Gödel," K. Gunderson's "The Imitation Game," H. Putnam's "Minds and Machines," P. Ziff's "The Feelings of Robots," J. J. C. Smart's "Professor Ziff on Robots," and N. Smart's "Robots Incorporated." A good anthology of essays on the philosophical distinctions between man and machine.

APTER, M. J. *The Computer Simulation of Behavior.* New York: Harper & Row, 1971. Pp. 180.

ARBIB, MICHAEL A. *Brains, Machines and Mathematics.* New York: McGraw-Hill, 1964. Pp. 152. A good introduction "to the common ground of brains, machines, and mathematics, where mathematics is used to exploit analogies between the working of brains and the control-computation-communication aspects of machines." Contains five chapters on neural nets, finite automata and Turing machines, structure and randomness, the correction of errors in communication, cybernetics, and Gödel's incompleteness theorem. Chap. 3 contains short sections on von Neumann's multiplexing scheme, Shannon's communication theory, the Cowan-Winograd theory of reliable automata, etc. Reviewed by J. Bronowski, *Scientific American* 210 (June 1964): 130-34.

――――. "Automata Theory." In R. E. Kalman, P. L. Falb and M. A. Arbib,

eds., *Topics in Mathematical System Theory.* New York: McGraw-Hill, 1969.

――――. *Theories of Abstract Automata.* Englewood Cliffs, N.J.: Prentice-Hall, 1969. Pp. 412. See also Michael A. Arbib, ed., *Algebraic Theory of Machines, Languages, and Semi-Groups* (New York: Academic, 1968).

ARMER, PAUL. "Attitudes toward Intelligent Machines." Pp. 389-405 in Edward A. Feigenbaum and Julian Feldman, eds., *Computers and Thought.* New York: McGraw-Hill, 1963. Originally published in E. J. Steele, ed., *Symposium on Bionics* (USAF Air Research and Development Command, 1960). Also published as *Attitudes toward Artificial Intelligence,* RAND report P-2114-2 (Santa Monica, Calif.: RAND Corp., June 1962).

ASHBY, W. ROSS. "Adaptiveness and Equilibrium," *Journal of Mental Science* 86 (1940): 478-83.

――――. "The Nervous System as a Physical Machine: With Special Reference to the Origin of Adaptive Behavior," *Mind* 56 (1947): 44-59.

――――. "The Cerebral Mechanism of Intelligent Action." In D. E. Richter, ed., *Perspectives in Neuropsychiatry.* London: H. K. Lewis, 1950.

――――. "Can a Mechanical Chess Player Outplay Its Designer?" *British Journal for the Philosophy of Science* 3 (May 1952): 44-47.

――――. *Design for a Brain.* New York: Wiley, 1954. Pp. 259. 2d rev. ed., 1960. Pp. 286. For an earlier account of the author's ideas on this subject, see "Design for a Brain," *Electronic Engineering* 20 (1948): 378-83.

――――. *An Introduction to Cybernetics.* New York: Wiley, 1956. Pp. 295.

――――. "What Is an Intelligent Machine?" *Proceedings of the Western Joint Computer Conference* (May 1961), pp. 275-80.

――――. "Principles of Self-organizing Systems." Pp. 255-78 in H. von Foerster and G. Zopf, eds., *Principles of Self-Organization.* Oxford: Pergamon, 1962.

"Automatic Control," *Scientific American,* vol. 187, no. 3 (September 1952). Contains: E. Nagel's "Automatic Control," A. Tustin's "Feedback," G. S. Brown and D. P. Campbell's "Control Systems," E. Ayres's "An Automatic Chemical Plant," W. Pease's "An Automatic Machine Tool," L. N. Ridenour's "The Role of the Computer," G. W. King's "Information," and W. Leontief's "Machines and Man." This collection can still serve as a good introduction to the subject. Also contains a brief bibliography on control technology.

AUWÄRTER, MAX. Voraussetzungen und Grundlagen für das technische Denken" [Preconditions and basis for technological thinking], *Technische Rundschau* (Bern) 54, no. 48 (1962): 1-2.

AXELOS, KOSTOS. *Einführung in ein künftiges Denken: über Marx und Heidegger* [Introduction to future thinking: concerning Marx and Heidegger]. Tubingen: Niemeyer, 1966. Pp. 104. "Both Heidegger and Marx—to some extent as the heirs of Hegel—are philosophers of the future. Axelos' book is written from the Heideggerian standpoint, centering on Heidegger's interpretation of the present age in terms of *Technik*, which is itself a movement in the mission *(Geschick)* of Being. The movements of the *Geschick* . . . are a world-play. The task of a thinking concerned with the future is to play along with the play. The future in Marxist *and* Heideggerian terms is a 'planetary' age, beyond all nationalism, ruled by a global or planetary technology. Thinking must decide whether and how such technology will be an instrument of liberation or of oppression. Axelos's confrontation raises many central and interesting questions: what is the relationship between Heidegger's *Gelassenheit* and Marx's *praxis?* What hope is there for a future governed by a play? etc."—J. D. Caputo, *Review of Metaphysics* 25, no. 2 (December 1971): 349.

BACH, K. "Denkformen der Technik" [Thought-forms of technology], *VDI-Nachrichten* 17, no. 15 (1963): 9.

BAKER, WILLIAM O. "Engineering and Science: a Sum and Not a Differ-ence," *American Scientist* 55, no. 1 (1967): 80-87.

BAR-HILLEL, YEHOSHUA. *Language and Information; Selected Essays on Their Theory and Application.* Reading, Mass.: Addison Wesley, 1964. Pp. 388. Reviewed by D. M. MacKay in *British Journal for the Philosophy of Science* 16 (November 1965): 253-55.

BENACERRAF, PAUL. "God, the Devil, and Gödel," *Monist* 51, no. 1 (January 1967): 9-32. A critique of J. R. Lucas and a detailed discussion of the implications of Gödel's theorem for a distinction between man and machine. For a reply, see J. R. Lucas, "Satan Stulified," *Monist* 52 (1968): 145-48.

BERKELEY, EDMUND CALLIS. *The Computer Revolution.* Garden City, N.Y.: Doubleday, 1962. Pp. 249. See also Berkeley's *Symbolic Logic and Intelligent Machines* (New York: Reinhold, 1959).

BERTALANFFY, LUDWIG VON. *Robots, Men, and Minds: Psychology in the Modern World.* New York: Braziller, 1967. Pp. 150. An attack on mechanism and behaviorism in the physical and social sciences, arguing for a general systems theory approach to understanding man. General systems theory should not be identified with cybernetics. The cybernated system is basically mechanistic and closed, but the behavior of general systems is determined by the interaction of many forces and variables in an open system. Good bibliography of general systems theory, and especially of the author's own work.

_____. *General Systems Theory; Foundations, Development, Applications.* New York: Braziller, 1968. Pp. 290. A collection of essays on systems theory by one of its founders. It "is predominantly a development in engineering science in the broad sense, necessitated by the complexity of 'systems' in modern technology, man-machine relations, programming and similar considerations which were not felt in yesteryear's technology but which have become imperative in the complex technological and social structures of the modern world. Systems theory, in this sense, is preeminently a mathematical field, offering partly novel and highly sophisticated techniques, closely linked with computer science. . . . Moreover, systems science, centered in computer technology, cybernetics, automation and systems engineering, appears to make the systems idea another—and indeed the ultimate—technique to shape man and society ever more into the 'megamachine' which Mumford (1967) has so impressively described. . . ." Also contains a good bibliography on Bertalanffy and on systems theory. The author has also edited, with A. Rapoport, a series of volumes under the title *General Systems* (Bedford, Mass.: Society for General Systems Research, 1956–68).

BLAKE, D. V., and A. M. UTTLEY, eds. *Proceedings of the Symposium on Mechanization of Thought Processes*. National Physical Laboratory Symposium, no. 10. 2 vols. London: Her Majesty's Stationery Office, 1959. Contains: D. M. MacKay's "Operational Aspects of Intellect," M. Minsky's "Some Methods of Artificial Intelligence and Heuristic Programming," etc.

BLENKE, HEINZ. "Zur Synthese von Wissenschaft und Technik" [On the synthesis of science and technology], *Mitteilungen der deutschen Forschungsgemeinschaft* 4 (1966): 2–26.

BLUMENBERG, H. Das Verhältnis von Natur und Technik als philosophisches Problem" [The relation of nature and technology as a philosophical problem], *Studium generale* 4, no. 8 (1951): 461–67.

———. "Lebenswelt und Technisierung unter Aspekten der Phänomenologie," *Filosofia* (Turin), vol. 14, no. 4, suppl. (November 1963).

BOBIK, JOSEPH, and KENNETH M. SAYRE. "Pattern Recognition Mechanisms and St. Thomas' Theory of Abstraction," *Revue philosophique de Louvain* 61 (February 1963): 24–43.

BODEN, MARGARET A. "Machine Perception," *Philosophical Quarterly* 19 (January 1969): 33–45. Do machines merit the term "percipient"? Two mutually irreducible forms of explanation are appropriate to machines: One refers to the mechanism, and one refers to the program. They are analogous to physiological and purposive explanations of behavior.

BOEHM, RUDOLF. "Pensée et technique; notes préliminaires pour une question touchant la problématique heideggerienne," *Revue internationale de philosophie* 14 (1960): 194–220.

BOIREL, RENÉ. *Science et technique.* Neuchatel: Editions du Griffon, 1955. Pp. 116. Boirel starts out by distinguishing between science and technique according as one aims at knowledge, the other at action. But "at the conclusion of this study, science and technique, grasped in their reciprocal combination, appear much more intimately linked than their divergent aims would have us suppose. While technique, in order to allow satisfaction of the needs of man living in the world, realizes a material effect by incarnating a bodily operation into matter, . . . science guarantees domination of these transforming operations upon matter by expressing them through a series of rational relations." Thus the argument of the book is that science and technique "are fundamentally two forms of human praxis" (p. 111).

BORGMANN, ALBERT. "Technology and Reality," *Man and World* 4, no. 1 (February 1971): 59–69. Technology procures the availability of things. "Where availability prevails things are present universally and instantly, without danger, subject to easy manipulation, and in a highly mobile manner. But what seem to be positive traits of presence are in fact indications of a decline in the presence of things. Advanced availability is identical with advanced evacuation of reality. . . . The more availability advances the more reality turns into a homogeneous horizon of possibilities" (p. 62). The only limit to technology, the only thing technology does not seem to be able to make available, is procurement itself, since this still "requires concentration of specialists in teams and of productive machinery in plants" (p. 68).

BORING, E. G. "Mind and Mechanism," *American Journal of Psychology* 59 (April 1946): 173–92.

BOULANGER, GEORGE R. "La révolution cybernétique," *Revue internationale de philosophie* 17, no. 2 (1963): 220–42.

BOYD, JOHN. "La révolution scientifique et technique," *Nouvelle revue internationale* 11, no. 6 (1968): 72–87. "The Brain and the Machine." Pp. 109–88 in Part 2 of Sidney Hook, ed., *Dimensions of Mind*. New York: New York Univ. Press, 1960. Reprinted, New York: Collier Books, 1961. Contents: N. Wiener's "The Brain and the Machine" (summary), M. Scriven's "The Compleat Robot: a Prolegomena to Androidology," S. Watanabe's "Comments on Key Issues," H. Putnam's "Minds and Machines," A. C. Danto's "On Consciousness in Machines," R. Lachman's "Machines, Brains, and Models," R. M. Martin's "On Computers and Semantical Rules," P. Weiss's "Love in a Machine Age," F. Heider's "On the Reduction Sentiment," and S. Hook's "A Pragmatic Note."

BRILLOUIN, LÉON. *Science and Information Theory*. New York: Academic, 1956. Pp. 320. 2d ed., 1962.

BROGLIE, C. "Uber die metaphysischen Hintergründe der Technik" [On the metaphysical background of technology], *Lehrerrundbrief* 14 (1959): 407–12.

BRONOWSKI, JACOB. "The Logic of Mind," *American Scientist* 54 (1966): 1–14. An informal discussion of how the theorems of K. Gödel, A. Church, A. Turing, and A. Tarski affect the concept of mind. Concludes that the common quality of imagination in science and literature can be traced to the logic of self-reference.

BRUN, JEAN. *Les conquêtes de l'homme et la séparation ontologique*. Paris: Presses Universitaires de France, 1961. Pp. 298.

——. *La main et l'esprit*. Bibliothèque philosophie contemporaine. See also the author's *La Main* (Paris: Delpire, 1968).

BRUNNER, AUGUST, S.J. "Perils of Technological Thought," *Philosophy Today* 1, no. 2 (June 1957): 114–17. From "Die Gefahren des technischen Denkens," *Stimmen der Zeit* 157 (February 1956): 335–46. A pedestrian essay about how the scientific-technological attitude toward things influences even those who are not dedicated to it. Another version of this essay appears in Walter Leifer, ed., *Man and Technology* (Munich: Hueber, 1963).

——. "Geist im technischen Zeitalter," *Stimmen der Zeit* 161 (1957–58): 161–72.

BUNGE, MARIO. "Do Computers Think?" *British Journal for the Philosophy of Science* 7 (August 1956): 139–48; and 7 (November 1956): 212–19. "What I propose to do here in order to ascertain whether machines think or not, is to examine succinctly the two main aspects of the question, namely (a) the nature of computers, and (b) the nature of mathematical thought. . . . Insofar as machines are the outcome of intelligent and purposive work, they cannot be put in the same class as natural inanimate objects; machines are matter intelligently organized by technology, and as such they stand on a level of their own. But, on the other hand, it should be kept in mind that artifacts, however complex, operate only with material objects, never with ideal, abstract objects, a sort of operation which is precisely one of the distinctive characterists of educated human beings. This elementary point is missed by most cyberneticians, and it seems to be the clue for the understanding of the whole question."

——. "Action." Pp. 121–50 in *Scientific Research*, Vol. 2: *The Search for Truth*. Berlin-Heidelberg-New York: Springer-Verlag, 1967. Technological theories are of two kinds and are as theory-laden as pure science. Substantive technological theories are applications of preexisting scientific theories. Operative technological theories are applications of the scientific method to man-machine complexes in nearly real situations; essentially they are scientific theories of action. There is a discussion of the differences between scientific laws and technological rules, between scientific predictions, technological forecasts, and expert prognoses. An earlier version of this chapter appeared as "Technology as Applied Science," *Technology and Culture* 7, no. 3 (Summer 1966): 329-47. Slightly revised version included in C. Mitcham and R. Mackey, eds., *Philosophy and Technology* (New York: Free Press, 1972).

―――. "Scientific Laws and Rules." Pp. 128-40 in R. Klibansky, ed., *Contemporary Philosophy; a Survey*, Vol. 2: *Philosophy of Science*. Florence: La Nuova Italia Editrice, 1968. The successful rules of modern technological action are grounded in a set of scientific laws, and this explains their effectiveness. This is the decisive difference between the rules of modern technology and the rules of art and craft which are found by trial and error and whose effectiveness is unexplained.

―――. "Is Scientific Metaphysics Possible?" *Journal of Philosophy* 68, no. 17 (September 2, 1971): 507-20. Automata theory, originally conceived of as a theoretical basis for computer science, has become so extremely general that it applies to all kinds of systems in all kinds of environments. "In particular, automata theory is applicable to neuron assemblies and to societies, not only to artifacts."

BURMEISTER, HANS. "Technologie — Mittler zwischen Theorie und Praxis" [Technology—intermediary between theory and practice], *Urania: Monatsschrift über Natur und Gesellschaft* (Jena) 5 (1968): 14-20.

CARPENTER, STANLEY ROBERT. "The Structure of Technological Action," Boston Univ., 1971. Pp. 198. A doctoral dissertation on the relationship of knowledge to technological action. Chaps. 1 and 2 characterize the philosophy of technology and introduce a working definition of technology: "those patterns of action by which man transforms knowledge of his environment into an instrument of control over that environment for the purpose of meeting human needs." In chap. 3, the author symbolizes a number of concepts derived from T. Kotarbinski's praxiology, and argues that a theoretical schema can be con-

structed which is compatible with the patterns of action by which man achieves control over his environment. "The method which has been presented amounts to an attempt, admittedly preliminary in nature, to derive a 'grammar of technological action.' "

CECCATO, SILVIO. "Cybernetics as a Discipline and an Interdiscipline,"*Diogenes* 53 (Spring 1966): 99-114.

CELLÉRIER, GUY, SEYMOUR PAPERT, and GILBERT VOYAT. *Cybernétique et épistémologie*. Paris: Presses Universitaires de France, 1968. Pp. 143. Chap. 1, "Modèles cybernétiques et adaptation," by G. Cellérier; chap. 2, "A propos du perception 'Qui a besion de l'épistémologie?' " by S. Papert and G. Voyat; chap. 3, "Note sur le perception et la motricité oculaire," by G. Cellérier; chap. 4, "McCulloch et la naissance de la cybernétique," by S. Papert. There is no index, but chap. 1 has a brief bibliography.

CHAPANIS, ALPHONSE. "Men, Machines, and Models," *American Psychologist* 16, no. 3 (March 1961): 115-16.

CHARI, C. T. K. "Further Comments on Minds, Machines and Gödel," *Philosophy* 38 (April 1963): 175-78. J. R. Lucas's argument based on Gödel's theorems should be considered significant in light of radical difficulties in classical mathematics involved with formalizing completely deductive systems.

CHURCHMAN, C. WEST. *Challenge to Reason*. New York: McGraw-Hill, 1968. Pp. 224. "Many present-day philosophers believe that our technological systems are more or less determined and that their particular destructive effects on man will happen whatever we do to change them. Yet most of us would not be inclined to arrive at so pessimistic a conclusion concerning our fate until we had tried to explore the more attractive assumption that we have a certain inherent ability to change the world for the better. . . . " In exploring this second possibility Churchman argues for the need to deal with the "ethics of whole systems" in a way that raises basic metaphysical and epistemological issues associated with operations research and systems theory, two disciplines closely allied with modern technology. Reviewed by A. McLaughlin, *Technology and Culture* 11, no. 1 (January 1970): 127–29. See also Churchman's *Prediction and Optimal Decision; Philosophical Issues of a Science of Values* (Englewood Cliffs, N.J.: Prentice-Hall, 1961).

———. *The Systems Approach.* New York: Delacorte, 1968. Pp. 243.

CHURCHMAN, C. WEST, RUSSELL L. ACKOFF, and E. LEONARD ARNOFF. *Introduction to Operations Research.* New York: Wiley, 1957.

CLACK, ROBERT J. "Can a Machine Be Conscious? Discussion of Dennis Thompson," *British Journal for the Philosophy of Science* 17 (November 1966): 232–34.

———. "The Myth of the Conscious Robot," *Personalist* 49, no. 3 (Summer 1968): 351–69.

CLARKE, J. J. Turing Machines and the Mind-Body Problem," *British Journal for the Philosophy of Science,* vol. 23 (February 1972).

CLARKE, W. NORRIS. "System: a New Category of Being," *Proceedings of the Twenty-third Annual Convention of the Jesuit Philosophical Association* (Woodstock, N.Y.: Woodstock College Press, 1961), pp. 5–17.

———. "Cybernetics and the Uniqueness of Man." Pp. 49–54 in *Proceedings of the Seventh Inter-American Congress of Philosophy.* Vol. 2. Quebec: Les Presses de l'Université Laval, 1968.

CODER, DAVID. "Gödel's Theorem and Mechanism," *Philosophy* 44 (July 1969): 234–37. "It is true that no minds can be explained as machines. But it is not true that Gödel's theorem proves this. At most, Gödel's theorem proves that not all minds can be explained as machines. Since this is so, Gödel's theorem cannot be expected to throw much light on why minds are different from machines. Lucas overestimates the importance of Gödel's theorem for the topic of mechanism, I believe, because he presumes falsely that being unable to follow any but mechanical procedures in mathematics makes something a machine."

Cf. J. R. Lucas's "Mechanism: a Rejoinder," *Philosophy* 45 (April 1970): 149–51.

COHEN, JONATHAN. "Can There Be Artificial Minds?" *Analysis* 16 (1955): 36–41.

COLLINS, N. L., and D. MICHIE, eds. *Machine Intelligence 1.* New York: American Elsevier, 1967. Pp. 278. Proceedings of a 1965 workshop in machine intelligence at the University of Edinburgh. Subsequent volumes have been published under various editors: E. Dale and D. Michie, eds., *Machine Intelligence 2* (1968); D. Michie, ed., *Machine Intelligence 3* (1968); B. Meltzer and D. Michie, eds., *Machine Intelligence 4* (1969); B. Meltzer and D. Michie, eds., *Machine Intelligence 5* (1970). Vols. 1–3 reviewed by M. A. Boden, *British Journal for the Philosophy of Science* 19 (1968): 271–74.

CONANT, JAMES BRYANT. *Modern Science and Modern Man.* New York: Columbia Univ. Press, 1952. Pp. 111. Reprinted, Garden City, N.Y.: Doubleday Anchor, 1952. Chaps. 1 and 3 especially offer a pragmatist interpretation of the nature of technology and its relation to science. Science defined as "a process of fabricating a web of interconnected concepts and conceptual schemes arising from experiments and observations and fruitful of further experiments and observations" (p. 62) turns out to be inherently active or technological. See also *On Understanding Science* (New Haven, Conn.: Yale Univ. Press, 1947) and *Science and Common Sense* (New Haven, Conn.: Yale Univ. Press, 1951).

———. "Induction and Deduction in Science and Technology." Pp. 1–31 in *Two Modes of Thought: My Encounters with Science and Education.* New York: Simon & Schuster, 1964. Science is "a complex combination of empirical-induction and theoretical-deduction. The practical arts, however, tend to be more empirical-inductive; and the transition from practical art to applied science "involves the introduction of concepts and conceptual schemes which were usually developed without regard for their use in improving a practical art" (p. 15). Chap. 2, "American Inventors of the Nineteenth Century," discusses a number of cases in which inventions were made with a minimum of theoretical guidance. See also Conant's "Die Verbindung zwischen der industriellen und der chemischen Revolution," *Humanismus und Technik* 1, nos. 3-4 (December 1, 1953): 135-44; and "Der Ingenier in der Zukunft," *Humanismus und Technik* 4, no. 3 (June 15, 1957): 157-65.

COTE, ALFRED J., JR. *The Search for the Robots.* New York: Basic, 1967. Pp. 243. A general introduction to bionics, the science of systems which function in a manner characteristic of, or resembling living systems.

COUFFIGNAL, LOUIS. *Les machines à calculer et la pensée humaine.* Brussels: Office de Publicité, 1953.

_____. "Science et technique de l'information," *Structure et évolution des techniques* (Paris), vol. 5, nos. 39-40 (July 1954-January 1955); and vol. 7, nos. 43-44 (September 1955-February 1956).

_____. "La cybernétique," *Structure et évolution des techniques* (Paris), vol. 8, nos. 51-52 (February-April 1957).

_____. *La cybernétique.* Paris: Presses Universitaires de France, 1963. Pp. 125.

_____. *Le concept d'information dans la science contemporaine.* Paris: Gauthier-Villars, 1965. Pp. 423. A collection of papers from a conference. Contributors include N. Wiener, L. Couffignal, B. Mandelbrot, G. Santillana, H. Greniewski, Goldmann, H. Frank, R. de Possell, A. Moles, and J. Zeman.

CRAIK, KENNETH J. W. *The Nature of Explanation.* Cambridge: Cambridge Univ. Press, 1943. "Among his contemporaries, Craik was legendary both for his mechanical skills and for his ability to conceptualize organic systems in mechanical terms. He was a master-builder of cybernetic models. Much of his published work, including a long chapter in *The Nature of Explanation*, is occupied with ingenious cybernetic designs for cognitive and behavioral functions."
—Merle B. Turner, *Realism and the*

Explanation of Behavior (New York: Appleton-Century-Crofts, 1971), p. 71.

CROSSON, FREDERICK J., and KENNETH M. SAYRE, eds. *Philosophy and Cybernetics.* Notre Dame, Ind.: Univ. Notre Dame Press, 1967. Pp. 271. Paperback, reprint, New York: Simon & Schuster. An important anthology of essays originally prepared for the Philosophic Institute for Artificial Intelligence at the University of Notre Dame. Contents: K. M. Sayre's "Philosophy and Cybernetics," "Choice, Decision, and the Origin of Information," "Toward a Quantitative Model of Pattern Formation," and "Instrumentation and Mechanical Agency"; F. J. Crosson's "Information Theory and Phenomenology," and "Memory, Models, and Meaning"; J. L. Massey's "Information, Machines, and Men"; and D. B. Burrell's "Obeying Rules and Following Instructions."

_____. "Information Theory and Phenomenology." Pp. 99-136 in F. J. Crosson and K. M. Sayre, eds., *Philosophy and Cybernetics.* Notre Dame, Ind.: Univ. Notre Dame, 1967. "There are striking parallels between the phenomenological account of the structure of perception and some analyses in a field which might seem to be unrelated, information theory. The aim of this essay, however, is not merely to draw attention to such parallelisms. It is to suggest that information-theoretic analysis possesses a conceptually interesting counterpart to meaning and also provides a formal model for the analysis of perception."

_____. "Phenomenology and Computer Simulation of Human Behavior." Pp. 160-69 in G. Murray, ed., *Philosophy and Science as Modes of Knowing.* New York: Appleton-Century-Crofts, 1968.

_____. "The Computer as Gadfly." Pp. 226-40 in R. S. Cohen and M. W. Wartofsky, eds., *Boston Studies in the Philosophy of Science.* Vol. 4. Dordrecht, Holland: Reidel, 1969. Discusses some recent work on handwriting and pattern recognition and

language translation and concludes that "both on the side of the discriminated perceptual whole and on the side of the subject, tacit elements appear to play an essential role in the human patterning of information."
———, ed. *Human and Artificial Intelligence.* New York: Appleton-Century-Crofts. 1970. Pp. 267. An important anthology containing H. A. Simon and A. Newell's "Information Processing in Computer and Man," D. E. Wooldridge's "Computers and the Brain," A. L. Samuel's "Some Studies in Machine Learning Using the Game of Checkers; II—Recent Progress," M. Scriven's "The Compleat Robot: a Prolegomena to Androidology," D. M. Mac-Kay's "The Use of Behavioural Language to Refer to Mechanical Processes," H. L. Dreyfus's "A Critique of Artificial Reason," H. Putnam's "Robots: Machines or Artificially Created Life?" R. Albritton's "Mere Robots and Others," U. Neisser's "The Imitation of Man by Machine," M. Polanyi's "The Logic of Tacit Inference," and M. Adler's "The Consequences for Action." Crosson's introduction is a good survey of the philosophical issues.

CULBERTSON, JAMES T. *The Minds of Robots; Sense Data, Memory Images and Behavior in Conscious Automata.* Urbana: Univ. Illinois Press, 1963. Pp. 466. "The purpose of this book is to show in detail how states of consciousness can be produced . . . in artificially constructed devices or robots." It is based on and uses the language and theory of "artificial nerve nets." Reviewed by D. M. MacKay, *British Journal for the Philosophy of Science,* vol. 16 (August 1965), and by T. Mischel, *Philosophy and Phenomenological Research,* vol. 26, no. 2 (December 1965).
"Cybernetics." Pp. 557-62 in *New Catholic Encyclopedia.* Vol. 4. New York: McGraw-Hill, 1967. The first part of this article, "Control Processes" by R. S. Ledley, gives a general description of the techniques of cybernetics; the second part, "Philosophical Implications" by W. A. Wallace, discusses general philosophical issues of man-

machine relationships and the problems of the nature of perception, memory, and abstraction, and offers a brief assessment from a Catholic perspective.
"La cybernétique," *Etudes philosophiques,* vol. 16, no. 2 (April-June 1961). Contains A. Rosenblueth's "Comportement, intention, téléologie," L. Couffignal's "La cybernétique comme méthodologie," R. Ruyer's "La cybernétique et al finalité," A. A. Moles's "La notion de quantité en cybernétique," L. Apostel's "Logique et cybernétique," P. Bertaux's "Les machines á traduire," J. Chaix-Ruy's "Actualité de Schopenhauer," and E. Przywara's "Una sancta."

DAEDALUS. "Pure Technology," *Technology Review* 72, no. 8 (June 1970): 38-45. The pseudonymous author is a columnist of *New Scientist.* A brief survey of the historical development, from classical time to the present, of pure as opposed to applied technology. "Pure technology is the building of machines for their own sake and for the pride or pleasure of accomplishment. It is a creative art form somewhere between art and science. Some examples of pure technology are the record-breaking vehicle, built purely to see if it will behave as intended; the chess-playing computer program, devised for the sheer entertainment of seeing how well it works out; and that masterpiece in miniature, *Scientific American's* Great International Paper Airplane Competition."

DANTO, ARTHUR C. "On Consciousness in Machines." Pp. 180-87 in Sidney Hook, ed., *Dimensions of Mind.* New York: New York Univ. Press, 1960.

DECHERT, CHARLES R. "Cybernetics and the Human Person," *International Philosophical Quarterly* 5, no. 1 (February 1965): 5-36. Discusses "a number of areas of correspondence between contemporary thought in the science of communication and control and traditional Aristotelian-Thomistic thought."

DESMONDE, WILLIAM H. "Gödel, Nondeterministic Systems, and Hermetic-Automata," *International Philosophi-*

cal Quarterly 11, no. 1 (March 1971): 49-74. Outlines a "model of a mixed mechanistic and non-mechanistic system" based on the theory of Turing machines. "The paper proceeds, in a non-technical way, to apply the proposed model to salvationistic theologies, such as that of Teilhard de Chardin, in which God redeems the natural world through an evolutionary process. The paper discusses the Hermetic tradition, in which the scientist and technologist are seen as outgrowths of the magician-alchemist who assists in the redemption of the cosmos. The inspiration of the scientist is viewed as a type of shamanistic ecstasy in which the individual contacts the non-deterministic part of the world system. The historical origin of axiomatization is regarded as a vision of the relation of the Many to the One. Automata are traced to the construction of icons. In making idols, early man sought to incarnate the Logos" (pp. 49-50). Note also the author's *Computers and Their Uses* (Englewood Cliffs, N.J.: Prentice-Hall, 1964).

DIAMOND, CORA. "The Interchangeability of Machines." Pp. 50-72 in J. J. MacIntosh and S. Coval, eds., *The Business of Reason*. London: Routledge & Kegan Paul, 1969.

DORE, P. "Natura e valore della 'tecnica' nel pensiero di S. Tommaso d'Aquino," *Rivista di filosofia neoscolastica* 56, nos. 3-4 (May-August 1964): 283-88.

DORFLES, G. "Technologia, oggettualità spaesamento alla XXXIII[a] Biennale," *Aut Aut; rivista di filosofia e di cultura* (Milan), vol. 95 (September 1966).

DREYFUS, HUBERT L. *Alchemy and Artificial Intelligence*. RAND Report P-3244. Santa Monica, Calif.: RAND Corp., December 1965. Behind research in artificial intelligence is the assumption that intelligent activities differ only in their degree of complexity. It has been assumed that the information processing which underlies intelligent behavior can be formulated in a program and simulated on a digital computer. This assumption is wrong because "the attempt to

analyse intelligent behavior in digital computer language systematically excludes three fundamental human forms of information processing (fringe consciousness, essence/accident discrimination, and ambiguity tolerance)." There is a highly critical account of current research on the ability of computers at game playing, problem solving, language translation and learning, and pattern recognition. He concludes that "we can then view recent work in artificial intelligence as a crucial experiment disconfirming the associationist assumption that all thinking can be analysed into discrete, determinate operations – the most important disconfirmation of this Humean hypothesis that has ever been produced" .(p. 85).

———. "Phenomenology and Artificial Intelligence." Pp. 31-47 in James Edie, ed., *Phenomenology in America*. Chicago: Quadrangle, 1967.

———. "Why Computers Must Have Bodies in Order to Be Intelligent," *Review of Metaphysics* 21, no. 2 (September 1967): 13-32. ". . . workers in AI [artificial intelligence] are necessarily committed to two basic assumptions: (1) an epistemological assumption that all intelligent behavior can be simulated by a device whose only mode of information processing is that of a detached, disembodied, objective observer. (2) The ontological assumption, related to logical atomism, that everything essential to intelligent behavior can in principle be understood in terms of a determinate set of independent elements." The author criticizes these two assumptions using arguments adapted from the phenomenologists. The general argument is that bodily skills and needs have an essential role in intelligent behavior. For a critical account of this essay, see Kenneth M. Sayre, "Intelligence, Bodies, and Digital Computers," *Review of Metaphysics* 21, no. 4 (June 1968): 714-23.

———. "Cybernetics as the Last Stage of Metaphysics." Pp. 493-99 in *Proceedings of the XIVth International Congress of Philosophy; Vienna, Sept. 2-9, 1968.*

Vol. 2. Vienna: Herder, 1968. The empirical arguments for artificial intelligence gain their plausibility only on the basis of an unjustified metaphysical axiom about the nature of language and intelligent human behavior, i.e., "that whatever people can in principle be formalized and processed by machine."

――――. "Mechanism and Phenomenology," *Noûs* 5, no. 1 (February 1971): 81-96.

――――. *What Computers Can't Do; a Critique of Artificial Reason.* New York: Harper & Row, 1972. Pp. 259. A detailed philosophical work on the limitations of research in artificial intelligence. Many arguments from the above articles are expanded and developed here. Current research in artificial intelligence (1957-67) is surveyed, and the work of Marvin Minsky is criticized in detail. Assumptions behind this research—biological, psychological, epistemological, and ontological—are critically related to arguments about the nature of perception and cognition in the works of Heidegger, Husserl, Merleau-Ponty, Polanyi, and Wittgenstein.

DRUCKER, PETER F. "Work and Tools," *Technology and Culture* 1, no. 1 (Winter 1959): 28-37. Technology is not about things: tools, processes, and products. It is about work: the specifically human activity by means of which man extends himself beyond the limitations of his animal nature. See also A. Zvorikine, "Concerning a Unified Concept of Technical Progress: Zvorikine on Drucker's 'Work and Tools,' " *Technology and Culture* 2, no. 3 (Summer 1961): 249-53, and Drucker's reply which follows (pp. 253-54). See also Drucker's "Technological Trends in the Twentieth Century" and "Technology and Society in the Twentieth Century," in Melvin Kranzberg and Carroll W. Pursell, Jr., eds., *Technology in Western Civilization* (New York: Oxford Univ. Press, 1967), 2:10-33. All of these essays are also reprinted in Drucker's *Technology, Management, and Society* (New York: Harper & Row, 1970).

DUCROCQ, ALBERT. *Découverte de la cybernétique* [Discovery of cybernetics]. Paris: Julliard, 1955. Pp. 279. A popular introduction. A German translation, *Die Entdeckung der Kybernetik; über Rechenanlagen, Regelungstechnik und Informationstheorie* (Frankfurt: Europäische Verlagsanstalt, 1959), has been influential. See also the author's *Logique de la vie* (Paris: Julliard, 1956) and *Logique générale des systèmes et des effets; introduction à une physique des effets fondements de l'intellectique* (Paris: Dunod, 1960).

EBACHER, ROGER. *La philosophie dans la cité technique; essai sur la philosophie bergsonienne des techniques.* Paris: Bloud & Gay, 1968; Quebec: Presses de l'Universite Laval, 1969. Pp. 242.

EVANS, C. R., and A. D. J. ROBERTSON, eds. *Cybernetics; Key Papers.* Baltimore: University Park Press, 1968. Pp. 289. A collection of papers by A. M. Turing, R. L. Gregory, W. S. McCulloch, K. J. W. Craik, C. Cherry, R. Thomson and W. Sluckin, etc.

FARRELL, B. A. "On the Design of a Conscious Device," *Mind* 79 (July 1970): 321-46. Also see Farrell's "Some Reflections on the Nature of Consciousness," in B. Rothblatt, ed., *Changing Perspectives on Man* (Chicago: Univ. Chicago Press, 1968).

FEIBLEMAN, JAMES K. "Pure Science, Applied Science, Technology, Engineering: an Attempt at Definitions," *Technology and Culture* 2, no. 4 (Fall 1961): 305-17. Distinguishes applied science and technology in terms of different approaches to practice. The applied scientist is guided by hypotheses deduced from theory, while the technologist employs skills derived from experience. However, the foundations of technology shifted from craft to science at the end of the 18th century, and the methods of technology and applied science are presently merging. Reprinted in an anthology of Feibleman's philosophical writings, Huntington Cairns, ed., *The Two-Story World* (New York: Holt, Rinehart & Winston, 1966), pp. 296-309. Included in C. Mitcham and R. Mackey, eds., *Philosophy and Technology* (New York: Free Press, 1972).

――――. "Technology as Skills," *Technology*

and *Culture* 7, no. 3 (Summer 1966): 318-28. "Skill" is defined as proficiency in the use of artifacts, and "artifact" is defined as a material object altered through human agency and intended for human use. "Every undertaking has its special technology, its tools, and the skills to use them. Technology is the material side of an enterprise, the discipline·which is equally necessary at every level. Thus both tools and skills are required for art, religion and philosophy as much as for economics and politics." See also "Artifactualism," *Philosophy and Phenomenological Research* 25, no. 4 (June 1965): 544-59, for the author's views on the role of material tools in human evolution.

FEIGENBAUM, EDWARD A., and JULIAN FELDMAN, eds. *Computers and Thought.* New York: McGraw-Hill, 1963. Pp. 535. "These research reports and discussions are concerned with the information processing activity that underlies intelligent behavior in human beings and computers." Of general interest are the major introductions to Part 1 on artificial intelligence and Part 2 on simulation of cognitive processes; P. Armer's "Attitudes toward Intelligent Machines"; M. Minsky's "Steps toward Intelligent Machines," and "A Selected Descriptor-indexed Bibliography to the Literature on Artificial Intelligence"; and A. M. Turing's "Computing Machinery and Intelligence." Also included are: A. Newell, J. C. Shaw, and H. A. Simon's "Chess-playing Programs and the Problem of Complexity"; A. L. Samuel's "Some Studies in Machine Learning Using the Game of Checkers"; A. Newell, J. C. Shaw, and H. A. Simon's "Empirical Explorations with the Logic Theory Machine: a Case Study in Heuristics"; H. Gelernter's "Realization of a Geometry-Theorem Proving Machine"; H. Gelernter, J. R. Hansen, and D. W. Loveland's "Empirical Explorations of the Geometry-Theorem Proving Machine"; F. M. Tonge's "Summary of a Heuristic Line Balancing Procedure"; J. R. Slagle's "A Heuristic Program That Solves Symbolic Integration Problems in Freshman Calculus"; B. F. Green, Jr., A. K. Wolf, C. Chomsky, and K. Laughery's "Baseball: an Automatic Question Answerer"; R. K. Lindsay's "Inferential Memory as the Basis of Machines Which Understand Natural Language"; O. G. Selfridge and U. Neisser's "Pattern Recognition by Machine"; L. Uhr and C. Vossler's "A Pattern-Recognition Program That Generates, Evaluates, and Adjusts Its Own Operators"; A. Newell and H. A. Simon's "GPS, a Program That Simulates Human Thought"; E. A. Feigenbaum's "The Simulation of Verbal Learning Behavior"; E. B. Hunt and C. I. Hovland's "Programming a Model of Human Concept Formulation"; J. Feldman's "Simulation of Behavior in the Binary Choice Experiment"; G. P. E. Clarkson's "A Model of the Trust Investment Process"; and J. T. Gullahorn and J. E. Gullahorn's "A Computer Model of Elementary Social Behavior."

———. "Artificial Intelligence: Themes in the Second Decade." *Proceedings of the 1968 International Federation for Information Processing Congress.* Amsterdam: North-Holland, 1968.

FELDMAN, JULIAN. "Computer Simulation of Cognitive Processes." Pp. 337-56 in H. Borko, ed., *Computer Applications in the Behavioral Sciences.* Englewood Cliffs, N.J.: Prentice-Hall, 1962.

FINK, DONALD G. *Computers and the Human Mind: an Introduction to Artificial Intelligence.* Garden City, N.Y.: Doubleday Anchor, 1966. Pp. 301.

FORES, MICHAEL. "Price, Technology and the Paper Model," *Technology and Culture* 12, no. 4 (October 1971): 621-27. D. J. de Solla Price's "paper model" for explaining scientific development by studying the relationships among scientific research papers has limited value when applied to technology. "I suggest that, in selecting 'research' to describe the main activity of technology, Price has, in some measure, prejudged the utility of the paper model in studying technology.... In fact, the main professional activities of technologists (design, development, and production) can quite easily be described as the opposite of research. ..."

FRANK, HELMAR G. *Kybernetik und Philosophie; Materialien und Grundriss zu einer Philosophie der Kybernetik.* Berlin: Duncker & Humblot, 1969. Pp. 190. Four chapters entitled: "The Relation between Cybernetics and Philosophy," "Cybernetics and Research on the Foundations of Philosophy," "Cybernetics and the Theory of Science," and "Cybernetics and Ideology." "This relatively small volume contains four basic findings or theses. (1) Cybernetics and philosophy cannot be traced back to each other, but they are related and complementary to each other. Thus, a philosophy of cybernetics is possible as well as a cybernetics of philosophy. (2) In cybernetics, through the analysis of the concept of information, a confrontation of the themes of consciousness with the methods of calculation and technical construction takes place, but basically without raising a metaphysical problem. (3) The principal significance of model representation for cybernetics lies in the heuristic function of the model ... (4) Through cybernetics, an extension of calculative thinking into the field of normative ideology, transcending cybernetics proper, becomes possible; what is to be considered 'ethical,' might, in the future, be determined by computers."—H. W. Brann, *International Philosophical Quarterly* 11, no. 1 (March 1971): 137–41.

———. *Kybernetische Grundlagen der Pädagogik; eine Einführung in die Pädagogistik für Analytiker, Planer und Techniker des didaktischen Informationsumsatzes in der Industriegesellschaft.* 2 vols. Baden-Baden: Agis-Verlag, 1969. Vol. 1, *Allgemeine Kybernetik;* vol. 2, *Angewandte kybernetische Pädagogik und Ideologie.* Vol. 2 contains a bibliography (pp. 245–64).

FRANK, PHILIPP. *Philosophy of Science: the Link between Science and Philosophy.* Englewood Cliffs, N.J.: Prentice-Hall, 1957. Pp. 394. Discusses technology very briefly in the introduction, section 5, "Technological and Philosophical Interest in Science."

———. "Contemporary Science and the Contemporary World View," *Daedalus* 87 (1958): 57–66.

FRENTZ, HANS-JÜRGEN. "Zur Situation der Technikphilosophie," *Physikalische Blätter* 14, no. 2 (1958): 67–71.

———. "Vom Wesen der Technik" [On the essence of technology], *VDI-Nachrichten* 17, no. 37 (1963): 9.

FUCHS, WALTER R. *Cybernetics for the Modern Mind.* New York: Macmillan, 1971. Pp. 357. Translated from *Exakte Geheimnisse: Knaurs Buch der Denkmaschinen, Informations-theorie und Kybernetik* (Munich: Droemersche Verlagsanstalt, 1968).

GABOR, DENNIS. What Have We Learned So Far from Cybernetics," *Methodos,* vol. 7 (1955).

GALEFFI, ROMANO. "A cibernética como problema filosófico," *Revista brasileira de filosofia* (São Paulo) 18 (1968): 161–72.

GARVIN, PAUL L. "Language and Machines," *International Science and Technology* 65 (May 1967): 63–76. Discussion of natural languages and their semantic ambiguities, syntax problems, synonomy of words, etc., and the attempt to simplify man's problems of communication with machines. Points out the current disagreement about whether development of a theory of language should precede the engineering of a reliable methodology for machine translation.

GAULD, ALAN. "Could a Machine Perceive?" *British Journal for the Philosophy of Science* 17 (May 1966): 44–58. Argues "that in order to perceive an object as an object of a certain sort one has in many instances to exercise concepts of a very advanced level; so advanced, in fact, that it is hard indeed to suppose that anyone could even in principle design a machine which might properly be said to 'possess' them" (p. 46). Cf. Bill Jones's "A Note on Alan Gauld's 'Could a Machine Perceive?'" *British Journal for the Phi-*

losophy of Science 20 (October 1969): 261-62.

GEORGE, F. H. "Could Machines Be Made to Think?" *Philosophy* 31 (July 1956): 244-52. The question of whether machines could think is easily reduced to meaninglessness, and the appropriate way of treating such matters is to carry through empirical projects to see what can in fact be done. Contains a discussion of A. M. Turing and a critique of W. Mays's "Can Machines Think?" *Philosophy* 27 (April 1952): 148-62. See also A. D. Richie's "Could Machines Be Made to Think?" *Philosophy* 32 (January 1957): 65-66, and George's reply in "Thinking and Machines," *Philosophy* 32 (April 1957): 168-69. The discussion continued in A. D. Richie's "Thinking and Machines," *Philosophy* 32 (July 1957): 258; W. Mays's "Thinking and Machines," *Philosophy* 32 (July 1957): 258-61; and F. H. George's "Finite Automata," *Philosophy* 33 (January 1958): 57-59.

————. "Machines and the Brain," *Science* 127 (May 30, 1958): 1269-74. Mathematical logic helps design complex nets whose arrangements resemble the structure of the brain.

————. *Automation, Cybernetics, and Society.* New York: Philosophical Library, 1959. Pp. 283.

————. *The Brain as a Computer.* New York-London: Pergamon, 1961. "The basis of the sort of work described is that anything, of which some observer is prepared to say that it behaves, can be regarded by him as having an input and an output, some identifiable characteristics which can be described in such a manner that the input expresses a set of arguments and the output the value of a function of those arguments which is computed by the behaving thing. It is in this sense that the brain is a computer, and the aim of this book is to use this fact to provide a conceptual framework for experimental psychology . . . a useful account of some of the contributions from cybernetics to the understanding of human perception, learning and thinking." — D. J. Stewart, *Philosophy* 38 (April 1963):

194. See also George's *Cognition* (London: Methuen, 1962).

————. "Minds, Machines and Gödel: Another Reply to Mr. Lucas," *Philosophy* 37 (January 1962): 62-63. "Cybernetics has been almost wholly concerned with what are called 'Inductive Systems,' or probabilistic machines that are capable of producing the axioms from which deductive operations start, and these are obviously beyond the range of being formal systems in the sense that makes Gödel's theorem applicable to them."

————. "Making Machines More Intelligent," *New Scientist* 34 (June 15, 1967): 656-58. Description of some of the progress that has been made in making a "truly intelligent machine," one which can solve problems, make plans, take decisions, pictorialize, symbolize, and conceptualize situations in such a way that it can describe them in a human-like language.

————. *Science and the Crisis of Society.* New York: Wiley-Interscience, 1970. Pp. 168.

GILL, ARTHUR. *Introduction to the Theory of Finite-State Machines.* New York: McGraw-Hill, 1962. Pp. 207.

GILSENBACH, REIMAR. "Technik contra Natur?" *Urania: Monatsschrift über Natur und Gesellschaft* (Jena) 29, no. 5 (1966): 12-20.

GINSBURG, ABRAHAM. *Algebraic Theory of Automata.* New York: Academic, 1968. Pp. 165.

GINSBURG, SEYMOUR. *An Introduction to Mathematical Machine Theory.* Reading, Mass.: Addison-Wesley, 1962. Pp. 148.

GOFF, ROBERT ALLEN. "Wittgenstein's Tools and Heidegger's Implements," *Man and World* 1, no. 3 (August 1968): 447-62. Not directly relevant. "In this paper the language of two philosophers *about* tools or implements becomes a means for seeing the language of each *as* a tool or implement."

GOLZ, WALTER. "Philosophisches Problembewusstsein und kybernetische Theorie," *Zeitschrift für Philosophische Forschung* 24, no. 2 (April-June 1970): 253-63. A discussion of Gotthard Günther, Karl Steinbuch, Helmar Frank, and Carl Friederich von Weizsäcker.

GOOD, IRVING JOHN. "Logic of Man and Machine," *New Scientist* 26 (April 15, 1965): 182-83.

_____. "Human and Machine Logic," *British Journal for the Philosophy of Science* 18 (August 1967): 144-47. The assertion that human logic can do some things that a Turing machine cannot do "cannot be proved by means of Gödel's theorem." See J. R. Lucas's "Human and Machine Logic: a Rejoinder," *British Journal for the Philosophy of Science* 19 (August 1968): 155-56; and see also I. J. Good's "Gödel's Theorem Is a Red Herring," *British Journal for the Philosophy of Science* 19 (February 1969): 357-58.

GOOD, IRVING JOHN, with ALAN JAMES MAYNE and JOHN MAYNARD SMITH. *The Scientist Speculates; an Anthology of Partly-baked Ideas.* New York: Basic, 1962. Pp. 413. A collection of short, half-serious comments solicited by the editors on a variety of topics. Of most relevance to this bibliography is chap. 3, "Minds, Meanings and Cybernetics," which contains pieces by A. Koestler, W. Mays, D. Michie, M. L. Minsky, G. Pask, M. Polanyi, etc.

GRAU, J. CORTS. "Sinn und Unsinn der Technik" [Sense and nonsense of technology], *Schweizerische Rundschau* 56, no. 1 (1956-57): 25-29.

GREGORY, R. L. "The Brain as an Engineering Problem." Pp. 307-30 in W. H. Thorpe and O. L. Zangwill, eds., *Current Problems in Animal Behavior.* Cambridge: Cambridge Univ. Press, 1961.

GRENIEWSKI, HENRYK. *Cybernetics without Mathematics.* Translated by O. Wojtasiewicz. New York: Pergamon, 1960. Pp. 201. From *Elementy cybernetyki sposobem niematematycznym wylozone* (Warsaw: Pánstwowe Wydawnictwo Naukowe, 1959).

GRUENBERGER, FRED. *Benchmarks in Artificial Intelligence.* RAND report P-2586. Santa Monica, Calif.: RAND Corp., June 1962.

GRUENDER, C. DAVID. "On Distinguishing Science and Technology," *Technology and Culture* 12, no. 3 (July 1971): 456-63. The chief distinction between applied science and technology "is in the scope or generality of the problem assigned. Those of broader scope we are inclined to think of as problems of 'applied' science; those that are closer to being specific and particular we think of as 'technology.'"

GRUNDER, KARLFRIED. "Heidegger's Critique of Science in Its Historical Background," *Philosophy Today* 7, no. 1 (Spring 1963): 15-32. From "M. Heidegger's Wissenschaftskritik in ihren geschichtlichen Zusammenhängen," *Archiv für Philosophie* 11, no. 3 (1962): 312-35.

GRUNFELD, JOSEPH. "Computers and Values," *Journal of Value Inquiry* 4, no. 1 (Spring 1970): 29-42. "The essential difference between man and computers is not that man can do things which a computer cannot do, but that a man constructs computers according to his values. The problem of whether and how far machines can 'think' is thus wrongly put. Processes that are expected to achieve something have a value that is inexplicable in terms of processes that have no such purpose. A 'problem' as well as a 'solution' is not in the machine as such but only in the conjunction of the machine and the human purpose."

GUILBAUD, GEORGES THÉODULE. *What Is Cybernetics?* Translated by Valerie MacKay. New York: Grove, 1960. Pp. 126. From *Le cybernétique* (Paris: Presses Universitaires de France, 1954).

GUNDERSON, KEITH. "Cybernetics." Pp. 280-84 in Paul Edwards, ed., *Encyclopedia of Philosophy.* Vol. 2. New York: Crowell-Collier and Macmillan, 1967.

_____. "Minds and Machines: a Survey." Pp. 416-25 in Raymond Klibansky, ed., *Contemporary Philosophy; a Survey,* Vol. 2: *Philosophy of Science.* Florence: La Nuova Italia Editrice, 1968. "The distinction between programme receptive features of mentality (well-defined tasks) and programme resistant features (basic capacities or non-tasks) if fully worked out should have an important bearing on CS [computer simulation of cognitive processes] research since it would involve a clarification of what it would make sense to simulate and what it would be absurd to attempt to simulate *by programming.* Some mental phenomena will no doubt have their analogues in machine software (or programming) whereas others may find counterparts only in hardware developments. This distinction may also have a bearing on the mind-body problem since it seems unlikely that the problem of explaining the relationship between the mind and the body is exactly the same problem for both classes of mentalistic phenomena."

―――. "Cybernetics and Mind-Body Problems," *Inquiry* 12 (1969): 406–19. To what extent do answers to such questions as "Can machines think?" and "Could robots have feelings?" yield insight into traditional mind-body questions? It is claimed that these three approaches to the first set of questions are mistaken: "(1) machines (and robots) obviously cannot think, feel, create, etc., since they do only what they are programmed to do; (2) on the basis of an analysis of the *meaning* of the words 'machine' ('robot', 'think', 'feel', etc.) we can see that *in principle* it would be impossible for machines (or robots) to think, feel, create, etc.; (3) machines (and robots) obviously can (or could) think, feel etc., since they *do* certain things which, if we were to do them, would require thought, feeling, etc. It is argued that, once it is seen why approach (2) is mistaken, it becomes desirable to decline 'in principle' approaches to the first set of questions and to favor 'piecemeal investigations' where attention is centered upon what is actually taking place in machine technology, the development of new programming techniques, etc."

―――. "Asymmetries & Mind-Body Perplexities." Pp. 273–309 in Michael Radner and Stephen Winokur, eds., *Minnesota Studies in Philosophy of Science.* Vol. 4. Minneapolis: Univ. Minnesota Press, 1970. ". . . I have at-tempted . . . to illustrate how certain arguments for a dualistic portrayal of the mental and the physical can as easily be generated for a purely mechanical pattern-recognizer. Such an illustration, if correct, of course, constitutes a *reductio* of that dualistic portrayal."

―――. *Mentality and Machines.* Garden City, N.Y.: Doubleday Anchor, 1971. Pp. 173. Chap. 1 contends "that Descartes formulated an effective argument against a certain type of strategy for proving that animals possessed a rich assortment of mental capacities and that Descartes' argument could also be employed against a certain type of modern strategy for proving that computers are intelligent." Originally published as "Descartes, La Mettrie, Language, and Machines," *Philosophy* 39 (1964): 193–222. Chap. 2 critically discusses the most influential argument (A. M. Turing) for believing that current computing machines have the capacity for thought. Originally published as "The Imitation Game," *Mind* 73 (April 1964): 234–45, and in A. R. Anderson, ed., *Minds and Machines* (Englewood Cliffs, N.J.: Prentice-Hall, 1964), pp. 60–71. Chap. 3 is a discussion of the claim (J. Cohen, M. Scriven, P. Ziff) that human mental life and behavior is not watch-like or programmed and that, if a subject's mental life or behavior is watch-like or programmed, then it cannot be human. Originally published as "Robots, Consciousness, and Programmed Behavior," *British Journal for the Philosophy of Science* 19 (1968): 109–22. Chap. 4 argues that computer simulation work on human problem solving has failed to explain the nonverbal features of mental processes and "until such nonverbal features are understood, it seems highly unlikely that the actual verbal capacities of human beings will be understood either. Thus there seems to be a sizeable gap between the current know-how of CS [computer simulation of cognitive processes] research and anything resembling that linguistically proficient mechanical man envisioned by La Mettrie." Originally published as "Philosophy and Computer Simulation," in George Pitcher and Oscar P. Wood, eds., *Ryle* (Garden City, N.Y.: Doubleday Anchor, 1970). Chap. 5 develops the distinction between *program-resistant* and *program-receptive* features of mentality and implies that "there is a host of features that are not susceptible to mechanization through anything like current programming techniques." See also "Interview with a Robot," *Analysis* 23 (1963): 136-42.

GÜNTHER, GOTTHARD. *Das Bewusstsein der Maschinen; eine Metaphysik der Kybernetik* [The consciousness of machines; a metaphysics of cybernetics]. 2d ed. Krefeld: Agis-Verlag, 1963. Pp. 213.

GUZZO, AUGUSTO. " 'Téchne' e tecnica," *Filosofia* (Turin) 12, no. 1 (January 1961): 87-108.

HAEGER, KLAUS-ALBRECHT. "Vom Wesen der Technik" [On the essence of technology], *Pädogogische Provinz* 13 (1959): 149-55.

HAMMING, R. W. "Intellectual Implications of the Computer Revolution," *American Mathematical Monthly* 70 (January 1963): 4-11. Reprinted in Zenon W. Pylyshyn, ed., *Perspectives on the Computer Revolution* (Englewood Cliffs, N.J.: Prentice-Hall, 1970).

HANSEN, FRIEDRICH. *Konstruktionssystematik; Grundlagen für eine allgemeine Konstruktionslehre* [Systematic contruction; foundations of a general theory of construction]. 2d ed. Berlin: Verlag Technik, 1966. Pp. 191.

HANSEN, W. H. "Mechanism and Gödel's Theorems," *British Journal for the Philosophy of Science* 22, no. 1 (February 1971): 9-16. Contains a discussion of Benacerraf.

HARRISON, MICHAEL A. *Introduction to Switching and Automata Theory.* New York: McGraw-Hill, 1965. Pp. 499. A bibliographic "Guide to the Literature" (pp. 473-90).

HARTMANN, OTTO JULIUS. "Von der Natur zur Unnatur: Stossen wir an die Grenzen unseres technischen Zeitalters?" [From nature to artificiality: are we reaching the limits of our technological age?], *Kommenden* 23, no. 18 (1969): 16-19.

HAUFFE, GERHARD. "Der Stand der Technik; seine Feststellung und Festlegung" [The position of technology; its establishment and definition], *Nachrichten für Dokumentation* (Frankfurt) 8, no. 3 (1957): 142-46.

HAUSTEIN, HEINZ-DIETER. "Zu einegen Problemen der Ausarbeitung wissenschaftlich-technischer Grundkonzeptionen" [On some problems of elaborating basic conceptions of science and technology], *Wissenschaftliche Zeitschrift der Hochschule für Ökonomie und Planung* (Berlin) 7 (1962): 324-31.

HAVERBECK, WERNER GEORG. "Die Evolution der Technik, ein Menschheitsphänomen; die Frage der Technik an das Christentum" [The evolution of technology, a phenomenon of mankind; the question of technology to christianity], *Kommenden* 15, no. 1 (1961): 9; 15, no. 2 (1961): 9-10; 15, no. 3 (1961): 9-10; 15, no. 4 (1961): 7; and 15, no. 8 (1961): 9.

_____. "Ist Technik tod- oder lebenbringend? Die Evolution der Technik als Menschheitsphänomen" [Is technology bringing life or death? The evolution of technology as a phenomenon of humanity], *Kommenden* 15, no. 11 (1961): 9.

_____. "Der Streit um die Technik; Kultur und Zivilisation," *Kommenden* 15, no. 16 (1961): 12.

_____. "Die Selbstverwirklichung des Menschen in der Technik" [The self-realization of man in technology], *Mitteilungen des Braunschweigischen Hochschulbundes* 2 (1962): 25-38.

_____. "Das Prinzip der Technik ist der Tod" [The principle of technology is death], *Frankfurter Heft* 20, no. 2 (1965): 115-20.

_____. *Das Ziel der Technik; die Menschwerdung der Erde* [The goal of technology; the anthropogenesis of the earth]. Olten: Walter-Verlag, 1965. Pp. 339. Bibliography (pp. 327-35).

HEIDEGGER, MARTIN. "Die Frage nach der Technik." In *Vorträge und Aufsätze.* Pfullingen: Neske, 1954. Also published in *Die Technik und die Kehre* (Pfullingen: Neske, 1963). Other works by Heidegger which are relevant to his analysis of technology are: "The Age of the World View," translated by M. Greene, *Measure* 2 (1951): 269–84; *The Question of Being*, translated by W. Kluback and J. T. Wilde (New Haven, Conn.: Twayne, 1958); "Memorial Address," in *Discourse on Thinking*, translated by J. M. Anderson and E. H. Freund (New York: Harper & Row, 1966); "Nietzches Wort 'Gott ist tot,'" in *Holzwege* (Frankfurt: Klostermann, 1957); and "The Principle of Identity," in *Identity and Difference*, translated by J. Stambough (New York: Harper & Row, 1969). Much of the other material in this section of the bibliography refers to Heidegger's analysis of modern technology as a "provoking setting-up disclosure of nature."

———. "La fin de la philosophie et la tâche de la pensée." In *Kierkegaard vivant*. Paris: Gallimard, 1966. Argues that philosophy has come to an end by being subsumed into the scientific world view, a world view that is essentially cybernetic or technological.

HEISENBERG, WERNER. *The Physicist's Conception of Nature.* Translated by A. J. Pomerans. New York: Harcourt, Brace & World, 1958. Pp. 192. From *Das Naturbild der heutigen Physik* (Hamburg: Rowohlt, 1955). The first chapter contains a brief discussion of technology.

HELBERG, WALTHER. "Beitrag zum Strukturverständnis des technischen Daseins" [Contribution to a structural understanding of technological existence], *VDI-Zeitschrift* 104 (1962): 669–74.

HERR, FRIEDRICH. "Die Technik als Ausdruck des schöpferischen Geistes" [Technology as expression of creative spirit], *VDI-Zeitschrift* 110, no. 8 (March 1968): 301–5. Today the engineering sciences are being liberated from ties to kings and ideological powers, and the engineer is becoming more aware that his true task is the internal and external liberation of man. This requires that the engineer himself be free and unconstrained in his actions.

HEYDE, JOHANNES ERICH. "Vom Wesen der Technik," *Forschungen und Fortschritte* 37, no. 9 (1963): 269–70.

HILTON, ALICE MARY. *Logic, Computing Machines, and Automation.* Washington, D.C.: Spartan, 1963. Pp. 427.

HOLLAK, M. J. "Hegel, Marx en de Cybernetica," *Tijdschrift voor filosofie,* vol. 25, no. 2 (June 1963).

———. "Von Causa sui zur Automation." Pp. 46–52 in *Proceedings of the XIVth International Congress of Philosophy; Vienna, Sept. 2–9, 1968.* Vol. 2. Vienna: Herder, 1968. Argues that the idea of self-causation *(causa sui)* as formulated from Descartes to Hegel, with its practical realization or ultimate objectification in the cybernetic ideal, ought to be explained by Marxism as a bourgeoise philosophy.

HOLZ, HANS HEINZ. "Nur die Philosophie kann das technische Weltbild kritisch deuten" [Only philosophy can critically interpret the technological world picture], *VDI-Nachrichten* 15, no. 45 (1961): 5–6.

HOOD, WEBSTER. "A Heideggerian Approach to the Problem of Technology." University Park: Pennsylvania State Univ., 1968. A doctoral dissertation which develops and compares the Heideggerian with the traditional (i.e., Aristotelian) view of technology, then argues for the superiority of the Heideggerian approach. Revised version included in C. Mitcham and R. Mackey, eds., *Philosophy and Technology* (New York: Free Press, 1972).

HUBBLE, EDWIN. *The Nature of Science and Other Lectures.* San Marino, Calif.: Huntington Library, 1954. In two brief lectures, a famous astronomer gives an overview of the nature of science and the relation of science to technology. Science as "the public domain of positive knowledge" is separate from values as "the private domain of personal convictions." Emphasizes "that technology is a by-product of pure research."

HÜBNER, KURT. "Von der Intentionalität der modernen Technik," *Sprache im technischen Zeitalter* 25 (1968): 27–48.

HUDSON, F. LYTH. "Scientific Method and the Nature of Technology," *Nature* (London) 196 (December 8, 1962): 933–35. "A scientist is one who applies scientific method in the fundamental natural sciences while a technologist is one who engages in the scientific study of the industrial arts" (p. 933). A technologist must therefore be a scientist, but a scientist need not be a technologist.

"Information," *Scientific American*, vol. 215, no. 3 (September 1966). Contents: J. McCarthy's "Information," D. C. Evans's "Computer Logic and Memory," I. E. Sutherland's "Computer Inputs and Outputs," C. Strachey's "System Analysis and Programming," R. M. Fano and F. J. Corbató's "Time Sharing on Computers," J. R. Pierce's "The Transmission of Computer Data," A. G. Oettinger's "The Uses of Computers in Science," S. A. Coons's "The Uses of Computers in Technology," M. Greenberger's "The Uses of Computers in Organizations," P. Suppes's "The Uses of Computers in Education," B.-A. Lipetz's "Information Storage and Retrieval," and M. L. Minsky's "Artificial Intelligence." The articles by Oettinger, Coons, and Minsky are most relevant to this bibliography.

JAKI, STANLEY L. *Brain, Mind, and Computers.* New York: Herder & Herder, 1969. Pp. 267. A critical survey of research in artificial intelligence which "consists in offering a detailed record of the claims and counterclaims. In a sense such an inquiry is equivalent to an investigation of the state of the art. Consequently, the views of scientists who have done the most creative work in the field and related areas shall be carefully registered. The record will show that, contrary to fashionable statements, reductionism and physicalism have no substantial support in the existence of electronic computers" (p. 12).

JARVIE, I. C. "The Social Character of Technological Problems: Comments on Skolimowski's Paper," *Technology and Culture* 7, no. 3 (Summer 1966): 384-90. It is the socially set problem posed to the technologist "which determines the character of thinking required." Efficiency is not the sole aim of technical thought and "whether the overriding concern is with accuracy, durability, 'efficiency,'

or what, is always dictated by the socially set problem and not the technological field." Included in C. Mitcham and R. Mackey, eds., *Philosophy and Technology* New York: Free Press, 1972).

———. *Technology and the Structure of Knowledge.* Dimension for Exploration Series. Oswego, N.Y.: Division of Industrial Arts and Technology, State Univ. College, 1967. Pp. 17. Logically, technology is only a part of the logical structure of knowledge. In this sense, it is "knowledge of what physicists call the 'initial conditions' " for the application of scientific laws. Anthropologically, technology "is coterminus with our attempts to come to terms with our world; that is, our culture and our society; and as such it contains within it both pure tools and all knowledge." In this sense, technology "includes under itself applied science, invention and the maintenance of the existing apparatus — these last two being something like planning and engineering." Included in C. Mitcham and R. Mackey, eds., *Philosophy and Technology* (New York: Free Press, 1972). See also "Is Technology Unnatural?" *Listener* 77 (March 9, 1967): 322-23, 333.

JASPERS, KARL. "Modern Technology." Pp. 100-125 in *The Origin and Goal of History,* translated by M. Bullock. New Haven, Conn.: Yale Univ. Press, 1953. From *Von Ursprung und Ziel der Geschichte* (Zurich: Artemis-Verlag, 1949). Argues that "technology is only a means, in itself it is neither good nor evil" (p. 125).

———. "Die Schicksalsfrage im technischen Zeitalter: den technischen Fortschritt zum Guten wenden?" [The fateful question in the technological age: turning technological progress to the good?], *Architekt BDA* (Essen) 9 (1960): 131-36.

———. "The Scientists and the 'New Way of Thinking.' " Pp. 187-208 in *The Future of Mankind.* Chicago: Univ. Chicago Press, 1961. From *Die Atombombe und die Zukunft des Menschen* (Munich: Piper, 1958). "We need an essentially new way of thinking if mankind is to survive. . . . Men must radically change their attitudes toward each other and their views of the future," writes Jaspers, quoting Einstein. Man must become better in order to face up to the responsibilities of his new technological powers. See also *Man in the Modern Age*, translated by E. and C. Paul (London: Routledge & Kegan Paul, 1951). First published, 1933. Reprinted, Garden City, N.Y.: Doubleday Anchor, 1957.

JEFFERSON, G. "The Mind of Mechanical Man," *British Medical Journal* 1 (June 25, 1949): 1105-21.

JESSEN, PALLE. *Indledning til en almen pragmatologi, med et specielt forsog pa afklaring af den systematiske forbindelse mellem behov og teknik* [Introduction to a general pragmatology, with a special attempt to clarify the systematic connection between need and technique]. Copenhagen, 1952. Pp. 19.

JONAS, HANS. "Cybernetics and Purpose: a Critique." Pp. 108-34 in *The Phenomenon of Life: Toward a Philosophical Biology*. New York: Harper & Row, 1966. An extended critique of A. Rosenblueth, N. Wiener, and J. Bigelow, "Behavior, Purpose and Teleology," *Philosophy of Science* 10, no. 1 (January 1943): 18-24, which argues that the cybernetical concept of purpose and teleology cannot be reduced to mechanical premises alone.

———. "The Practical Uses of Theory." Pp. 188-210 in *The Pheomenon of Life: Toward a Philosophical Biology*. New York: Harper & Row, 1966. An examination of the epistemological and metaphysical presuppositions and implications of modern practical theory, that is, technology. Included in C. Mitcham and R. Mackey, eds., *Philosophy and Technology* (New York: Free Press, 1972). Some of the author's other essays in this volume are also valuable. See "Is God a Mathematician?" and "Gnosticism, Existentialism, and Nihilism."

———. "The Scientific and Technological Revolutions: Their History and Meaning," *Philosophy Today* 15, no. 2 (Summer 1971): 76-101. On the metaphysical unity between the scientific revolution of the 17th century and the technological revolution of the 18th century.

JÜNGER, ERNST. *Der Arbeiter*. Hamburg: Hanseatische Verlagsanstalt, 1932. Pp. 300. Included in the author's *Werke*, vol. 6 (Stuttgart: E. Klett, n.d.). Argues that nihilism is simply the first stage in the development of a new kind of humanity, for whom the worker serves as the best illustration. Technology is not just a neutral means but a total language, a metaphysics. "Technology is the mobilization of the world through the *Gestalt* of the worker." Sections 44-57 translated in C. Mitcham and R. Mackey, eds., *Philosophy and Technology* (New York: Free Press, 1972). This book grew out of the author's *Total Mobilmachung* (1930); and for a full appreciation of Jünger's ideas on the subject, his later essay "Über die Linie" (1950) ought to be examined. Heidegger's commentary on this essay in *The Question of Being*, translated by J. T. Wilde and W. Kluback (New York: Twayne, 1958), might also be consulted.

KANE, R. H. "Turing Machines and Mental Reports," *Australasian Journal of Philosophy* 44, no. 3 (December 1966): 344-52.

KAPP, REGINALD O. "Living and Lifeless Machines," *British Journal for the Philosophy of Science* 5 (August 1954): 91-103. "The available facts point to a more basic resemblance between living and lifeless machines than has generally been recognized on either side of the dispute."

KATTSOFF, L. O. "Brains, Thinking and Machines," *Methodos* 6 (1954): 279-86.

———. "Some Philosophical Issues and Automata," *Filosofia* (Turin), vol. 18, no. 4, suppl. (November 1967).

KEMENY, J. G. "Man Viewed as a Machine," *Scientific American* 192 (April 1955): 58-67. A survey of the possibilities and limitations of computers with special reference to theories about the nature of machines in the works of J. von Neumann, A. M. Turing, W. S. McCulloch, and W. A. Pitts. Based on von Neumann's detailed comparison of human and mechanical "brains" in *The Computer and the Brain* (New Haven, Conn.: Yale Univ. Press, 1958).

KESSELRING, FRITZ. "Grenzen des Technischen" [Limits of the technological], *VDI-Zeitschrift* 97, no. 26 (1955): 916-20.

KISSER, PETER. "Die Voraussetzungen der Technik" [The preconditions of technology], *Universum: Monatszeitschrift für Natur, Technik and Wirtschaft* (Vienna) 22, no. 1 (1967): 1-5.

KLAUDY, P. "Gedanken zur Bedeutung der Technik" [Thoughts on the meaning of technology], *Physikalische Blätter* 23, no. 2 (1967): 49-61.

KLÖPPEL, KURT. "Die Entwicklung der Ingenieurwissenschaften" [The development of the science of engineering], *VDI-Zeitschrift* 103, no. 23 (1961): 1145-53.

KNAYER, M. "Zum systematischen Erfinden und Beurteilen in der Technik" [On systematic invention and evaluation in technology], *Technische Rundschau* (Bern) 47, no. 13 (1955): 25-29.

KNIEHAHN, WERNER. "Im Blickfeld industrieller Praxis" [From the perspective of industrial practice]. *Humanismus und Technik* 5, no. 3 (August 15, 1958): 151-59.

———. "Vom tieferen Sinn des technischen Denkens" [On the basic meaning of technological thinking], *Humanismus und Technik* 7, no. 3 (September 29, 1960): 114-27.

KOCHEN, M., D. M. MACKAY, M. E. MARON, M. SCRIVEN, and L. UHR. *Computers and Comprehension.* RAND Memo. RM-4065-PR. Santa Monica, Calif.: RAND Corp., April 1964.

KOCKELMANS, JOSEPH J. "Physical Science and Technology." Pp. 170-75 in *Phenomenology and Physical Science.* Pittsburgh: Duquesne Univ. Press, 1966. Argues for the Heideggerian understanding of technology as a means for the disclosure of Being.

KOTARBIŃSKI, TADEUSZ. "I principi di un'etica indipendente," *Rivista di filosofia* (Turin), vol. 49, no. 4 (October 1958).

———. "The Concept of Action," *Journal of Philosophy* 57, no. 7 (March 31, 1960): 215-22. "...a form of causal relationship is needed for the purpose of defining action: an elementary process of action is an individual case of bringing about an effect by a cause.... Elementary action includes simple pressure.... To put it briefly, the initial event of an elementary action consists, strictly speaking, not in purposeful pressure, but in a purposeful change in pressure...."

———. "Praxiological Sentences and How They Are Proved." Pp. 211-23 in E. Nagel, P. Suppes, and A. Tarski, eds., *Logic, Methodology and Philosophy of Science.* Stanford, Calif.: Stanford Univ. Press, 1962.

———. "Practical Error," *Danish Year-Book of Philosophy* (Copenhagen) 1 (1964): 65-71. Observing the various kinds of practical errors can help in formulating and systematizing the methods used in practical skills.

———. "Les problèmes de la praxiologie ou théorie générale de l'activité efficace," *Revue philosophique de la France et de l'étranger* 154 (1964): 453-72.

———. *Praxiology: an Introduction to the Sciences of Efficient Action.* Translated by O. Wojtasiewicz. New York: Pergamon, 1965. Pp. 219. The scope of Praxiology is the general theory of efficient action—the technique of good, efficient work as such. Its principal concern is the formulation and justification of standards of efficient work. For epistemological background, see the author's *Gnosiology: the Scientific Approach to the Theory of Knowledge,* translated by O. Wojtasiewicz, 2d ed. (New York: Pergamon, 1966). For a brief history and discussion of the scope and nature of the subject, see Henryk Skolimowski's "Praxiology" in *Polish Analytical Philosophy* (New York: Humanities Press, 1967). This originally appeared as "Praxiology—the Science of Accomplished Acting," *Personalist,* vol. 46, no. 3 (Summer 1965).

———. "Investigating the Scientific Creation" (in Russian), *Voprosy filosofii* 22, no. 6 (1968): 43-52.

———. "La philosophie de la technique de Dupréel," *Revue internationale de philosophie* 22 (1968): 156-66.

———. "Les formes positives et négatives de la coopération," *Revue du métaphysique et de morale,* vol. 75, no. 3 (July-September 1970).

————. "The Methodology of Practical Skills: Concepts and Issues," *Metaphilosophy* 2, no. 2 (April 1971): 158–70.

KRANZBERG, MELVIN. "The Spectrum of Science-Technology," *Journal of the Scientific Laboratories* 48 (December 1967): 47–58.

————. "The Unity of Science-Technology," *American Scientist* 55 (March 1967): 48–66. "The pressures of social need and the intellectual needs of the practitioners . . . are bringing science and technology into even closer association" (p. 49). This review of three hundred years of science and technology was a Sigma Xi-RESA Lecture in 1966.

————. "The Disunity of Science-Technology," *American Scientist* 56 (Spring 1968): 21–34. Argues that there is at present a spectrum of interrelationships between science and technology which runs from "the lone inventor working in his basement" to the "scientist working with a graduate student . . . on a small problem at some frontier of his own choosing." A Sigma Xi-RESA Lecture in 1967.

KRISHNA, D. " 'Lying' and the Compleat Robot," *British Journal for the Philosophy of Science* 12 (August 1961): 146–49.

KUHLENKAMP, ALFRED. "Experiment und Erfahrung in der Technik" [Experiment and experience in technology]. Pp. 69–85 in Walter Stolz, ed., *Experiment und Erfahrung in Wissenschaft und Kunst.* Freiburg, 1963.

LACEY, A. R. "Men and Robots," *Philosophical Quarterly* 10 (January 1960): 61–72. ". . . provided we stay on a theoretical level and do not involve ourselves too far in technological details, it is difficult to state any fundamental difference between minds and machines in terms of what they can do . . ." (p. 69). He concludes "that the difference between men and machines lies primarily not, as usually seems to have been thought, in what they can do but in what they can suffer or experience." A survey of arguments in W. Mays, M. Polanyi, D. M. MacKay.

LAER, P. HENRY VAN. *Philosophy of Science.* Vol. 2. Pittsburgh: Duquesne Univ. Press, 1956. Pp. 342. Chaps. 4 and 7 contain an Aristotelian-Thomistic analysis of technology.

LALANDE, ANDRÉ. "Technique et science," *Journal de psychologie normale et pathologique* 41 (1948): 79–88.

LANDGREBE, LUDWIG. "Knowledge in Action." Chap. 6 of *Major Problems in Contemporary European Philosophy.* Translated by K. F. Reinhardt. New York: Ungar, 1966. From *Philosophie der Gegenwart* (Frankfurt: Ullstein, 1957). A good survey of European thought on the subject of technological action.

LANGAN, THOMAS. "The Notion of Technique." Pp. 191–200 in *The Meaning of Heidegger.* New York: Columbia Univ. Press, 1959.

LASZLO, ERVIN. *Introduction to Systems Philosophy; toward a New Paradigm of Contemporary Thought.* New York: Gordon & Breach, 1971. For a popular essay based on this volume, see Laszlo's "Systems Philosophy," *Main Currents in Modern Thought* 28, no. 2 (November–December 1971): 55–60, a general defense of synthetic against analytic philosophy culminating in the argument that a "general theory is emerging these days from the workshops of the system scientists. Their disciplines—general systems theory, cybernetics, information and game theory, etc.—come up with theories applicable to a wide range of empirical phenomena." This new general theory looks at "the world as organization." As such, systems philosophy may be viewed both as a metaphysical outgrowth and as a metaphysics of technology. Note: Laszlo, in conjunction with L. von Bertalanffy and S. C. Pepper, is editing a series called *The International Library of Systems Theory and Philosophy* (New York: Braziller). Vol. 1 by Laszlo, is *The Systems View of the World: the Natural Philosophy of the New Developments in the Sciences* (1972). Vol. 2, *The Relevance of General Systems Theory* is a seventieth birthday Festschrift for Bertalanffy with essays by A. Rapoport, Pattee, Rosen, Zerbst, K. Boulding, Thayer,

Gray, Rizzo, Livesey, Clark, and Bertalanffy.

LATIL, PIERRE DE. *Thinking by Machine; a Study of Cybernetics.* Translated by Y. M. Golla. New York: Houghton Mifflin, 1957. Pp. 355. From *La pensée artificielle* (Paris: Gallimard, 1956). An early popular introduction to the subject with historical information which still remains of some value. Weakened by the absence of an index or bibliography.

LEMAIRE, JOSSE. "La cybernétique, nouveau mode penser?" In *Deuxième congrès international de cybernétique.* Namur: Association Internationale de Cybernétique, 1960.

LEWIS, DAVID. "Lucas against Mechanism," *Philosophy* 44 (July 1969): 231-33.

LICKLIDER, J. C. R. "Man-Computer Symbiosis," *IRE Transactions on Human Factors in Electronics,* HFE-1 (March 1960), pp. 4-11.

LINERA, ANTONIO ALVAREZ DE. "Cybernetics as Seen by the Philosophers," *Philosophy Today* 1, no. 3 (Fall 1957): 202-6. From "La cibernética a la luz de la filosofía," *Rivista de filosofía* (Madrid), vol. 15 (January-March 1956).

LINGIS, A. F. "On the Essence of Technique." Pp. 126-38 in Manfred S. Frings, ed., *Heidegger and the Quest for Truth.* Chicago: Quadrangle, 1968.

LUCAS, J. R. "Minds, Machines and Gödel," *Philosophy* 36 (April-July 1961): 112-27. Gödel's theorems imply that a mechanical model of the mind is, in principle, impossible. These theorems refute mechanisms, establish a basis for distinguishing conscious and unconscious beings, and indicate a unique kind of self-consciousness in man not attributable to machines. Reprinted in A. R. Anderson, ed., *Minds and Machines* (Englewood Cliffs, N.J.: Prentice-Hall, 1964), and in K. M. Sayre and F. J. Crosson, eds., *The Modeling of Mind* (Notre Dame, Ind.: Univ. Notre Dame Press, 1963). Lucas defends himself against various critics in the following papers: "Human and Machine Logic: a Rejoinder," *British Journal for the Philosophy of Science* 19 (August 1968): 155-56; "Satan Stultified," *Monist* 52 (1968): 145-58; "Mechanism: a Rejoinder," *Philosophy* 45 (April 1970): 149-51; and "Metamathematics and the Philosophy of Mind: a Rejoinder," *Philosophy of Science* 38 (1971): 310-13.

_____. *The Freedom of the Will.* Oxford: Oxford Univ. Press, 1970. Pp. 181. Lucas's main work on the place of Gödel's argument within the philosophical problem of free will and determinism. In the first two-thirds of the book there is a discussion of the general problem and this leads, in the last one-third, to the author's primary argument that "Kant was right in thinking freedom to be an unavoidable problem set by human reason, though wrong in despairing of the power of reason to solve it. For the very fact that freedom is a problem of reason is itself a reason for believing that we are free. Many men have sensed this, but the argument is extremely difficult to articulate and assess. I have attempted to articulate it . . . and to reformulate it in formal terms with the aid of a profound theorem of mathematical logic discovered by Gödel in 1929. It is a controversial argument, and I have tried to meet objections raised against it, for I believe it to be ultimately a decisive argument which will refute the one sort of determinism – physical determinism – which seriously worries men today" (p. 2). Contains bibliography of literature on Gödel's theorem and its application.

LUDEWIG, WALTER. Die Ingenieursarbeit – eine Kunst oder eine Wissenschaft?" [The work of the engineer – an art or a science?], *Fridericiana; Zeitschrift der Universität Karlsruhe* 1, no. 1 (1967): 23-27.

LUTZ, THEO. "Kybernetik, Portrait einer Wissenschaft. Part 1, Der Begriff und seine Geschichte," *VDI-Zeitschrift* 112, no. 7 (April 1970): 413-17; "Part 2, Die Grundbegriffe der Kybernetik," 112, no. 15 (August 1970): 1009-14; "Part 3, Die Theorie von der Information," 112, no. 20 (October 1970): 1341–46; and "Part 4, Kybernetische Maschinen," 112, no. 24 (December 1970): 1613–18.

McCLINTOCK, ROBERT. "Machines and Vitalism; Reflections on the Ideology of Cybernetics," *American Scholar* 35, no. 2 (Spring 1966): 249–57. Argues that "cybernetic materialism needs to be complemented by a biological vitalism."

McCULLOCH, WARREN S. *Embodiments of Mind.* Cambridge, Mass.: M.I.T. Press, 1965. A collection of McCulloch's essays, primarily on the physiological theory of knowledge. Two essays coauthored with Walter A. Pitts have been especially influential in automata theory. They are: "A Logical Calculus of the Ideas Imminent in Nervous Activity," *Bulletin of Mathematical Biophysics* 5 (1943): 115–33, and "How We Know Universals: the Perception of Auditory and Visual Forms," *Bulletin of Mathematical Biophysics* 9 (1947): 127–47. Some other McCulloch essays contained in this work are: "Why the Mind Is in the Head," in L. A. Jeffress, ed., *Cerebral Mechanisms in Behavior; the Hixon Symposium* (New York: Wiley, 1951); "Through the Den of the Metaphysician," *British Journal of the Philosophy of Science* 5 (May 1954): 18–31, translated from "Dans l'antre du métaphysicien," *Thales* 7 (1951): 37–49; "Mysterium Iniquitatis—of Sinful Man Inquiring into the Place of God," *Scientific Monthly* 80, no. 1 (January 1955): 35–39, and in P. G. Frank, ed., *The Validation of Scientific Theories* (Boston: Beacon, 1956); "Toward Some Circuitry of Ethical Robots or an Observation Science of the Genesis of Social Evaluation in the Mind-like Behavior of Artifacts," *Acta Biotheoretica* 11 (1956): 147–56. "In their early work McCulloch and Pitts established what has become one of the most significant theorems in all of the theory of automata. Making certain assumptions as to the functions of neurons, for any input-output behavior of a finite automata, there is a possible construction of a nerve-net which will simulate that behav-

ior."—Merle B. Turner, *Realism and the Explanation of Behavior* (New York: Appleton-Century-Crofts, 1971), p. 60. For a popular discussion of McCulloch, see Nilo Lindgren's "The Birth of Cybernetics—an End to the Old World," *Innovation* 6 (October 1969): 13–25.

MacKAY, D. M. "Mind-like Behaviour in Artefacts," *British Journal for the Philosophy of Science* 2 (August 1951): 105–21. "Our principal conclusion is that we have failed to find any distinction in principle between the observable behavior of a human brain and the behavior possible in a suitably designed artefact. . . . In such an artefact analogues of concepts such as emotion, judgment, originality, consciousness, and self-consciousness appear."

——. "On Comparing the Brain with Machines," *Advancement of Science* 40 (March 1954): 402–6. Also in *American Scientist* 42 (1954): 261–68, and *Annual Report of the Smithsonian Institution* (1954), pp. 231–40.

——. "Information Theory and Human Information Systems," *Impact of Science on Society* 8, no. 2 (1957): 86–101.

——. "Man as a Mechanism," *Faith and Thought* 91 (1960): 145–57. Revised in D. M. MacKay, ed., *Christianity in a Mechanistic Universe* (London: Tyndale, 1965).

——. "The Use of Behavioural Language to Refer to Mechanical Processes," *British Journal for the Philosophy of Science* 13 (August 1962): 89–103. Reprinted in F. J. Crosson, ed., *Human and Artificial Intelligence* (New York: Appleton-Century-Crofts, 1970). Cf. P. J. Fozzy's "Professor MacKay on Machines," *British Journal for the Philosophy of Science* 14 (August 1963): 154–57, and D. M. MacKay's "Consciousness and Mechanism: a Reply to Miss Fozzy," *British Journal for the Philosophy of Science* 14 (August 1963): 157–59.

——. "From Mechanism to Mind." Pp. 163–91 in J. R. Smythies, ed., *Brain and Mind; Modern Concepts of the Nature of Mind.* New York: Humanities Press, 1965. This essay is based on two essays originally published as "Mentality in Machines," *Proceedings of the Aristotelian Society,* suppl. vol. 26 (1952): 61–86, and "From Mechanism to Mind," *Transactions of the Victorian Institute* 85 (1953): 17–32. ". . . any attempt to 'maintain the dignity of man' by searching for limits to the information-processing powers of artefacts is misguided and foredoomed. This is no prophecy but a deduction from the demonstrable fact that to specify exactly a behavioral test of information-processing capability amounts in principle to specifying a mechanism that can meet it. We have left open the question whether we could ever enunciate an adequate test for mentality in the full human sense. Indeed our plea would be for more open-mindedness in facing an issue on which it is difficult to conceive of the kind of evidence that would be adequate. The view here offered is that these developments only illuminate and in no way controvert the Christian doctrine of Man."

———. "Cerebral Organization and the Conscious Control of Action." In J. C. Eccles, ed., *Brain and Conscious Experience.* New York: Springer-Verlag, 1966.

———. "What Is Cybernetics?" Pp. 417–28 in Webster P. True, ed., *Smithsonian Treasury of 20th Century Science.* New York: Simon & Schuster, 1966.

———. *Freedom of Action in a Mechanistic Universe.* Cambridge: Cambridge Univ. Press, 1967. Pp. 37. The Leslie Stephens Lecture, 1967.

———. *Information, Mechanism and Meaning.* Cambridge, Mass.: M.I.T. Press, 1969. A collection of essays on information theory. Part 1 consists of three introductory broadcast talks on "Measuring Information," "Meaning and Mechanism," and on "What Makes a Question" (pp. 1–38). Part 2 contains more technical essays. An explanatory survey of the terminology of information theory, prepared for the 1950 London Symposium on Information Theory, has been added as an appendix.

McNAUGHTON, R. "The Theory of Automata." Pp. 379–421 in F. L. Alt, ed., *Advances in Computers.* Vol. 2. New York: Academic, 1961. A good survey.

MACOMBER, W. B. "Science and Technology: Mathematics and Manipulation." Pp. 198–208 in *The Anatomy of Disillusion; Martin Heidegger's Notion of Truth.* Evanston, Ill.: Northwestern Univ. Press, 1967.

MALCOLM, NORMAN. "Conceivability of Mechanism," *Philosophical Review* 77, no. 1 (January 1968): 45–73.

MARITAIN, JACQUES. *The Degrees of Knowledge.* Translated by G. B. Phelan. New York: Scribner's, 1959. An attempt to extend Thomist epistemology to take in modern natural science. See especially chaps. 2 and 4.

MARTIN, MICHAEL. "On the Conceivability of Mechanism," *Philosophy of Science* 31, no. 1 (March 1971): 79–86. A discussion of N. Malcolm's article of the same title.

MAYBERRY, THOMAS C. "Consciousness and Robots," *Personalist* 51, no. 2 (Spring 1970): 222–36.

MAYS, WOLFE. "The Hypothesis of Cybernetics," *British Journal for the Philosophy of Science* 2 (November 1951): 249–50.

———. "Can Machines Think?" *Philosophy* 27 (April 1952): 148–62. A critique of A. M. Turing's "Computing Machines and Intelligence," *Mind,* vol. 59 (October 1950). "The basic assumption in applying the calculating machine analogy to the mind is that thinking operates in the form of an atomic system. . . . There is a good deal of psychological evidence that we think and perceive in terms of 'gestalten,' which are not merely the algebraic sums of the elements into which they may be analysed." See also Mays's "Thinking and Machines," *Philosophy* 32 (July 1957): 258–61, and "Minds and Machines," *Listener* (December 27, 1956). Cf. F. H. George's "Could Machines Be Made to Think?" *Philosophy* 31 (July 1956): 244–52.

———. "The Use and Misuse of Logical Principles in Cybernetic Discussions." Pp. 528–33 in *Proceedings of the XIVth*

International Congress of Philosophy; Vienna, Sept. 2-9, 1968. Vol. 2 Vienna: Herder, 1968. ". . . it is doubtful whether philosophical arguments in terms of the logical possibility of constructing devices for simulating conscious activities, in any way strengthen the case for or against their practical possibility. As far as logic is concerned, there is nothing to choose between the propositions 'Machines are conscious' and 'Machines are not conscious'; neither is self-contradictory. The question whether the former proposition could be true in some future state of the world, is something about which logic tells us nothing."

MAZLISH, BRUCE. "The Fourth Discontinuity," *Technology and Culture* 8, no. 1 (January 1967): 1-15. Examines man's relationship with tools and machines and suggests that man is now on the threshold of breaking past the discontinuity between himself and machines. Cf. also Theodore A. Wertime, "Culture and Continuity: a Commentary on Mazlish and Mumford," *Technology and Culture* 9, no. 2 (April 1968): 203-12.

MAZUR, MARIAN. "Concept of Autonomous System and Problem of Equivalence of Machine to Man." Pp. 534-39 in *Proceedings of the XIVth International Congress of Philosophy; Vienna, Sept. 2-9, 1968.* Vol. 2. Vienna: Herder, 1968.

MESAROVIĆ, MIHAJLO D., ed. *Views on General Systems Theory.* New York: Wiley, 1964. Pp. 178. Proceedings of the Second Systems Symposium at Case Institute of Technology. Contains M. D. Mesarovic's "Foundations for a General Systems Theory," K. E. Boulding's "General Systems as a Point of View," R. L. Ackoff's "General System Theory and Systems Research: Constrasting Conceptions of Systems Science," H. Putnam's "The Complete Conversationalist: a 'Systems Approach' to the Philosophy of Language," J. Myhill's "The Abstract Theory of Self-Reproduction," A. Rapoport's "Remarks on General Systems Theory," C. West Churchman's "An Approach to General Systems Theory," etc.

METZGER, ARNOLD. *Automation und Autonomie; das Problem des freien Einzelnen im gegenwärtigen Zeitalter.* Pfullingen: Neske, 1964. Pp. 61. See also Oscar Oppenheimer's "Auseinandersetzung mit Arnold Metzgers Gedanken uber 'Automation und Autonomie,'" *Zeitschrift für Philosophische Forschung* 19, no. 3 (1965): 480-88; and Metzger's reply which follows (pp. 488-92).

MILES, T. R. "On the Difference between Men and Machines," *British Journal of the Philosophy of Science* 7 (February 1957): 277-92. Even if all distinguishing marks of appearance, physical make-up, and behavior were removed, men differ from machines "in having awareness in relation to the body-schema, i.e., in being able to recognize houses and trees as external to the body-schema, and pains, aches, and tickles as internal. . . . Men can relate their perceptions to a body-schema, but in the case of machines there is no body-schema, to which perceptions could be related."

MILLER, GEORGE A. "Thinking Machines: Myth and Actualities," *Public Interest* 2 (Winter 1966): 92-112. A brief history of computer development, followed by the question of man-machine relations, discussed in both theroretical and practical terms. Cybernetics is the way to an answer. "I expect that the division of labour between man and machines, described in the most general terms, will ultimately correspond to a division between finding problems and solving them, but exactly what I mean by that distinction is not yet clear, even to me."

MILLER, GEORGE A., EUGENE GALANTER, and KARL H. PRIBRAM. *Plans and the Structure of Behavior.* New York: Holt, Rinehart & Winston, 1960. Pp. 226.

MINSKY, MARVIN L. "Steps toward Artificial Intelligence," *Proceedings of the Institute of Radio Engineers* 49 (January 1961); 8-30. Included in E. Feigenbaum and J. Feldman, eds., *Computers and Thought* (New York: McGraw-Hill, 1963), pp. 406-50.

———. "Artificial Intelligence," *Scientific American* 215, no. 3 (September

1966): 247-60. A description of some programs that enable a computer to behave in ways that probably everyone would agree show intelligence: set up goals, make plans, consider hypothesis, recognize analogies, etc.

———. *Computation: Finite and Infinite Machines.* Englewood Cliffs, N.J.: Prentice-Hall, 1967. Pp. 317. A major work on the mathematical theory of computers by a leading researcher in artificial intelligence who has spoken optimistically about its future. "Within a generation . . . I am convinced, few compartments of intellect will remain outside the machine's realm — the problem of creating artificial intelligence' will be substantially solved."

MITCHAM, CARL, and ROBERT MACKEY. "Jacques Ellul and the Technological Society," *Philosophy Today* 15, no. 2 (Summer 1971): 102-21. Exposition of Ellul's thought followed by a critique of its three basic ideas: the distinction between technical operation and human action, technical operations and technical phenomena, and the ancient and modern forms of the technical phenomena. Conclusion: Ellul has an inadequate concept of technology based on questionable metaphysical foundations.

MOLES, ABRAHAM A. *Information Theory and Esthetic Perception.* Translated by J. E. Cohen. Urbana: Univ. Illinois Press, 1966. Pp. 217. See also Moles's "Cybernétique et oeuvre l'art," *Revue d'esthétique*, vol. 18, no. 1 (January-March 1965).

MOLES, ABRAHAM A., and ANDRÉ NOIRAY. "La pensée technique." Pp. 496-525 in *La Philosophie.* Paris: Culture, Art, Loisirs, 1969.

MORAY, NEVILLE. *Cybernetics.* New York: Hawthorn, 1963. Pp. 125. An informal introduction from a Catholic perspective to some concepts of cybernetics: feedback, system, behavior matrix, etc. Primarily an inquiry into the ultimate limitations of cybernetics and machine construction with special reference to the theories of A. M. Turing, and W. S. McCulloch and W. Pitts.

———. "Superman and Supermachine,"

Modern Schoolman 47, no. 3 (March 1970): 339-46. An extended discussion of K. M. Sayre's *Recognition: a Study in the Philosophy of Artificial Intelligence* (Notre Dame, Ind.: Univ. Notre Dame Press, 1965).

MOSIER, RICHARD D. "The Theory of Technology." Pp. 299-302 in *The American Temper.* Berkeley: Univ. California Press, 1952. The argument is that Dewey's instrumental theory of knowledge is a philosophy of technology in the sense of being a technological epistemology. As such, "the instrumental theory unites the traditional American interests in mechanism and idealism. . . . The subject and object are drawn together in the active, causal relation of knowing, and the activity of knowing is seen to be a participant in what is finally known. Ideas are embodied in matter by way of technique, and every advance in technology, in perfected methods of inquiry and production, brings the possibility of realizing secure values ever closer. Such is the philosophy in which the long history of American ideas culminates; for the instrumental theory is a technological theory both of knowledge and value."

MOTHES, HANS. "Vom Sinn und Unsinn der Technik" [On sense and nonsense of technology], *Naturlehre; Zeitschrift für die experimentellen Unterricht der Volksschulen in Physik und Chemie* (Frankenberg-Eden) 3 (1954): 41-47.

MÜHLE, H. "Heidegger und das Wesen der Technik," *Handelsblatt* (Düsseldorf) (May 5, 1953).

NADLER, GERALD. "An Investigation of Design Methodology," *Management Science* 13, no. 10 (June 1967): B643-B655. The argument is that recent works on design or engineering employ the methodology of research, but that to be really efficient, design requires a methodology that is significantly different from the methodology of research. Some relevant works on design engineering: John R. M. Alger and Carl V. Hays's *Creative Synthesis in Design* (Englewood Cliffs, N.J.: Prentice-Hall, 1962), Morris Asimow's *Introduction to Design* (Englewood Cliffs, N.J.: Prentice-Hall,

1962), Harold W. Buhl's *Creative Engineering Design* (Ames: Iowa State Univ. Press, 1960), John R. Dixon's *Design Engineering: Inventiveness, Analysis, and Decision Making* (New York: McGraw-Hill, 1966), J. Christopher Jones and D. G. Thornley, eds., *Conference on Design Methods* (New York: Macmillan, 1963), J. Christopher Jones's *Design Methods: Seeds of Human Futures* (New York: Wiley, 1970), E. V. Krick's *An Introduction to Engineering and Engineering Design* (New York: Wiley, 1965), Gerald Nadler's *Work Design: a Systems Concept* (Homewood, Ill.: Irving, 1963; rev. ed., 1970), Gerald Nadler's *Work Systems Design: the IDEALS Concept* (Homewood, Ill.: Irving, 1967), and Thomas T. Woodson's *Introduction to Engineering Design* (New York: McGraw-Hill, 1966).

NAGEL, ERNEST, and JAMES R. NEWMAN. *Gödel's Proof.* New York: New York Univ. Press, 1958. Pp. 118. "Gödel's conclusions bear on the question whether a calculating machine can be constructed that would match the human brain in mathematical intelligence. Today's calculating machines have a fixed set of directives built into them; these directives correspond to the fixed rules of inference of formalized axiomatic procedure. . . . But, as Gödel showed in his incompleteness theorem, there are innumerable problems in elementary number theory that fall outside the scope of a fixed axiomatic method, and that such engines are incapable of answering, however intricate and ingenious their built in mechanisms may be and however rapid their operations. . . . [Gödel's theorem] does indicate that the structure and power of the human mind are far more complex and subtle than any non-living machine yet envisaged." See the review by H. Putnam, *Philosophy of Science* 27 (April 1960): 205-7, and Nagel and Newman's "Putnam's review of *Gödel's Proof*," *Philosophy of Science* 28 (April 1961): 209-11.

NANIWADA, HARUO. *Die Logik der technischen Welt: eine sozialwissenschaftliche Studie.* The Science Council of Japan,

Division of Economics, Commerce and Business Administration, Economic Series no. 47. Tokyo, 1970. Pp. 34. Despite the publisher, this is a metaphysical commentary with reference to Artistotle, Kant, Dessauer, Heidegger, etc.

NEGLEY, GLENN. "Cybernetics and Theories of Mind," *Journal of Philosophy* 48, no. 19 (September 13, 1951): 574-82.

NEISSER, ULRIC. "The Imitation of Man by Machine," *Science* 139 (January 18, 1963): 193-97. Reprinted in F. J. Crosson, ed., *Human and Artificial Intelligence* (New York: Appleton-Century-Crofts, 1970).

NEUMANN, JOHN VON. "The General and Logical Theory of Automata." In L. A. Jeffress, ed., *Cerebral Mechanisms in Behavior; the Hixon Symposium.* New York: Wiley, 1951. Also in J. R. Newman, ed., *The World of Mathematics,* vol. 4 (New York: Simon & Schuster, 1956); J. von Neumann, *Collected Works,* vol. 5 (New York: Macmillan, 1963); and Z. W. Pylyshyn, ed., *Perspectives on the Computer Revolution* (Englewood Cliffs, N.J.: Prentice-Hall, 1971). For an introductory discussion of von Neumann's work on the theory of machines and on the idea of an automaton capable of reproducing itself, see J. G. Kemeny's "Man Viewed as a Machine," *Scientific American,* vol. 192 (April 1955).

———. *The Computer and the Brain.* New Haven, Conn.: Yale Univ. Press, 1958. A nontechnical elaboration of ideas first introduced in "The General and Logical Theory of Automata." The material was originally prepared as the Silliman Lecture but was not presented because of von Neumann's death in 1957; a classic despite its preliminary nature.

———. *Theory of Self-reproducing Automata.* Edited and completed by Arthur W. Burks. Urbana: Univ. Illinois Press, 1966. Pp. 388.

NEWELL, ALLEN. "Some Problems of Basic Organization in Problem-solving Programs." In Marshall C. Yovits, George T. Jacobi, and Gordon D. Goldstein, eds., *Self-organizing Systems.* Washington, D.C.: Spartan, 1962. Also published as *Some Problems of Basic Organization in Problem-solving Programs*, RAND report RM-3283-PR (Santa Monica, Calif.: RAND Corp., December 1962).

———. *Learning, Generality and Problem-Solving*. RAND report RM-3285-1-PR. Santa Monica, Calif.: RAND Corp., February 1963.

NEWELL, ALLEN, J. C. SHAW, and HERBERT A. SIMON. "The Elements of a Theory of Human Problem Solving," *Psychological Review* 65 (March 1958): 151-66.

NEWELL, ALLEN, J. C. SHAW, and HERBERT A. SIMON. "The Processes of Creative Thinking." Pp. 63-119 in Howard E. Gruber, Glenn Terrell, and Michael Wertheimer, eds., *Contemporary Approaches to Creative Thinking*. New York: Atherton, 1962. "... we would have a satisfactory theory of creative thought if we could design and build some mechanisms that could think creatively (exhibit behavior just like that of a human carrying on creative activity), and if we could state the general principles on which the mechanisms were built and operated.... The success already achieved in synthesizing mechanisms that solve difficult problems in the same manner as humans is beginning to provide a theory of problem-solving that is highly specific and operational. The purpose of this paper is to draw out some of the implications of this theory for creative thinking. To do so is to assume that creative thinking is simply a special kind of problem-solving behavior. This seems to us a useful working hypothesis." Also published as *The Processes of Creative Thinking*, RAND report P-1320. (Santa Monica, Calif.: RAND Corp., September 16, 1958).

NEWELL, ALLEN, and HERBERT A. SIMON. "Computer Simulation of Human Thinking," *Science* 134 (December 22, 1961): 2011-17. "A digital computer is a general-purpose symbol-manipulating device. If appropriate programs are written for it, it can be made to produce symbolic output that can be compared with the stream of verbalizations of a human being who is thinking aloud while solving problems." Also published as *Computer Simulation of Human Thinking*, RAND report P-2276 (Santa Monica, Calif.: RAND Corp., April 20, 1961). See also Newell and Simon's *Computer Simulation of Human Thinking and Problem Solving*, RAND report P-2312 (Santa Monica, Calif.: RAND Corp., May 29, 1961).

NEWELL, ALLEN, and H. A. SIMON. "Computers in Psychology." In R. D. Luce, R. R. Bush, and E. Galanter, eds., *Handbook of Mathematical Psychology*. Vol. 1. New York: Wiley, 1963.

NORTHROP, F. C. S. "The Neurological and Behavioristic Psychological Basis of the Ordering of Society by Means of Ideas," *Science* 107 (April 1948): 411-16. Discusses the underlying assumptions of cybernetics as a model for interpreting all forms of human mentality, with particular reference to Rosenblueth, Weiner, and Bigelow and to McCulloch and Pitts.

ORTEGA Y. GASSET, JOSÉ. "Man the Technician." Pp. 87-161 in *Toward a Philosophy of History*. Translated by H. Weyl. New York: Norton, 1941. From "Meditacion de la tecnica," in *Ensimismamiento y alteracion; obras completas*, vol. 5 (Madrid: Revista de Occidente, 1939); first published, 1933. Revised translation included in C. Mitcham and R. Mackey, eds., *Philosophy and Technology* (New York: Free Press, 1972). Defines technology as the system of activities through which man endeavors to realize the extranatural project that is himself. See also *The Dehumanization of Art* (1925) (Princeton, N.J.: Princeton Univ. Press, 1948) and *The Revolt of the Masses* (1929) (New York: Norton, 1932).

OSTROWSKI, JEAN J. "Essai d'une typologie métapraxéologique," *Sophia*, vol. 37, nos. 1-2 (1969). Also published in *Proceedings of the XIV International Congress of Philosophy; Vienna, Sept. 2-9, 1968* (Vienna: Herder, 1968), 2:540-47.

OTTO, GÜNTER. "Technik als Denkweise und Handlungsform" [Technology as a way of thinking and a form of acting]. Pp. 162-77 in Hartmut Sellin and Bodo Wessels, eds., *Beiträge zur*

Didaktik der technischen Bildung: Beiträge zum Werkunterricht. Vol. 2. Weinheim-Berlin-Basel: Beltz, 1970.

PALMAERS, MARTIN. "The Technics of Rational Civilization," *Diogenes* 61 (Spring 1968): 16-31.

PACOTTE, J. "Idée de science de la technique," *Revue philosophique de la France et de l'étranger* 117 (1934): 226-47.

PARSEGIAN, V. L. *This Cybernetic World of Men, Machines, and Earth Systems.* Garden City, N.Y.: Doubleday, 1972. Pp. 217.

PASK, GORDON. *An Approach to Cybernetics.* New York: Harper & Row, 1961. Pp. 128. A general survey with some technical material. Includes brief bibliography and index.

PEDELTY, MICHAEL J. *An Approach to Machine Intelligence.* Washington, D.C.: Spartan, 1963. Pp. 125. A fairly short but detailed examination of the relations between cybernetic systems (including the abstract notion of systems representation) and psychological and physiological systems. Much of the book was originally written as class notes for the course on automata theory and neuromimes (that is, devices which imitate the properties of neurons) sponsored first by the U.S. Department of Agriculture Graduate School and later by the Aeronautical Systems Division of the Air Force Systems Command. See especially the introduction and chap. 2, "A Philosophy of Machine Intelligence." See also Pedelty's *Machine Intelligence and Its Implications for Design Philosophy* (Dayton, Ohio: National Aerospace Electronics Conference, 1962), and L. O. Gilstrap, R. J. Lee, and M. J. Pedelty's "Learning Automata and Artificial Intelligence," in E. M. Bennett, J. Degan, and J. Spiegel, eds., *Human Factors in Technology* (New York: McGraw-Hill, 1962).

PEJOVIĆ, DANILO. "Technology and Metaphysics," *Praxis* 2, nos. 1-2 (1966): 202-16. Technology is not just "a heap of instruments" but part of a specific attitude toward the world as a whole; as such it must be understood against the background of modern philosophy. A comparison of Aristotle's and Descartes's attempts to elucidate this fundamental background after which technology is described in terms of three characteristics: (1) its machines, (2) its progress, and (3) its metaphysical dominance of modern life. This dominance can be overcome only by recognizing the reality of something "that in its essence is not at all technical, although it makes technology itself possible and lends it a super-technical meaning"—that is, the *"natura-humanum* proportions and limits and thus points to the place where technology has its beginning and end."

PINSKY, L. "Do Machines Think about Machines Thinking?" *Mind* 60 (1951): 397-98.

POLANYI, MICHAEL. "Note on the Hypothesis of Cybernetics," *British Journal for the Philosophy of Science* 2 (February 1952): 312-15. The operations of a formalized deductive system cannot be equivalent to those of a mind because the semantic operations of knowing, understanding, and acknowledging the truth of axioms, etc., are not formalized, and if they were formalized, further informal semantic operations would then be needed, so that there would always be an originating mind at the end of the chain. Revised as a section in *Personal Knowledge.*

———. "Problem Solving," *British Journal for the Philosophy of Science* 8 (August 1957): 89-103.

———. *Personal Knowledge: toward a Post-critical Philosophy.* Chicago: Univ. Chicago Press, 1958. "He divides science up into pure science, technically justified science, and systematic technology.... Pure science is concerned with revealing ultimate reality. Technically justified science is concerned with that part of pure science which can bring great utilitarian benefits, e.g. the study of coal, but has little relevance to the main body of pure science. Systematic technology is the study of a technology which has developed systematic ideas of its own, e.g., electronics.... [F]or Polanyi applied science consists of technically justified science and systematic technology ... which differ from pure science by the fact that the main reason for their pursuit is their utility whereas pure science is pursued in order to apprehend ultimate reality."—R. J. Brownhill, "Toward a Philosophy of Technology," *Scientia*, vol. 104, no. 7 (November–December 1969).

———. *The Tacit Dimension*. Garden City, N.Y.: Doubleday, 1966; Doubleday Anchor, 1967. Pp. 108.

———. "The Logic of Tacit Inference," *Philosophy* 41 (January 1966): 1–18. Reprinted in F. J. Crosson, ed., *Human and Artificial Intelligence* (New York: Appleton-Century-Crofts, 1970).

POLYA, GYORGY. *How to Solve It*. Princeton, N.J.: Princeton Univ. Press, 1945. Reprinted, Garden City, N.Y.: Doubleday Anchor, 1957. See also *Mathematics and Plausible Reasoning*, Vol. 1: *Induction and Analogy in Mathematics;* Vol. 2: *Patterns of Plausible Inference*. Princeton, N.J.: Princeton Univ. Press, 1954.

POPPER, KARL. "Naturgesetze und Wirklichkeit." Pp. 43–60 in Simon Moser, ed., *Gesetz und Wirklichkeit*. Innsbruck: Tyrolia-Verlag, 1949. ". . . in applied science, unlike pure science, the problem of deducibility is to find initial conditions which, together with given theories, yield conditions specified by practical considerations. This is, indeed, how Popper characterizes technology at large. . . ."—Joseph Agassi, "The Confusion between Science and Technology in the Standard Philosophies of Science," *Technology and Culture*, vol. 7, no. 3 (Summer 1966).

———. *Of Clouds and Clocks; an Approach to the Problem of Rationality and the Freedom of Man*. Saint Louis: Washington Univ., 1966. Pp. 38.

PRICE, DEREK J. DE SOLLA. "Is Technology Historically Independent of Science? A Study in Statistical Historiography," *Technology and Culture* 6, no. 4 (Fall 1965): 553–68. In this same issue of *Technology and Culture*, cf. John J. Beer's "The Historical Relations of Science and Technology; Introduction" (pp. 547–52), Carl W. Condit's "Comment: Stages in the Relationships between Science and Technology" (pp. 587–90), and Robert E. Schofield's "Comment: On the Equilibrium of a Heterogeneous Social System" (pp. 591–95).

———. *The Difference between Science and Technology*. Detroit: Thomas Alva Edison Foundation, February 1968. Pp. 16. An examination of the similarities, differences, and parallels between science and technology which emphasizes the difference more than the similarities. "We have the position then, that in normal growth, science begets more science, and technology begets more technology. The pyramid-like exponential growths parallel each other, and there exists what the modern physicist would call a weak interaction . . . that serves just to keep the two largely independent growths in phase."

———. "The Structures of Publication in Science and Technology." Pp. 91–104 in W. H. Gruber and D. G. Marquis, eds., *Factors in the Transfer of Technology*. Cambridge, Mass.: M.I.T. Press, 1969. ". . . I now hypothesize that though it cannot be diagnosed by papers, technology has a structure that is formally identical with that of science. We shall define technology as that research where the main product is not a paper, but instead a machine, a drug, a product, or a process of some sort. In the same way as before there will be a highly competitive rat race, a similar exponential rate of growth, an archive of past technology, and a research front of the current state of the art. . . . With these definitions and a hypothesis we now have conjured up a pair of similar and parallel systems for science and technology." See also Price's "Networks of Scientific Papers," *Science* 149 (July 1965): 510–15; "Statistical Studies of Networks of Scientific Papers," *Symposium on Statistical Association Methods for Mechanized Documentation* (Washington, D.C.: National Bureau of Standards, June 1964); and "Citation Measures of Hard Science, Soft Science, Technology, and Non-science," paper presented at a conference on Communication among Scientists and Technologists, Johns Hopkins Univ., October 28–30,

1969; "Issledovanie o' issledovanii [Study about study], *Voprosy istorii estestvoznaniia i tekhniki*, no. 2 (1970), pp. 30–39. For a critique of this last paper see I. Ia. Konfederatov, "O zakonomernostiakh razvitiia nauki i tekhniki na sovremennom etape" [On regularities in the development of science and technology on the present-day state], *Voprosy istorii nauki i tekhniki*, no. 2 (1970), pp. 40–49.

PUCCETTI, ROLAND. "On Thinking Machines and Feeling Machines," *British Journal for the Philosophy of Science* 18 (May 1967): 39–51. Concludes that "... we seem to have reached a logical limit to machine technology. A machine can think, be intelligent, arrive at original and unforeseeable solutions to problems, modify its responses in accord with experience, assimilate human language—including the language of sensations and emotions—and even be self-reproducing. But so long as it is a *machine* it will not have feelings, precisely because its components belong to another order of nature than the organic. This is why the phrase 'feeling machine' seems permanently deviant, while 'thinking machine' is not" (p. 50). Discusses F. H. George, H. Putnam, and M. Scriven.

PUTNAM, HILARY. "Minds and Machines." Pp. 148–79 in Sidney Hook, ed., *Dimensions of Mind*. New York: New York Univ. Press, 1960. Reprinted, New York: Collier Books, 1961. The conceptual issues surrounding the traditional mind-body problem have nothing to do with the special character of human subjective experience, but arise for any highly complex computing system, in particular for any computing system able to construct theories concerning its own nature. Reprinted in Alan R. Anderson, ed., *Minds and Machines* (Englewood Cliffs, N.J.: Prentice-Hall, 1964), pp. 72–97.

———. "Robots: Machines or Artificially Created Life?" *Journal of Philosophy* 61, no. 21 (November 12, 1964): 668–91. "What I hope to persuade you is that the problem of the Minds of Machines will prove, at least for awhile,

to afford an exciting new way to approach quite traditional issues in the philosophy of mind. ... I suggest the question: Are robots conscious? calls for a decision, on our part, to treat robots as fellow members of our linguistic community, or not to so treat them. As long as we leave this decision unmade, the statement that robots (of the kind described) are conscious has no truth value." Reprinted in Stuart Hampshire, ed., *Philosophy of Mind* (New York: Harper & Row, 1966), and in F. J. Crosson, ed., *Human and Artificial Intelligence* (New York: Appleton-Century-Crofts, 1970).

———. "Brains and Behavior." Pp. 211–35 in R. J. Butler, ed., *Analytical Philosophy*. Oxford: Blackwell & Mott, 1965.

———. "The Mental Life of Some Machines." Pp. 177–200 in H. N. Castaneda, ed., *Intentionality, Minds, and Perception*. Detroit: Wayne State Univ. Press, 1967. Discusses the concepts of preferring, believing, and feeling in terms of a machine analog. "I hope to show by considering the use of these words in connection with a machine analog that the traditional alternatives—materialism, dualism, logical behaviorism—are incorrect, even in the case of these machines. ... What I claim is that seeing why it is that the analogs of materialism, dualism, and logical behaviorism are false in the case of these Turing Machines will enable us to see why the theories are incorrect in the case of human beings..." See also A. Plantinga's "Comments," ibid. (pp. 201–6), and the author's "Rejoinder," ibid. (pp. 206–13).

RASHEVSKY, N. "Is the Concept of an Organism as a Machine a Useful One?" *Scientific Monthly* 80 (January 1955): 32–35.

REAGAN, MICHAEL D. "Basic and Applied Research: a Meaningful Distinction?" *Science* 155 (March 17, 1967): 1383–86. Current debate over federal expenditures on basic or applied research and development assumes that an operational difference exists between the two. Includes a number of author's thoughts on the difference between basic and applied science.

REITMAN, WALTER R. *Cognition and Thought; an Information-processing Approach.* New York: Wiley, 1965. Pp. 312.

RENAUD, P. "L'invention rationnelle," *Structure et évolution des techniques,* vol. 8, nos. 53-54 (May-July 1957).

REUTER, H. K. "Zu einigen Begriffen auf technologischem Gebiet" [Some concepts in the domain of technology], *Fertigungstechnik,* new series, 16, no. 2 (1966): 88-90.

RICHARDSON, WILLIAM J. "Heidegger's Critique of Science," *New Scholasticism* 42, no. 4 (Fall 1968): 511-36. "What Heidegger offers is not a philosophy of science but a humanism for a scientific age."

RIERDIJK, CORNELIS WILLEM. *Een filosofie voor her cybernerisch-biotechnische tijdperk; 91 stellingen* [A philosophy for the cybernetic-biotechnical era; 91 theses]. Assen: Van Gorcum, 1967. Pp. 116.

ROBERTS, H. R. T. "Thinking and Machines," *Philosophy* 33 (October 1958): 356.

"Le robot peut-il servir a la connaissance de l'homme?" *Dialectica,* vol. 10 (December 1956). Contains P. de Latil's "De la machine considérée comme un moyen de connaître l'homme," A. Doyen and L. Liaigre's "Méthodologie comparée du biomécanisme et de la mecanique comparée, "L. Couffignal's "Quelques réflexions et suggestions," and P. Vernotte's "Le domaine de la machine et celui de l'homme."

"Le robot peut-il servir a la connaissance de l'homme?" *Dialectica* 13, nos. 3-4 (1959): 262-349. Contains an introduction and J. Sauvan's "Le facteurs de connaissance," A. David's "Réflexions pour un nouveau schéma de l'homme: but et moyens," P. Vernott's "A propos de la connaissance," F. Bonsack's "Connaissance et vérité," and F. Gonseth's "La connaissance de l'homme par l'intermédiaire du robot."

ROMBACH, H. *Substanz System Struktur: die Ontologie des Funktionalismus und der philosophische Hintergrund der modernen*

Wissenschaften. 2 vols. Freiburg-Munich: Alber, 1966. A historico-philosophical analysis of the movement from substance to structure or organization in Western ontological development. "This is a monumental work. . . . Authentic philosophical thinking has always been ontological, and structure no less than substance is a form or *species* of being. System too is a *species* of being which leads from substance to structure. Structure is only an articulation and intensification of substance. The concept of structure is the central notion of contemporary thought; it is at the focus of the natural as well as of the social sciences and it alone can bridge the gap between the 'two cultures.'"—M. J. Vetö, *Review of Metaphysics* 23, no. 1 (September 1969): 137.

RORTY, AMELIE O. "Slaves and Machines," *Analysis* 22 (1962): 118-20.

RORTY, RICHARD. "Functionalism, Machines, and Incorrigibility," *Journal of Philosophy* 69, no. 8 (April 20, 1972): 203-20. A reply to H. Putnam's "Mind and Machines," in S. Hook, ed., *Dimensions of Mind* (New York: New York Univ. Press, 1960).

ROSE, J., ed. *Survey of Cybernetics: a Tribute to Norbert Wiener.* London: Iliffe, 1970. Pp. 391. Contents: G. R. Boulanger's "Prologue: What Is Cybernetics?" S. Demczynski's "The Tools of the Cybernetic Revolution," J. F. Schuh's "What a Robot Can and Cannot Do," V. M. Glushkov's "Contemporary Cybernetics," F. H. George's "Behavioral Cybernetics," W. G. Walter's "Neurocybernetics," J. H. Clark's "Medical Cybernetics," A. M. Rosie's "Cybernetics and Information," G. Pask's "Learning and Teaching Systems," M. J. Apter's "Models of Development," C. T. Leondes and J. M. Mendel's "Artificial Intelligence Control," V. Strejc's "Cybernetics and Process Control," D. B. Foster's "Cybernetics and Industrial Processes," D. A. Bell's "Cybernetics and Ergonomics," N. S. Prywes's "Information and Retrieval," A. Crawford's "Management Cybernetics," R. W. Revans's "The Structure of Disorder," J. Rose's

"Cybernetics; Technological Barriers to Education," and S. Beer's "Epilogue: Prospects of the Cybernetic Age."

RUMPF, HANS. "Wissenschaft und Technik." In Ernst Oldemeyer, ed., *Die Philosophie und die Wissenschaften: Simon Moser Zum 65, Geburtstag.* Meisenheim: Hain, 1967.

——. "Gedanken zur Wissenschaftstheorie der Technik-Wissenschaften" [Thoughts on the scientific theory of the technology-sciences], *VDI-Zeitschrift* 111, no. 1 (1969): 2-10. Argues that there are no sharp limits between technical science and natural science with regard to the range of disciplines, formulation of problems, methodology, or structure of statements. The only distinction is that in technical science manufacturability is the central issue. Technical science (Tecknik-wissenschaften) is not, however, the same thing as technology (Technik) since knowledge of manufacturability is not identical to the manufacturing itself.

RUSSO, FRANÇOIS. "Une perspective fondamentale de la théorie de l'information et de l'action technique: la dualité de l'action concrete et du signal," *Dialectica*, vol. 12, no. 1 (March 1958).

——. "La création scientifique et technique," *Etudes* 305 (1960): 28-40.

RUYER, RAYMOND. "The Mystery of Reproduction and the Limits of Automatism," *Diogenes* 48 (Winter 1964-65): 53 ff.

SALLIS, JOHN. "Towards the Movement of Reversal: Science, Technology, and the Language of Homecoming." Pp. 138-68 in John Sallis, ed., *Heidegger and the Path of Thinking.* Pittsburgh: Duquesne Univ. Press, 1970.

SAYRE, KENNETH M., and FREDERICK J. CROSSON, eds. *The Modeling of Mind; Computers and Intelligence.* Notre Dame, Ind.: Univ. Notre Dame Press, 1963. Paperback reprint, New York: Simon & Schuster, 1968. Pp. 275. An anthology designed for "two groups of persons concerned with the impact of computer technology upon our conception of the human mind. . . . The two fields to which we refer may be described as (*a*) the philosophy of mental acts, and (*b*) the computer-oriented technology of the simulation of mental behavior. For convenience of reference, we will dub these two fields simply (*a*) 'philosophy' and (*b*) 'technology.'" Contains F. Crosson and K. Sayre's "Modeling: Simulation and Replication," A. Rapoort's "Technological Models of the Nervous System," L. Hiller and L. Isaacson's "Experimental Music," A. Newell's "The Chess Machine," H. Wang's "Toward Mechanical Mathematics," L. Wittgenstein's "Remarks on Mechanical Mathematics," G. Ryle's "Sensation and Observation," K. Sayre's "Human and Mechanical Recognition," N. Sutherland's "Stimulus Analysing Mechanisms," A. Gurwitsch's "On the Conceptual Consciousness," M. Polanyi's "Experience and the Perception of Pattern," D. M. MacKay's "Mindlike Behaviour in Artefacts," M. Scriven's "The Mechanical Concept of Mind," and J. R. Lucas's "Minds, Machines and Gödel."

——. *Recognition: a Study in the Philosophy of Artificial Intelligence.* Notre Dame, Ind.: Univ. Notre Dame Press, 1965. Pp. 312. Current research in artificial pattern recognition is mistaken in its assumption that recognition is identical to classification. In fact, logical analysis of the terms "classify" and "recognize" indicate that the two performances which they describe are not identical; thus, human beings recognize while machines classify. For a critical discussion, see N. Moray's "Superman and Supermachine," *Modern Schoolman* 47, no. 3 (March 1970): 339-46.

——. "Philosopny and Cybernetics." Pp. 3-33 in F. J. Crosson and K. M. Sayre, eds., *Philosophy and Cybernetics.* Notre Dame, Ind.: Univ. Notre Dame Press, 1967. ". . . a review of the current impact of cybernetics upon philosophy with (1) a survey of philosophic issues stemming from information theory . . . (2) a discussion of the mind-machine problem from the point of view of computing science . . . [and] (3) an estimation of the relevance of these topics for the tradi-

tional problems of mechanism, physicalism and determinism." Among the topics discussed are: the technical concept of information and its relation to meaning and semantics, entropy of natural language, game-playing and theorem-proving mechanisms, pattern recognition, and creativity.
——. "Intelligence, Bodies, and Digital Computers," *Review of Metaphysics* 21, no. 4 (June 1968): 714-23. "My basic disagreement with Dreyfus' analysis [Hubert L. Dreyfus's "Why Computers Must Have Bodies in Order to Be Intelligent," *Review of Metaphysics* 21, no. 2 (September 1967): 13-32] is with his argument that machines are intrinsically incapable of the sort of information processing exemplified in human pattern recognition.... I do not believe that a successful argument against the possibility of intelligent behavior in machines can be mounted on the fact that digital computers process information in the form of determinate machine states. In point of fact, we simply do not understand the basis of *human* intelligent behavior well enough to argue definitively either for or against the possibility of machine intelligence. My own present estimation, however, is that when we finally achieve an adequate understanding of human pattern recognition and problem solving, we will be able to simulate these functions on an entirely digital machine. Indeed, our being able to do so would itself provide evidence of an adequate understanding of these human capacities."
——. *Consciousness: a Philosophic Study of Minds and Machines.* New York: Random House, 1969. Pp. 237. "... an appropriate way to test and to extend one's understanding of a mental function is to attempt to describe in unequivocal terms how that function might possibly be approximated in a mechanical system or, if it cannot be so approximated, to explain in unequivocal terms why this is the case. The present essay thus begins and ends with a discussion of what I call 'the question of machine consciousness'" (p. vii). The closing

chapters develop the basic outlines of an "Information-processing Theory of Consciousness."
——. "Teaching Ourselves by Learning Machines," *Journal of Philosophy* 67, no. 21 (November 5, 1970): 908-18.
——. "Kybernetic." Pp. 1266-67 and 1271-86 in C. D. Kernig, ed., *Sowjetsystem und Demokratische Gesellschaft.* Freiburg: Herder, 1971.
——. "Information Processing and Mind-Brain Identity," *Kubenetes*, vol. 2 (April 1972).
SCHADEWALDT, WOLFGANG. *Die Anforderungen der Technik an die Geisteswissenschaften* [The demands of technology on the humanities]. Göttingen: Musterschmidt, 1957. Pp. 48.
——. *Natur, Technik, Kunst; drei Beiträge zum Selbstverständnis der Technik in unserer Zeit* [Nature, technology, art; three essays toward an understanding of technology in our time]. Göttingen: Musterschmidt, 1960. Pp. 64. The second essay is a valuable historico-philosophical study of the concepts of nature and technique among the Greeks which examines also the etymology of the modern terms "nature" and "technique."
——. "Humanität und Technik," *VDI-Zeitschrift* 104 (September 11, 1962): 1327. Analysis of the relationship between humanity and technology by a professor of classical philology at the University of Tübingen.
——. "Der Mensch in der technischen Welt," *Universitas: Zeitschrift für Wissenschaft, Kunst und Literatur* 19, no. 5 (1964): 449-67. Related, more popular articles: "Technik und menschliche Existenz," *Bulletin des Presse- und Informationsamtes der Bundesregieurung* (Bonn) 107 (1957): 974; "Der Mensch und die Technik," *Physikalische Blätter* 15 (1959): 337-42; and "Technology and Man," in Walter Leifer, ed., *Man and Technology* (Munich: Hueber, 1963). See also "Technik und Humanität sind einander zugeordnet," *Berichte und Informationen des Österreichischen Forschungsinstitutes für Wirtschaft und Politik* (Salzburg) 19 (1964): 15-16.
——. "Die Welt der modernen Tech-

nik und die altgriechische Kulturider" [The world of modern technology and the ancient Greek idea of culture], *Hellenika* 1 (1965): 14-21. Also included in *Hellas und Hesperion*, vol. 2 (Zurich: Artemis, 1960), along with "Die Anforderungen der Technik an die Geisteswissenschaften," and the three essays from *Natur-Technik- Kunst* in a section entitled "Griechentum und moderne Technik."

SCHAEFFER, RICHARD. "Martin Heidegger und die Frage nach der Technik," *Zeitschrift für Philosophische Forschung* 9 (1955): 116-27.

SCHEELE, WALTER. "Gedanken über Technik und Wissenschaft," *Kautschuk und Gummi*, vol. 8, no. 8 addition (1955).

SCHISCHKOFF, GEORGI. "Philosophie und kybernetik; zur Kritik am kybernetischen Positivismus," *Zeitschrift für Philosophische Forschung* 19, no 2 (1965): 248-78.

————. "Wissenschaftstheoretische Betrachtungen zum Informationsbegriff; Philosophisch-Anthropologische Aspekte, entbehrliche Sprachbereiche und Innenweltinformation" [Theoretical scientific reflections on the concept of information; philosophical-anthropological aspects, unnecessary spheres of language and information of the inner world], *Zeitschrift für Philosophische Forschung* 25, no. 1 (January-March 1971): 60-88.

SCHMIDT, HERMANN. "Die Regelungstechnik als technisches und biologisches Grundproblem" [Control technology as a fundamental problem of technology and biology], *VDI-Zeitschrift* 85, no. 4 (1941): 81-88.

————. "Der Mensch in der technischen Welt," *Physikalische Blätter* 7 (1953): 289 ff.

————. "Die Entwicklung der Technik als Phase der Wandlung des Menschen" [The development of technology as a phase of the transformation of man], *VDI-Zeitschrift* 96, no. 5 (1954): 118-22. The author sees automation as the final fulfillment of technology; he argues that the control loop is a "physical shadow" of man's somatic circuits and the circle of human action; and, finally, he has a vision of life coming to full self-awareness through seeing itself objectified in modern control loop technology.

SCHMITT, RICHARD. "Heidegger's Analysis of 'Tool,'" *Monist* 49, no. 1 (January 1965): 70-86. An attempt to restate in analytic terms Heidegger's distinction between two different senses of "thing." The same material is available in chap. 2 of Schmitt's *Martin Heidegger on Being Human* (New York: Random House, 1969).

SCHON, DONALD A. *Displacement of Concepts.* London: Tavistock, 1963. Pp. 208. Reprinted as *Invention and the Evolution of Ideas* (London: Associated, 1967). "The germ of this book was the notion that the evolution of theories is very much like processes of invention and product development as they occur in industry; that the two kinds of development can be seen as embodiments of a single underlying process, which I call the displacement of concepts." There is a discussion of the use of tools as a source of concept displacement to theories of decision, with particular reference to the pragmatic tradition (J. Dewey, C. Peirce, C. Stevenson). Concludes that, "if we see deciding in terms of the use of tools, it too becomes a manipulation of instruments to achieve an anticipated goal. All problems become a species of technical problem."

SCHORSTEIN, JOSEPH. "The Metaphysics of the Atom Bomb," *Philosophical Journal; Transactions of the Royal Philosophical Society of Glasgow* 1, no. 1 (January 1964): 33-46.

"Science and Engineering," Symposium in *Technology and Culture* 2, no. 4 (Fall 1961): 305-99. Contains J. K. Feibleman's "Pure Science, Applied Science, Technology, Engineering: an Attempt at Definitions," J. K. Finch's "Engineering and Science: a Historical Review and Appraisal," A. R. Hall's "Engineering and the Scientific Revolution," P. F. Drucker's "The Technological Revolution: Notes on the Relations of Technology, Science, and Culture," H. M. Leicester's

"Chemistry, Chemical Technology, and Scientific Progress," C. S. Smith's "The Interaction of Science and Practice in the History of Metallurgy," F. Kohlmeyer and F. L. Herum's "Science and Engineering in Agriculture: a Historical Perspective," M. Kerker's "Science and the Steam Engine," and J. B. Rae's "Science and Engineering in the History of Aviation."

SCRIVEN, MICHAEL. "The Mechanical Concept of Mind," *Mind* 62 (April 1953): 230–40. Also included in K. M. Sayre and F. J. Crosson, eds., *The Modeling of Mind* (Notre Dame, Ind.: Univ. Notre Dame Press, 1963) and in A. R. Anderson, ed., *Minds and Machines* (Englewood Cliffs, N.J.: Prentice-Hall, 1964).

———. "The Compleat Robot: a Prolegomena to Androidology." Pp. 118–42 in S. Hook, ed., *Dimensions of Mind.* New York: New York Univ. Press, 1960. Reprinted in F. J. Crosson, ed., *Human and Artificial Intelligence* (New York: Appleton-Century-Crofts, 1970).

———. "Philosophy of Education: Learning Theory and Teaching Machines," *Journal of Philosophy* 67, no. 21 (November 5, 1970): 896–908.

SEUNTJENS, HUBERT. Definition cybernétique de l'homme," *Cybernética* 5, no. 2 (1962): 103–15. A paper delivered at the IIId International Congress of Cybernetics, Namur, Belgium, September 11–15, 1961.

SHANNON, CLAUDE E. "Computers and Automata," *Proceedings of the Institute of Radio Engineers* (1953). A discussion of Turing machines, logic machines, learning machines, game-playing machines, and self-reproducing machines. Also in Z. W. Pylyshyn, ed., *Perspectives on the Computer Revolution* (Englewood Cliffs, N.J.: Prentice-Hall, 1971), pp. 114–27.

———. "Von Neumann's Contributions to Automata Theory," *Bulletin of the American Mathematical Society* 64 (1958): 123–29.

SHANNON, CLAUDE E., and JOHN MCCARTHY, eds. *Automata Studies.* Annals of Mathematics Studies, no. 34. Princeton, N.J.: Princeton Univ. Press, 1956. Contains D. M. MacKay's "The Epistemological Problem for Automata," J. McCarthy's "The Inversion of Functions by Turing Machines," J. von Neumann's "Probabilistic Logics and the Synthesis of Reliable Organisms from Unreliable Components," W. Ross Ashby's "Design for an Intelligence-Amplifier," J. T. Culbertson's "Some Uneconomical Robots," A. M. Uttley's "Patterns in a Conditional Probability Machine," S. C. Kleene's "Representation of Events in Nerve Nets and Finite Automata," etc.

SHANNON, CLAUDE E., and WARREN WEAVER. *The Mathematical Theory of Communication.* Urbana: Univ. Illinois Press, 1949. Pp. 117. Contains Shannon's "The Mathematical Theory of Communication," and Weaver's "Recent Contributions to the Mathematical Theory of Communication."

SHAW, R., T. HALWES, and J. JENKINS. "The Organism as a Mimicking Automation." Mimeographed. Minneapolis: Center for Research in Human Learning, Univ. Minnesota, 1966.

SHRADER, KRISTIN. "Cybernetics and Materialism." Notre Dame, Ind.: Univ. Notre Dame, 1972. Pp. 270. An unpublished dissertation which "is an attempt to determine whether adherence to the informational theories of cybernetics thereby commits one to materialism."

SIMON, HERBERT A. *Modeling Human Mental Processes.* RAND report P-2221. Santa Monica, Calif.: RAND Corp., February 20, 1961.

———. "The Logic of Rational Decision," *British Journal for the Philosophy of Science* 16 (November 1965): 169–86.

———. "Thinking by Computers" and "Scientific Discovery and the Psychology of Problem Solving." Pp. 3–40 in R. G. Colodny, ed., *Mind and Cosmos.* Pittsburgh: Univ. Pittsburgh Press, 1966. In the first essay, a theory of human problem solving is advanced with reference to some evidence for its validity. The theory "has been formalized and tested by incorporating it in programs for digital computers and studying the behavior of these programs when they are confronted with problem-solving tasks." In the second essay, his thesis is that "scientific discovery is a form of problem solving and . . . the processes whereby science is carried on can be explained in the terms that have been used to explain the processes of problem solving."

——. "The Logic of Heuristic Decision Making." Pp. 1–20 in Nicholas Rescher, ed., *The Logic of Decision and Action*. Pittsburgh: Univ. Pittsburgh Press, 1967. See also comments by R. Binkley and N. D. Belnap, Jr., with a reply by the author (pp. 21–35).

——. *The Sciences of the Artificial*. Cambridge, Mass.: M.I.T. Press, 1969. Pp. 123. The three Karl Taylor Compton Lectures at M.I.T. in the spring of 1968 in which the author makes explicit and "develops at some length a thesis that has been central to much of my research, at first in organization theory, later in management science, and most recently in psychology. The thesis is that certain phenomena are 'artificial' in a very specific sense: They are as they are only because of a system's being molded, by goals or purposes, to the environment in which it lives. If natural phenomena have an air of 'necessity' about them in their subservience to natural law, artificial phenomena have an air of 'contingency' in their malleability by environment. . . . Engineering, medicine, business, architecture, and painting are concerned not with the necessary but with the contingent—not with how things are but with how they might be—in short, with design. The possibility of creating a science or sciences of design is exactly as great as the possibility of creating any science of the artificial. The two possibilities stand or fall together. These essays, then, attempt to explain how a science of the artificial is possible. . . . I have taken as my main examples—in the second and third lectures respectively—the fields of the psychology of cognition and engineering design." And because the topics of artificiality and complexity are closely interwoven, the essay "The Architecture of Complexity" (originally published in *Proceedings of the*

American Philosophical Society 106, no. 6 [December 1962]: 462–82), is also included in this volume. Reviewed by A. C. Michalos, *Technology and Culture* 11, no. 1 (January 1970): 118–20.

SIMON, HERBERT A., and ALLEN NEWELL. "Information Processing in Computer and Man," *American Scientist* 52 (September 1964): 281–300. Also in Z. W. Pylyshyn, ed., *Perspectives on the Computer Revolution* (Englewood Cliffs, N.J.: Prentice-Hall, 1970), pp. 256–73, and in F. J. Crosson, ed., *Human and Artificial Intelligence* (New York: Appleton-Century-Crofts, 1970).

SIMON, MICHAEL ARTHUR. "Could There Be a Conscious Automaton?" *American Philosophical Quarterly* 6, no. 1 (January 1969): 71–78. Mentalistic language is the most natural, efficient, and logically appropriate kind of description to use for a machine whose behavior is indistinguishable from human behavior. It is impossible to use this mental terminology for machines and then to withhold ascription of consciousness.

SIMONDON, GILBERT. *Du mode d'existence des objets techniques*. Paris: Montaigne-Aubier, 1958. Pp. 269.

SINGH, JAGJIT. *Great Ideas in Information Theory, Language, and Cybernetics*. New York: Dover, 1966. Pp. 338. One of the best general introductions to the field. For a good discussion, see reviews by Rubin Gotesky and Hugh G. Petrie, with a reply from the author in *Philosophy Forum* 7, no. 4 (June 1969): 67–86.

SLUCKIN, W. *Minds and Machines*. Baltimore: Penguin, 1954. Pp. 223. Revised ed., 1960. Pp. 239. Still a good introduction by an engineer and psychologist. Each chapter is followed by a bibliography.

SMART, J. J. C. "Gödel's Theorem, Church's Theorem and Mechanism," *Synthèse* 13, no. 2 (1961): 105–10. Also see the author's *Between Science and Philosophy: an Introduction to the Philosophy of Science* (New York: Random House, 1968), chap. 9, "Determinism, Free Will, and Intelligence" (pp. 291–330).

SPENGLER, OSWALD. *Man and Technics; a Contribution to a Philosophy of Life*.

Translated by C. F. Atkinson. New York: Knopf, 1932. Pp. 104. From *Der Mensch und die Technik; Beitrag zu einer Philosophie des Lebens* (Munich: Beck, 1931). The argument is that "Technics is the tactics of living; it is the inner form of which the procedure of conflict—the conflict that is identical with Life itself—is the outward expression." Man is neither *Homo sapiens* nor *Homo faber*, but inventive carnivore. See also the distinction between "Faustian" and "Apollonian nature-knowledge" in *Decline of the West*, translated by C. F. Atkinson, 2 vols. (New York: Knopf, 1926-28), chap. 10; and chap. 21, "The Form-World of Economic Life: the Machine."

SPILSBURY, R. J. "Mentality in Machines," *Proceedings of the Aristotelian Society*, suppl. vol. 26 (1952): 27-60.

SPORN, PHILIP. *Foundations of Engineering*. New York: Macmillan, 1964. Pp. 143. By a leading electrical engineer. See chap. 1, "Philosophy of Engineering" (pp. 7-28), for a discussion of differences among the roles of the scientist, technician, technologist, and engineer. "The engineer is the key figure in the material progress of the world. It is his engineering that makes a reality of the potential value of science by translating scientific knowledge into tools, resources, energy, and labor to bring them into the service of man. Engineering goes a great deal beyond technical know-how, beyond the work of the technologist, skilled in the technology of a particular field" (pp. 22-23). Reviewed by J. K. Finch, *Technology and Culture* 6, no. 2 (Spring 1965): 263-66. See also Sporn's *Technology, Engineering, and Economics* (Cambridge, Mass.: M.I.T. Press, 1969).

SRZEDNICKI, JAN. "Could Machines Talk?" *Analysis* 22 (1962): 113-17. "It is the purpose of this article to find out what a machine would have to do if we were to say that it uses a language."

STACHOWIAK, HERBERT. *Denken und Erkennen im kybernetischen Modell* [Thought and perception in cybernetic model]. Vienna-New York: Springer-Verlag, 1965. Pp. 247.

STÄGER, H. J. "Noglichkeiten und Grenzen der technischen Entwicklung" [Possibilities and limits of technological development], *Schweizerische Technische Zeitschrift* 54, no. 40 (1957): 833-38.

STEEL, T. B., JR. "Artificial Intelligence Research," *Computers and Automation*, vol. 16 (January 1967). The contemporary controversy over artificial intelligence is resolving into two kinds of approaches. One believes that the creation of artificial intelligence is still the ultimate goal, but recognizes that it will be a long, slow process of development. The other group believes that the ultimate goal is more practical, namely, to perform certain operations more accurately or more cheaply than is possible for human operators. For this group, the goal is a division of labor between men and machines.

STEGMÜLLER, WOLFGANG. *Wissenschaftliche Erklärung und Begründung.* Vol. 1: *Probleme und Resultate der Wissenschaftstheorie und Analytischen Philosophie.* Berlin-New York: Springer-Verlag, 1969. Pp. 812. Contains a discussion (pp. 570-623) of automata, self-regulating systems, learning machines, Turing machines, etc.

STEINBUCH, KARL. *Automat und Mensch; über menschliche und maschinelle Intelligenz* [Automata and man; on human and machine intelligence]. Berlin-New York. Springer-Verlag, 1961. Pp. 253. 3d ed., 1965. A technical but comprehensive survey. Includes good bibliography and index.

———. "Künstliche Intelligenz," *Studium generale* 14, no. 7 (1961): 400-408.

———. "Zwölf Fragen zur Kybernetik" [Twelve questions on cybernetics], *Studium generale* 14, no. 10 (1961): 592-600.

———. "Kybernetik, Weg zu einer neuen Einheit der Wissenschaften" [Cybernetics, path to a new unity in the sciences], *VDI-Zeitschrift* 104, no. 26 (1962): 1307-14.

———. "Der Ingenieur und die Zukunft," *VDI-Zeitschrift* 111, no. 9 (May 1969): 549-52.

SZASZ, THOMAS S. "Minds and Machines," *British Journal for the Philoso-*

phy of Science 8 (February 1958): 310-16.

TAUBE, MORTIMER. *Computers and Common Sense; the Myth of Thinking Machines.* New York: Columbia Univ. Press, 1961. Pp. 136. A critique of research in artificial intelligence which, although valuable, is marred by some factual errors. Critically reviewed by Richard Laing in *Behavioral Science* 7 (April 1962): 238-40, and by W. R. Reitman in *Science* 135 (March 2, 1962).

"Technology and Causality." *Technology and Culture* 14, no. 1 (January 1973): 1-27. A symposium containing the following articles: M. McLuhan and B. Nevitt's "The Argument: Causality in the Electric World," J. Owen's "Comment: Effects Precede Causes," and F. Wilhelmsen's "Comment: Through a Rearview Mirror—Darkly."

THOMPSON, DENNIS. "Can a Machine Be Conscious?" *British Journal for the Philosophy of Science* 16 (May 1965): 33-43. Also in John R. Burr and Milton Goldinger, eds., *Philosophy and Contemporary Issues* (New York: Macmillan, 1972), pp. 283-91. "If we refuse to regard as conscious a robot which can do everything a human can, we (1) strengthen a certain kind of solipsism, and (2) commit ourselves more fully to Epiphenomenalism than we would if we rejected the Machine Theory." Cf. Robert J. Clack's "Can a Machine Be Conscious? Discussion of Dennis Thompson," *British Journal for the Philosophy of Science* 17, no. 3 (November 1966): 232-34.

THOMSON, R., and W. SLUCKIN. "Cybernetics and Mental Functioning," *British Journal for the Philosophy of Science* 4 (August 1953): 130-46. "The principal worth of cybernetics is to be found in those writings which confine themselves to what is at the root of the discussion: the negative feed-back hypothesis of neurophysiology." One function of metaphysical statements in cybernetics is to "express the analogy of 'man as something resembling a robot' in order to proclaim confidence in the cybernetic hypothesis as the most rational and comprehensive starting point for attacking the problem of explaining human thought and purpose."

TIMMERMANN, GERHARD. "Handwerk, Volkstechnik und Ingenieurtechnik und ihre gegenseitungen Beziehungen" [Crafts, folk technology, and engineering and their relationships], *Technikgeschichte* 37, no. 2 (1970): 130-45.

TÖRNEBOHM, HAKAN. "Kybernetik," *Studium generale* 11, no. 5 (1957): 283-91.

TOULMIN, STEPHEN. "Innovation and the Problem of Utilization." Pp. 24-38 in W. H. Gruber and D. G. Marquis, eds., *Factors in the Transfer of Technology.* Cambridge, Mass.: M.I.T. Press, 1969. "The development, testing, and spread of new machines, techniques, and industrial processes is only one species of innovation." The general process of innovation consists of three linked phases and "these three aspects of the process of innovation may be conveniently referred to by using names borrowed from zoology: (1) the phase of *mutation*, (2) that of *selection*, and (3) that of *diffusion* and eventual *dominance*." This "evolutionary" analysis applies to the internal processes of development of innovation within either natural science or basic technology or productive industry. For a critical discussion of Toulmin's model in this same book, cf. Tom Burns's "Models, Images, and Myths" (pp. 11-23). See also Toulmin's "Conceptual Revolutions in Science," in R. S. Cohen and M. Wartofsky, eds., *Boston Studies in the Philosophy of Science,* vol. 3 (New York: Humanities, 1967); and "The Complexity of Scientific Choice, 1: A Stocktaking," *Minerva* 2, no. 3 (Spring 1964): 343-59; and "The Complexity of Scientific Choice, 2: Culture, Overheads or Tertiary Industry?" *Minerva,* 4, no. 2 (Winter 1966): 155-69.

TROLL, JOHN H. "The Thinking of Men and Machines," *Atlantic Monthly* 194 (July 1954): 62-65. Also in John R. Burr and Milton Goldinger, eds., *Philosophy and Contemporary Issues* (New York: Macmillan, 1972), pp. 277-83.

TURING, A. M. "On Computable Numbers, with an Application to the *Entscheidungsproblem*," *Proceedings of the London Mathematical Society* 42 (1937): 230-65. "This is the classic paper on computability in which the Turing machine was first introduced. The *Entscheidungsproblem* was proposed by the Hilbert school of mathematicians as the central problem of logic. It is the problem of deciding whether a statement made in the symbolism of formal logic follows from the axioms of the logic—i.e., whether it is a theorem of the logic. It was shown by Turing to be an uncomputable problem."—Z. W. Pylyshyn, *Perspectives on the Computer Revolution* (Englewood Cliffs, N.J.: Prentice-Hall, 1971). Reprinted in M. Davis, ed., *The Undecidable* (Hewlett, N.Y.: Raven, 1965).

———. "Computing Machinery and Intelligence," *Mind* 59 (October 1950): 433-60. A classical statement of the view that computing machines can be constructed to simulate human conscious activities. Also in Alan R. Anderson, ed., *Minds and Machines* (Englewood Cliffs, N.J.: Prentice-Hall, 1964); E. A. Feigenbaum and J. Feldman, eds., *Computers and Thought* (New York: McGraw-Hill, 1963); and Z. W. Pylyshyn, ed., *Perspectives on the Computer Revolution* (Englewood Cliffs, N.J.: Prentice-Hall, 1971). Reprinted under the title "Can a Machine Think? in James R. Newman, ed., *The World of Mathematics*, vol. 4 (New York: Simon & Schuster, 1956).

[NOTE: Many philosophers have discussed the significance of Turing's work on the theory of automata; a number are referred to in this section. Two good introductions are B. A. Trakhtenbrot, *Algorithms and Automatic Computing Machines* (Boston: Heath, 1963), and Merle B. Turner, *Realism and the Explanation of Behavior* (New York: Appleton-Century-Crofts, 1971).]

TURNER, MERLE B. *Realism and the Explanation of Behavior.* New York: Appleton-Century-Crofts, 1971. Pp. 257. A defense of reductionism in psychology with chapters on how this general thesis is affected by Gödel's

theorem, the cybernetic hypothesis, the theory of automata, and computer simulation of human behavior. Contains a good survey of the philosophical literature surrounding Gödel's theorem. (J. R. Lucas, P. Benacerraf, J. J. C. Smart, etc.) and a discussion of the work of A. M. Turing, J. von Neumann, W. S. McCulloch, and W. A. Pitts on the theory of automata. Also surveys Anglo-American philosophical literature on the distinctions between men and machines (H. Putnam, F. H. George, M. Scriven, etc.). Good bibliography.

ULMER, KARL. "Die Wandlung des naturwissenschaftlichen Denkens zur Beginn der Neuzeit bei Galilei," *Jahrbuch für Philosophie* (1949), pp. 289-351. Concerned with the question of whether modern technology arose out of modern natural science or whether modern science arose out of technology.

VATE, DWIGHT VAN DE, JR. "The Problem of Robot Consciousness," *Philosophy and Phenomenological Research* 32, no. 2 (December 1971): 149-65. "...the production of a conscious being is always the collective achievement of society as a whole, nothing smaller having sufficient power. If one assumes that conscious flesh-and-bone persons are actual social achievements, is then the conscious machine a possible social achievement? This is the version of the problem of robot consciousness I propose to explore."

WALENTYNOWICZ, BOHDAN. "On Methodology of Engineering Design." Pp. 587-90 in *Proceedings of the XIV International Congress of Philosophy; Vienna, Sept. 2-9, 1968.* Vol. 2. Vienna: Herder, 1968. Identifies technology and engineering. "The thesis of this paper is that, seeing the differences between pure science and engineering as to their aims, and owing to the fact that the scientific cognition is based upon analysis and induction, whereas the engineering attitude upon synthetic and creative approach—the usefulness of the general scientific methodology for the engineering design is considerably limited." The methodology of engineering design and of engineering as a whole requires the concepts of such general theories as praxiology, general systems theory, cybernetics, and the problem-formulating and problem-solving procedures found in such new disciplines as decision theory, operations research, and game theory.

WALTER, W. GREY. *The Living Brain.* New York: Norton, 1953. Pp. 311. By an electroencephalographer. But see especially the chapter on automata, "Totems, Toys and Tools"(pp.114-32).

WARTOFSKY, MARX W. *Conceptual Foundations of Scientific Thought.* New York: Macmillan, 1968. Pp. 560. See the section, "Legislative Rules, Technical Maxims, and Normative Laws" (pp. 57-62), where three prescientific modes of knowing are considered: (1) anthropomorphic explanation, (2) inductive generalization, and (3) technical rules. "In each case certain fundamental patterns of explanation become formulated in their earliest way...." Technical rules are prescriptive or normative formulations of laws "in which these laws themselves are considered to have a certain necessity and a divine or extrahuman status, an objectivity transcending local human purposes." Natural necessity and determinism, the wider concepts of a developed science, are contained in germinal form in that prescientific way of knowing represented by technical rules.

WATANABE, SATOSI. "Epistemological Implications of Cybernetics." Pp. 594-600 in *Proceedings of the XIVth International Congress of Philosophy; Vienna, Sept. 2-9, 1968.* Vol. 2. Vienna: Herder, 1968. Cybernetics brings about a revolution in philosophical outlook, two salient features of which are: "First, the world in cybernetics is no longer an isolated system which is developing by itself independently of an observer and outside the reach of an agent. On the contrary, it is something which is constantly acted upon by the observer-agent and which is reacting on the observer-agent.... Second, knowl-

edge in cybernetics is not an unedited all-including collection of facts or its apathetic theorization. Knowledge is an intentionally selected collection of facts and a purposively structured theory about those selected facts." Contains a brief account of how this new cybernetic viewpoint affects the epistemological problems of causality, deductive logic, inductive inference, time, and concept formation. Concludes that the cybernetic outlook is a "humanization of our outlook."

WEBB, JUDSON. "Metamathematics and the Philosophy of Mind," *Philosophy of Science* 35, no. 2 (June 1968): 156-78. Argues that it is a mistake to apply the theorems of Gödel and Church to the philosophy of mind as evidence against mechanism. These theorems are presented as purely mathematical theorems, and various applications are discussed critically, in particular the use of Gödel's theorem by J. R. Lucas to distinguish conscious and unconscious beings. Concludes that general attempts to extract philosophy from metamathematics involve only dramatizations of the constructivity problem in foundations. More specifically, philosophical extrapolations from metamathematics involve premature extensions of Church's thesis. Cf. J. R. Lucas's "Metamathematics and the Philosophy of Mind: a Rejoinder," *Philosophy of Science* 38 (1971): 310-13.

WEBER, WERNER. "Vom Wesen und Sinn des technischen Schaffens" [On the essence and meaning of technological production], *Technische Rundschau* (Bern) 55, no. 45 (1963): 1-3.

WEIZÄCKER, CARL F. VON. "The Experiment." In *The World View of Physics,* translated by M. Grene. Chicago: Univ. Chicago Press, 1952. First published as a separate essay, "Das Experiment," *Studium generale,* vol. 1, no. 1 (1947-48). Argues that the experiment is a dialogue with nature. See also his essay "I-Thou and I-It in the Contemporary Natural Sciences," in P. A. Schilpp and M. Friedman, eds., *The Philosophy of Martin Buber* (LaSalle, Ill.: Open Court, 1967).

WHITELEY, C. H. "Minds, Machines and

Gödel: a Reply to Mr. Lucas," *Philosophy* 37 (January 1962): 61-62.

WIESER, WOLFGANG. "Organismen und Maschinen," *Merkur* 11, no. 5 (1957): 425-40.

WILKES, M. V. "Can Machines Think?" *Proceedings of the IRE* 41 (1953): 1230-34. See also Wilkes's *Automatic Digital Computers* (New York: Wiley, 1957).

WISDOM, J. O. "The Hypothesis of Cybernetics," *British Journal for the Philosophy of Science* 2 (May 1951): 1-24. "The basic hypothesis of cybernetics is that the chief mechanism of the central nervous system is one of negative feedback. . . . Secondly, cybernetics makes the hypothesis that the negative feedback mechanism explains 'purposive' and 'adaptive' behavior. Broadly speaking, what the cybernetic model does for our outlook is to make us understand how purposive behavior can be manifested by a machine, for 'purposive' can now be defined in terms of negative feedback." Studies the development of these concepts in the works of W. R. Ashby, N. Wiener, D. M. MacKay, W. S. McCulloch, W. A. Pitts, and L. von Bertalanffy. For comments in this same journal, cf. B. M. Adkins's "The Homeostat," p. 248; W. Mays's "The Hypothesis of Cybernetics," pp. 249-50; F. M. R. Walshe's "The Hypothesis of Cybernetics," pp. 161-63; and M. Polanyi's "The Hypothesis of Cybernetics," pp. 313-15.

————. "A New Model for the Mind-Body Relationship," *British Journal for the Philosophy of Science* 2 (February 1952): 295-301.

————. "Mentality in Machines," *Proceedings of the Aristotelian Society,* suppl. 26 (1952): 1-26.

————. "The Need for Corroboration; Comments on Agassi's Paper," *Technology and Culture* 7, no. 3 (Summer 1966): 367-70. "The application of Newtonian mechanics to resisting media is applied science; if the medium is highly specific, so that we take a special interest in it (such as water we want to fire torpedoes in it), we move into technology. The one is concerned with understanding and extending knowledge, the other with using it. . . . Applied science, though a step on the way to do something, is itself an extension of understanding. Applied science has sometimes been described as concerning with doing, which seems to me to be wrong in linking it with technology rather than with (pure) science."

WRIGHT, M. A. "Can Machines Be Intelligent?" *Process Control and Automation* 6 (1959): 2-6.

YOUNG, JOHN F. *Cybernetics.* New York: American Elsevier, 1969. Pp. 139.

YOVITS, MARSHALL C., and SCOTT CAMERON, eds. *Self-organizing Systems.* New York: Pergamon, 1960. Pp. 322. Contains H. von Foerster's "On Self-organizing Systems and Their Environments"; A. Newell, J. C. Shaw, and H. A. Simon's "A Variety of Intelligent Learning in a General Problem Solver"; G. Pask's "The Natural History of Networks"; W. S. McCulloch's "The Reliability of Biological Systems"; A. W. Burk's "Computation, Behavior, and Structure in Fixed and Growing Automata"; A. M. Uttley's "The Mechanization of Thought Processes"; etc. Also contains several useful panel discussions.

ZAPPONE, DOMENICO G. "Autonomia del autonomia e crisi del sapere filosofico," *Sapienza,* vol. 21, nos. 1-2 (January-June 1968).

ZIELENIEWSKI, JAN. "Why 'Cybernetics and the Philosophy of Technical Science' Only? Some Comments." Pp. 601-8 in *Proceedings on the XIVth International Congress of Philosophy; Vienna, Sept. 2-9, 1968.* Vol. 2. Vienna: Herder, 1968. A general account of the relations between cybernetics, philosophy of technology, praxiology, and general systems theory. Argues that the philosophy of action includes all these other fields.

[GENERAL NOTE: This section of the bibliography selectively includes both technical works on artificial intelligence, automata theory, cybernetics, and information theory, along with some general introductions to these same fields. This has been done primarily on the basis of those works most generally referred to by philosophers writing on these topics. For example, although the line is not a sharp one, Michael A. Arbib, W. Ross Ashby, Arthur Gill, Abraham Ginsburg, Seymour Ginsburg, Friedrich Hansen, A. Harrison, Yehoshua Bar-Hillel, Warren S. McCulloch, Marvin Minsky, John von Neumann, Gyorgy Polya, Claude E. Shannon, and W. Grey Walter (in English); Leon Brillouin and Louis Couffignal (in French); and Herbert Stachowiak and Karl Steinbuch (in German) are all primarily concerned with technical problems and only indirectly with philosophical issues. The same is true with the volumes edited by Franz L. Alt, D. V. Blake and A. M. Uttley, N. L. Collins and D. Michie, Claude E. Shannon and John McCarthy, and Marshall C. Yovits and Scott Cameron. Works listed under the following authors and titles can, however, serve as more general introductions to these subjects: "Automatic Control," Alfred J. Cote, Jr., Albert Ducrocq, Donald G. Fink, Walter R. Fuchs, Georges Theodule Guilbaud; "Information," Pierre de Latil, Neville Moray, V. L. Parsegian, Gordon Pask, Jagjit Singh, W. Slucken, and John F. Young.]

V. APPENDIX

A. CLASSICAL DOCUMENTS

ADAMS, HENRY (1838–1918). "The Dynamo and the Virgin." Chap. 25 in *The Education of Henry Adams.* Boston: Houghton Mifflin, 1918. A classic essay arguing that the dynamo symbolizes for modern man what the Virgin Mary did for medieval man. Cf. Lynn White, jr., "Virgin and Dynamo Reconsidered," *American Scholar* (Spring 1958), and Harvey Cox, "The Virgin and the Dynamo Revisited: an Essay on the Symbolism of Technology," *Soundings* 44, no. 2 (Summer 1971): 125–46.

BABBAGE, CHARLES (1792–1871). *On the Economy of Machinery and Manufactures.* 4th enlarged ed. London: Knight, 1841. 1st ed., 1832. See also

Babbage's *Reflections on the Decline of Science in England, and on Some of Its Causes* (London: Fellowes, 1830), and *The Exposition of 1851; or Views of the Industry, the Science, and the Government* (London: Murray, 1851). Cf. also Maboth Moseley, *Irascible Genius: the Life of Charles Babbage* (Chicago: Regnery, 1970).

BACON, FRANCIS (1561–1626). *Works.* The standard edition of the *Works* is that edited by Spedding, Ellis, and Heath, in 14 vols. (London: 1857–74). See especially the Preface and "Plan of the Work" which introduce *The Great Instauration* (1620). Two studies: Paolo Rossi, *Francis Bacon: from Magic to Science* (Chicago: Univ. Chicago Press, 1968), and Benjamin Farrington, *The Philosophy of Francis Bacon* (Chicago: Univ. Chicago Press, 1967).

BECKMANN, JOHANN (1739–1811). *Anleitung zur Technologie, oder zur Kentniss der Handwerke, Fabriken und Manufacturen* [Instruction on technology, or on knowledge of handicraft, management, and manufacture]. Gottingen, 1777. Pp. 460. 6th ed., 1809. The book that founded technology as a university subject. Cf. also Beckmann's *A History of Invention and Discoveries,* translated by W. Johnston (London: Bell, 1797). See also Wilhelm Franz Exner's *Johann Beckmann; Begründer der technologischen Wissenschaften* (Vienna: C. Gerold, 1878). Pp. 59.

COUDENHOVE-KALERGI, RICHARD NIKOLAUS (1894–). *Apologie der Technik.* Der Neue Geist 41/44. Leipzig: Neue Geist Verlag, 1922. Pp. 71. Largely popular sociology, although there is a sort of code-word use of some philosophical terms. Rather pontifical in tone. Dangers of technology treated in such cliches as "technology without ethics will just as surely lead to catastrophes as ethics without technology."—R. J. R.

———. *Revolution durch Technik.* Vienna–Leipzig: Paneuropa-Verlag, 1932. Pp. 101. A revised version of *Apologie der Technik* (1922) by an essayist and diplomat. Draws a Europe-Asia dichotomy, arguing that "Asia's significance lies in its ethics. Europe's significance lies in its technology." Technology is war with a nonhuman object. "The Asian spirit threatens the technological world revolution not in Asia but in Europe." Slightly more awareness of the dangers of technology, but the most significant change from the first draft of the book is the altered view of Soviet communism; it is no longer an experiment or fiasco, but a power to reckon with and to include in the world technological revolution. – R. J. R. For other articles, see Dessauer's bibliography.

D'ALEMBERT, JEAN (1717-83). *Preliminary Discourse to the Encyclopedia of Diderot.* First published, 1751. A good recent translation is that by Richard N. Schwab and Walter E. Rex (Indianapolis: Bobbs-Merrill, 1963). A number of articles in the *Encyclopedia* (1751-72) might also be consulted. See, e.g., those on "Art," "Automaton," "Commerce," "Invention," "Luxury," and "Man."

DuBOIS-REYMOND, A. *Erfindung und Erfinder* [Invention and inventor]. Berlin, 1906. Pp. 291. Partly philosophical, with biological analogies.

ENGELHARDT, VIKTOR (1891-). *Weltanschauung und Technik.* Leipzig: Meiner, 1922. Pp. 88. Short work with chapter titles such as "Darwinism and Technology," "Monism and Technology," "Positivist Empiricism and Technology," "Vaihinger and the Concept of Truth in Technology," "Pragmatism and Technology," "Will and Personality in Technology," "The Idea of Technology," "Ethics and Technology." Although one may be slightly suspicious when one sees this last topic, for instance, treated in three pages, the author makes no claims to exhaustiveness. The book's greatest value is probably historical, having first raised a number of questions which were not taken seriously until thirty or forty years later. – R. J. R. See also Engelhardt's *An der Wende des Zeitalters* (Berlin: Arbeitjugend-Verlag, 1925).

ESPINAS, ALFRED VICTOR (1844-1922). *Les origines de la technologie.* Paris: Al-

can, 1897. Pp. 290. An attempt "to give the history of technology in general or praxiology. The philosophy of knowledge has had its historians; it is perhaps not inappropriate to undertake the history of the philosophy of action." Chap. 1 of this book was first published as "Technologie physico-théologique," *Revue philosophique* (Paris: Alcan) 30, no. 2 (August 1890): 112-35; and 30, no. 3 (September 1890): 295-314. See also the author's *Etudes sur l'historie de la philosophie de l'action* (Paris: Bossard, 1925).

EYTH, MAX VON (1836-1906). *Lebendige Krafte; Sieben Vortrage aus dem Gebiete der Technik* [Living powers; seven lectures from the domain of technology]. Berlin: Julius Springer, 1903. Pp. 262. Lectures delivered between 1893 and 1903; mostly popular science. But the last lecture, "On the Philosophy of Invention" – although it deals primarily with the history of inventions – is of some historical interest. – R. J. R. Discussed by van Riessen in *Filosofie en Techniek* (Kampen: Kok, 1949). For a number of other articles by this author, see Dessauer's bibliography.

FORD, HENRY (1863-1947). *My Life and Work.* In collaboration with Samuel Crowther. Garden City, N.Y.: Doubleday, 1922. Pp. 289. The great industrialist assesses the significance of his own career and describes the ideals which he believes guided him. A German translation of this autobiography had great influence (indeed, Ford was decorated by the German government in 1936), even among intellectuals. See also *My Philosophy of Industry,* an authorized interview with Fay Leone Faurote (New York: Coward-McCann, 1929), which contains "Machinery, the New Messiah," "My Philosophy of Industry," "Success," and "Why I Believe in Progress."

GEISSLER, KURT W. "Ist Philosophie der Technik möglich?" [Is philosophy of technology possible?], *Technik und Kultur* (Verband deutscher diplomingenieure, Berlin) 16 (1925): 108.

HANFFSTENGEL, GEORG VON (1874-). *Technisches Denken und Schaffen* [Technological thinking and creation]. Berlin: Julius Springer, 1919. Pp. 212. 4th ed., 1927. Pp. 228. Discussed by van Riessen in *Filosofie en Techniek* (Kampen: Kok, 1949).

HARDENSETT, HEINRICH (1899–1947). *Der kapitalistische und der technische Mensch.* Munich: Oldenbourg, 1932. Pp. 128. Discussed by van Riessen in *Filosofie en Techniek* (Kampen: Kok, 1949). For a large number of other works by this author from the 1920s and 1930s, see Dessauer's bibliography.

HUXLEY, THOMAS HENRY (1825–95). "On the Hypothesis That Animals Are Automata, and Its History." In *Essays*, Vol. 1: *Methods and Results.* New York: Appleton, 1893. First published 1874.

KEOWN, R. MCARDLE. *Mechanism.* New York: McGraw-Hill, 1912. Pp. 169. 3d revised ed. by V. M. Faires. New York: McGraw-Hill, 1931. Pp. 242.

KIMBALL, D. S. *Elements of Machine Design.* New York: Wiley, 1909. Pp. 446. 2d revised enlarged ed. New York: Wiley, 1923.

KOREVAAR, ARIE. "Filosofie der Techniek," *Polytechnisch Weekblad* 26 (1932): 261–64.

———. *Techniek en Wereld-beschouwing.* Volkuniversiteits Bibliotheek, no. 59. Haarlem, 1934. Pp. 267. By a Dutch engineer. Discussed by van Riessen in *Filosofie en Techniek* (Kampen: Kok, 1949).

LA METTRIE, JULIEN OFFRAY DE (1709–51). *L'homme machine* (1747). Argues that although thought depends on matter and man is a machine, he is so complicated a machine he may never be understood. Numerous translations.

MACH, ERNST (1838–1916). *The Science of Mechanics; a Critical and Historical Exposition of Its Principles.* Translated by T. J. McCormack. 5th revised English ed. LaSalle, Ill.: Open Court, 1942. A translation of *Die Mechanik in ihrer Entwicklung historisch-kritisch Dargestellt,* first published in 1883.

MCKAY, ROBERT FERRIER. *The Theory of Machines.* London: E. Arnold, 1915. Pp. 440. Contents: Part I, "Mechanics"; Part II, "Kinematic of Ma-

chines." See also *Principles of Machine Design* (New York: Longmans, 1924).

MARX, KARL (1818–83). *Notes on Machines.* Translated by Ben Brewster. Leicester: Sublation, Students' Union, Leicester Univ., 1966. Pp. 17. A translation of a small section from *Grundrisse der Kritik der politischen Okonomie* written by Marx in 1857–58. Analyzes machinery as it affects labor and as "the most adequate form of *fixed capital.*" Conclusion: "Nature does not construct machines. . . . They are products of human industry: natural material transformed into organs of human power over nature or of its activity in nature. *They are organs of the human brain created by the human hand;* the power of knowledge objectified." Cf. also *Capital,* chap. 15, "Machinery and Modern Industry," where Marx distinguishes between tools and machines and argues at length that machines are a means for producing surplus-value in a capitalist economy.

MASON, OTIS T. (1838–1908). *The Origins of Invention: a Study of Industry among Primitive Peoples.* Cambridge, Mass.: M.I.T. Press, 1966. Pp. 448. First published, 1895. A classic study on the origin of modern industry which shows how man interacting with his environment learned the art of inventing.

MAXWELL, CLERK. "On Governors," *Proceedings of the Royal Society* 16 (1868): 270.

MORRIS, WILLIAM (1834–96). *Works.* See especially "News from Nowhere," "How We Live and How We Might Live," "Useful Work versus Useless Toil," and "A Factory as It Might Be."

REULEAUX, FRANZ (1829–1905). *The Kinematics of Machinery; Outlines of a Theory of Machines.* Translated and edited Alex B. W. Kennedy. London: Macmillan, 1876. Pp. 662. Reprinted, New York: Dover, 1963. From *Theoretische Kinematik; Grundzuge einer Theorie des Machinenwesens* (Braunschweig: F. Vieweg und Sohn, 1875). Attempts "to determine the conditions which are common to all machines, in order to decide what it is, among its great variety of forms, that essentially constitutes a machine" (p. 1). Chap. 1 examines the "machine-problem" and arrives at a tentative solution. Then, following a detailed examination of different types of machines and their combinations, Reuleaux arrives in chap. 13 at a confirmed definition: "the complete machine is a closed kinematic chain" (p. 502). And, "A machine is a combination of resistant bodies so arranged that by their means the mechanical forces of nature can be compelled to do work accompanied by certain determinate motions" (p. 503). See also O. Bottema, "F. Reuleaux, Filosoof der Techniek," *Algemeen Nederlands Tijdschrift voor Wijsbegeerte en Psychologie* 59, no. 1 (April 1967): 28-32.

_____. *Cultur und Technik; Vortrag gehalten im Niederösterreichischen Gewerbevereine am 14. November 1884.* Special issue of the "Wochenschriften des Niederösterreichischen Gewerbevereines." Vienna: Verlag des Niederösterreichischen Gewerbevereines, 1884. Pp. 37.

_____. "Technology and Civilization." Pp. 705-19 in Smithsonian Institute Annual Report for 1890. Translated from *Prometheus* (Berlin) 1 (1890): 625, 641, 666.

ROUSSEAU, JEAN JACQUES (1712-78). "A Discourse on the Arts and Sciences" (1750). A recent translation is by G. D. H. Cole in *The Social Contract and Discourses* (New York: Dutton, 1950). Argues that the arts and sciences have caused not just material progress but moral decay as well.

RUSKIN, JOHN (1819-1900). *The Two Paths; Being Lectures on Art and Its Application to Decoration and Manufacture.* New York: Wiley, 1859. Lectures delivered in 1858-59. See also Ruskin's *For Clavigera: Letters to the Workmen and Labourers of Great Britain* (New York: Wiley, 1871).

SCHNEIDER, MAX. "Uber Technik, technisches Denken und technische Wirkungen" [On technology, technological thinking and technological operations]. Erlangen, 1912. An unpublished dissertation.

SIEMENS, G. "Zur Philosophie der Technik," *Hochland,* vol. 24 (1926).

TAYLOR, FREDERICK W. (1856-1915). *Principles of Scientific Management.* New York: Harper, 1911. Pp. 144. Classical exposition of that theory of industrial management which has become known as Taylorism—crucial features of which are time-and-motion studies and systems of pay incentives.

URE, ANDREW (1778-1857). *The Philosophy of Manufactures: or, An Exposition of the Scientific, Moral, and Commercial Economy of the Factory System of Great Britain.* London: Knight, 1835. Pp. 480.

VEBLEN, THORSTEIN (1857-1929). *The Instinct of Workmanship and the State of the Industrial Arts.* New York: Sentry Press, Macmillan, 1914. Pp. 335. Contrasts elements of craft-oriented technology with industrial technology. Some of Veblen's other works might also be consulted for his analysis of industrial society and the conflict between technology and institutions. See, e.g., *The Theory of Business Enterprise* (New York: Scribner's, 1904), *The Place of Science in Modern Civilization* (New York: Huebsch, 1919), and *The Engineers and the Price System* (New York: Huebsch, 1921).

WENDT, ULRICH. *Die Technik als Kulturmacht.* Berlin: Reimer, 1906. Pp. 322. Discussed by van Riessen in *Filosofie en Techniek* (Kampen: Kok, 1949).

WIESNER, JULIUS VON. *Natur-Geist-Technik.* Leipzig: Engelmann, 1910. Pp. 428.

WILLIS, ROBERT (1800-1875). *Principles of Mechanism; Designed for the Use of Students in the Universities, and for the Engineering Students Generally.* London: Parker; Cambridge: Deighton, 1841. Pp. 446. 2d ed. London: Longmans, 1870.

[GENERAL NOTE 1: There are a number of other figures—particularly German historians, economists, and engineers—whose works might have been cited in this section. Among them: Bernard Bavink, Hans Drischel, R. Dvorak, Julius Goldstein, Friedrich von Gottl-Ottlilienfeld, Johann Gröttrup, Erich von Holst, Paul Oesterreich, Josef Popp, Julius Schenck, Franz Schnabel, Werner Sombart, E. Spranger, and Max Maria von Weber. For works by these and other figures, see Dessauer's bibliography along with the first section of Herlitzius's. Many are discussed by Dessauer in *Striet um die Technik* (Frankfurt: Knecht, 1956). But cf. too A. Spielhoff, *Die Technik als Problem der Kulturwissenschaft* (n. p., 1949).]

[GENERAL NOTE 2: For a good index to discussions relevant to the philosophy of technology in classical philosophical literature, see *Great Books of the Western World* (Chicago: Encyclopaedia Britannica, 1952), vols. 2 and 3, *The Great Ideas; a Syntopicon of Great Books of the Western World*, under the headings of "Art," "Knowledge," "Physics," "Progress," "Science," and "Mechanics."]

B. BACKGROUND MATERIALS

ALFORD, J. "Problems of a Humanistic Art in a Mechanistic Culture," *Journal of Aesthetics and Art Criticism*, vol. 20, no. 1 (Fall 1961).

ARMYTAGE, W. H. G. "Origins and Philosophy of Technology," *Nature* (London) 182 (November 15, 1958): 1349. Brief review of two books: *The History and Philosophy of Science and Technology* (London: Education and Training Department of the Electricity Council, 1958), and Sir Henry Self, *Some Implications of Modern Science* (London: Electricity Council, 1958). The first is a 139-page collection of six popular lectures; the second is an 18-page pamphlet.

———. *The Rise of the Technocrats; a Social History*. London: Routledge & Kegan Paul, 1965. Pp. 448. Cf. also this author's *A Social History of Engineering*, 2d ed. (London: Faber; and Cambridge, Mass.: M.I.T. Press, 1966).

ARON, RAYMOND. "Technology and History." Pp. 208–22 in *Progress and Disillusion*. New York: Praeger, 1968.

"L'art dans le mode de la technique," *Praxis* 2, no. 3 (1966): 267–324. Papers from a symposium of the same

title held at Varazdin December 6–9, 1965. Contains D. Grlic's "Wozu Kunst?" I. Focht's "Kunst in der Welt der Technik," D. Pejovic's "L'art et l'esthétique," and M. Kangra's "Philosophie und Kunst." For a brief report on the symposium, see *Praxis* 2, no. 1 (1966): 257–58.

AYRES, CLARENCE E. "Technology and Progress," *Antioch Review* 3, no. 1 (March 1943): 6–20. Argues for an understanding of progress in terms of technological advancement. Dated. Also in *Theory of Economic Progress* (Chapel Hill: Univ. North Carolina Press, 1944), pp. 105–24. Another chapter of interest in this book is "Technology and Institutions" (pp. 177–202).

BABINI, JOSÉ. *Ciencia y tecnologia: breve historia*. Buenos Aires: Columba, 1969. Pp. 68.

BEDINI, S. A. "The Role of Automata in the History of Technology," *Technology and Culture* 5, no. 1 (Winter 1964): 24–42.

BERNAL, J. D. *The Social Function of Science*. Cambridge, Mass.: M.I.T. Press, 1967. First published, 1939. Pp. 482. Chapters include "Historical Description of Science," "The Existing Organization of Scientific Research in Britain," "Science in Education," "The Application of Science," "Scientific Communication," "Science and Social Transformation."

———. "Science, Invention, and Social Applications of Technology." In Roy Wood Sellers, V. J. McGill, and Marvin Farber, eds., *Philosophy for the Future: the Quest of Modern Materialism*. New York: Macmillan, 1949.

———. *Science and Industry in the 19th Century*. London: Routledge & Kegan Paul, 1953. Pp. 230. Two essays: one on the relations of science and technology in the 19th century; the other analyzes in detail one interaction—the discovery of molecular asymmetry by Pasteur in 1848.

———. *Science in History*. 3d revised and illustrated ed. 4 vols. Cambridge, Mass.: M.I.T. Press, 1971. A Marxist analysis, originally the subject of the Charles Beard Lectures at Oxford in 1948. First published in 1954. 2d ed.,

1957. Vol. 1, *The Emergence of Science;* vol. 2, *The Scientific and Industrial Revolutions;* vol. 3, *The Natural Sciences in Our Time;* vol. 4, *The Social Sciences: Conclusion.*

BERNSTEIN, JEREMY. *The Analytical Engine: Computers—Past, Present and Future.* New York: Random House, 1964. Pp. 113. Paperback reprint, New York: Vintage, 1966.

BERNSTEIN, RICHARD J. *Praxis and Action: Contemporary Philosophies of Human Activity.* Philadelphia: Univ. Pennsylvania Press, 1971. Pp. 360. A study of Marxist, existentialist, pragmatist, and analytic theories of the nature of human activity. Provides a good background for considering technology as a type of human *praxis* or action, although this subject is not directly discussed.

BRUMBAUGH, ROBERT S. *Ancient Greek Gadgets and Machines.* New York: Crowell, 1966. Pp. 152. By a well-known Plato scholar and philosopher.

BURLINGAME, ROGER. *Backgrounds of Power; the Human Story of Mass Production.* New York: Scribner's, 1949. Pp. 372.

————. *Machines That Built America.* New York: Harcourt, Brace & World, 1953. Pp. 241. Popular history.

BURY, J. B. *The Idea of Progress; an Inquiry into Its Origin and Growth.* New York: Macmillan, 1932. Pp. 357. Classic study of the development of the modern faith in progress.

CADDEN, JOHN J., and PATRICK R. BROSTOWIN, eds. *Science and Literature: a Reader.* Boston: Heath, 1964. Pp. 310. Essays on the relation between science and culture, followed by essays on the relation between science and literature since the Renaissance, and an anthology of relevant literature. Valuable background material.

CALDER, RITCHIE. *After the Seventh Day: the World Man Created.* New York: Simon & Schuster, 1961. Pp. 448. Published in England as *The Inheritors: the Story of Man and the World He Made* (London: Heinemann, 1961).

————. *The Evolution of the Machine.* Princeton, N.J.: Van Nostrand, 1968.

Pp. 160. Published for *American Heritage,* in association with the Smithsonian Institution.

CAMPBELL, BLAIR. "La Mettrie: the Robot and the Automaton," *Journal of the History of Ideas* 31, no. 4 (October-December 1970): 555-72.

CARDWELL, DONALD S. L. "Power Technologies and the Advance of Science," *Technology and Culture* 6, no. 2 (Spring 1965): 188-207.

————. *Turning Points in Western Technology; Technology, Science and History.* New York: Neale Watson Academic Publications, 1972. Pp. 264.

CHAPUIS, ALFRED. *Les automates dans les oeuvres d'imagination.* Neuchâtel: Editions du Griffon, 1947.

CHAPUIS, ALFRED, and EDMOND DROZ. *Automata: a Historical and Technological Study.* Translated by A. Reid. Neuchâtel: Editions du Griffon, 1958. Pp. 408. See also A. Chapuis and Édouard Gélis's *Le monde des automates; étude historique et technique* (Paris, 1928).

CHERRY, C. E. "A History of the Theory of Information," *Methodos,* vol. 8 (1956). First published in a *Symposium on Information Theory* (London: Ministry of Supply, 1950). Reprinted, 1953.

CHILD, ARTHUR HENRY. *Making and Knowing in Hobbes, Vico, and Dewey.* University of California Publications in Philosophy, 16, no. 13: 271-310. Los Angeles: Univ. California Press, 1953.

CLAGETT, MARSHALL. *The Science of Mechanics in the Middle Ages.* Madison: Univ. Wisconsin Press, 1959. Pp. 771.

COHEN, JOHN. *Human Robots in Myth and Science.* Cranbury, N.J.: Barnes, 1967. Pp. 156.

CONZE, WERNER. "Die Strukturgeschichte des technisch-industriellen Zeitalters als Aufgabe für Forschung und unterricht" [The structural history of the technological-industrial age as task for research and study], *Arbeitsgemeinschaft für Forschung des Landes Nord-rhein-Westfalen; Geisteswissenschaften; im Auftrag des Ministerpräsidenten* (Cologne-Opladen) 66 (1957): 5-43.

————. "Technik und Gesellschaft in der

Geschichte," *Elektrotechnischer Zeitschrift,* Ausgabe A, 83, no. 25 (1962): 831-37.

COUTTS-SMITH, KENNETH. *The Dream of Icarus; Art and Society in the Twentieth Century.* New York: Braziller, 1970. Pp. 228. "Technology gives man power. Power gives man options. Parallel to the history of technology is the history of the options it has brought. The extraordinary thing striking anyone engaged in an exchange of views with an artist is that sooner or later he hears him making a statement that asserts or implies that the development of technology has *reduced* our options."—opening paragraph of a negative review by J. Le Corbeiller which tries to suggest why Coutts-Smith might feel this way, *Technology and Culture* 12, no. 2 (April 1971): 378.

CROMBIE, A. C. "Technics and Science in the Middle Ages." Chap. 4 in *Medieval and Early Modern Science.* Vol. 1. Garden City, N.Y.: Doubleday Anchor, 1959. This book is a revised edition of *Augustine to Galileo: the History of Science A.D. 400-1650* (Cambridge, Mass.: Harvard Univ. Press, 1953).

DANIELS, GEORGE H. "The Big Questions in the History of American Technology," *Technology and Culture* 11, no. 1 (January 1970): 1-21. From a symposium on the Historiography of American Technology. Cf. also in this same issue of *Technology and Culture,* J. G. Burke's "Comment: the Complex Nature of Explanations in the Historiography of Technology" (pp. 22-26), E. Layton's "Comment: the Interaction of Technology and Society" (pp. 27-31), and Daniel's "The Reply: Differences and Agreements."

DAUMAS, MAURICE. "Le myth de la révolution technique," *Revue d'histoire des sciences et leurs applications,* vol. 16, no. 4 (October-December 1963).

———. *A History of Technology and Invention: Progress through the Ages.* Translated by Eileen B. Hennessy. 2 vols. New York: Crown, 1969. Vol. 1, pp. 596; vol. 2, pp. 694.

———. "L'histoire des techniques: son objet, ses limites, ses méthodes,"

Revue d'histoire des sciences de leurs applications, vol. 22, no. 1 (January-March 1969).

DORN, HAROLD. "Technology and History," *Vector* 28 (January 1964): 16 ff. The importance of the history of technology in the evolution of culture.

DOUGHERTY, JUDE P. "Lessons from the History of Science and Technology." Pp. 34-50 in J. K. Ryan, ed., *Studies in Philosophy and the History of Philosophy.* Vol. 4. Washington, D.C.: Catholic Univ. America Press, 1969. Brief review of historical research of the last forty years which "firmly establishes the historical continuity of scientific thought and attests to the contributions which Christian theology and philosophy made by their teachings to the advancement of science and technology."

DRACHMANN, A. G. *The Mechanical Technology of Greek and Roman Antiquity.* Madison: Univ. Wisconsin Press, 1963. Pp. 218.

DUROCHER, AURELE ADELARD. *Verbal Opposition to Industrialism in American Magazines, 1830-1860.* Ann Arbor, Mich.: University Microfilms, 1955.

EASTWOOD, W., ed. *A Book of Science Verse: the Poetic Relations of Science and Technology.* London: Macmillan, 1961. Pp. 279.

EFRON, ARTHUR. "Technology and the Future of Art," *Massachusetts Review* 7, no. 4 (Fall 1966): 677-710. Begins with McLuhan but goes on to develop a more general analysis. "Technology, like violence, is too much a part of this age not to show up in a thousand forms in art. But also, like violence, it makes all the difference whether the artist treats it say, in the manner of William Faulkner, or in the manner of Mickey Spillane" (p. 689).

EISELEY, LOREN. *Francis Bacon and the Modern Dilemma.* Montgomery Lectures on Contemporary Civilization. Lincoln: Univ. Nebraska Press, 1962. Pp. 98. A rather literary meditation on Bacon's relationship to the modern technological world. After chapters on Bacon's prophetic vision and his work as scientist and educator, Eiseley argues that "for all his cynicism and knowledge of human fraility" Bacon "still believed in man." Moreover, while rejecting dogmatism in favor of a scientific doubt which led to certainty, "he failed to see that science, the doubter, might end in a metaphysical doubt itself — doubt in the rationality of man, doubt as to the improvability of man" (p. 70). There is also a suggestive discussion of the problem of human choice in the face of the ambiguous multiplication of possibilities from modern science.

EKIRCH, ARTHUR A. *Man and Nature in America.* New York: Columbia Univ. Press, 1963. Pp. 231.

ELIADE, MIRCEA. *The Forge and the Crucible.* New York: Harper & Row, 1962. Pp. 208. An analysis of primitive technology as a type of magic or alchemy in which matter is transformed into a higher form through the artisan's imitation of the divine creative act. The technologist completes divine creation and brings about the salvation of the cosmos.

ELSNER, HENRY, JR. *The Technocrats: Prophets of Automation.* Syracuse, N.Y.: Syracuse Univ. Press, 1967. Pp. 252. Traces the technocracy movement from 1919 to the present and concludes with a chapter on a sociological-political interpretation of technocracy.

EURICH, NELL. *Science in Utopia: a Mighty Design.* Cambridge, Mass.: Harvard Univ. Press, 1967. Pp. 332. Focusing on the 17th century, the author discusses, among others, Campanella, Andreae, Bacon, Hartib, Cowley, and Glanvill, pointing out the uses that they made of the science of their day in the ideal societies which they described.

FEUER, LEWIS S. *The Scientific Intellectual: the Psychologial and Sociological Origins of Modern Science.* New York: Basic, 1963. Pp. 441. Affirms, against the theories of Weber, Koestler, Merton, et al. that "the spirit of scientific research was born of hedonist-libertarian values" rather than a Protestant asceticism.

FINCH, JAMES K. *The Story of Engineering.* Garden City: N.Y.: Doubleday, 1960. Pp. 528.

FINLEY, M. I. "Technical Innovation and Economic Progress in the Ancient World," *Economic History Review,* 2d series, 18, no. 1 (August 1965): 29-45.

FLECKENSTEIN, JOACHIM O. "Die Einheit von Technik, Forschung und Philosophie im Wissenschaftsideal des Barock" [The unity of technology, research and philosophy in the scientific ideal of the Baroque Age], *Technik-Geschichte* 32, no. 1 (1965): 19-30.

FORBES, R. J. "Technology and Society," *Impact of Science on Society* 2, no. 1 (January-March 1951): 7-9. See also in this issue the annotated bibliography on technology and society.

———. *Man the Maker: a History of Technology and Engineering.* London: Abelard-Schuman, 1958. Pp. 365.

———. *The Conquest of Nature: Technology and Its Consequences.* New York: Praeger, 1968. Pp. 142. A generally positive account which the author contrasts to the conclusions of J. Ellul. "It is . . . the burden of the . . . preceding exposition that technology does *not* have such internal dynamism [as Ellul thinks] and is wholly incapable of setting its own rules on the basis of its own logic within a completely closed circle" (p. 110).

FORBES, R. J., and E. J. DIJKSTERHUIS. *A History of Science and Technology.* 2 vols. Baltimore: Penguin, 1963.

FORTI, UMBERTO. "Perché la storia della técnica," *Le macchine: bollettino dell' Instituto Italian per la Storia della Técnica* 1 (1967-68): 11-26. A survey of the philosophy of the history of technology.

FRANKLIN, H. BRUCE. *Future Perfect: American Science Fiction of the Nineteenth Century.* New York: Oxford Univ. Press, 1966. Pp. 402. The relegation of science fiction to the category of "sub-literature" as a result of the rise of realism at the end of the 19th century "appears at least in part, to be motivated by . . . the virulent anti-scientism and anti-rationalism so fashionable in literary circles today." A study of the work of Melville, Poe, Hawthorne, Twain, Bierce, Bellamy, etc. See, too, Kingsley Amis, *New Maps of Hell* (New York: Harcourt,

Brace, 1960), for a good survey of contemporary science fiction. Then cf. Dennis Livingston, "Science Fiction as a Source of Forecast Material," *Futures* 1, no. 3 (March 1969): 232-38.

FREYRE, GILBERTO. "Time, Leisure and the Arts: Reflections of a Latin-American on Automation," *Diogenes* 54 (Summer 1966): 104-15.

FRITSCH, VILMA. "Technik, Wissenschaft und Magie," *Technische Rundschau* (Bern) 58, no. 40 (1966): 1-2.

FURNAS, C. C., JOE MCCARTHY, and the editors of *Life. The Engineer.* New York: Time, Inc., 1966. Pp. 200. A good journalistic picture book introduction to the conceptual world of the engineer.

GANDILLAC, MAURICE DE. "The Role and Significance of Technique in the Medieval World," *Diogenes* 47 (Fall 1964): 125-39.

GELLNER, ERNST. *Thought and Change.* Chicago: Univ. Chicago Press, 1965. Pp. 224. Older philosophical theories of social change failed because they explained change as the result of Divine Will or an inexorable progress toward a material utopia. The author advances a new theory of change centered around man whose essence "resides in his capacity to contribute to and profit from industrial society."

GERICKE, HELMUTH. "Naturwissenschaften und Technik in früheren Zeiten" [Natural science and technology in earlier periods], *Humanismus und Technik* 9, no. 1 (December 19, 1963): 1-12.

GIEDION, SIEGFRIED. *Mechanization Takes Command.* New York: Oxford Univ. Press, 1948. Pp. 743. Documents the concrete processes by which the modern experience of space and time has become dominated by the characteristics of machine efficiency. For a comprehensive account of an attempt to bridge the gap that opened as a result between artistic conception and commercial production, see Hans M. Wingler, ed., *Bauhaus,* translated by W. Jabs and B. Gilbert (Cambridge, Mass.: M.I.T. Press, 1969).

GILLE, BERTRAND. *Esprit et civilisation techniques au Moyen Age* [Technical spirit and civilization in the Middle Ages]. Paris: Université, 1952. Pp. 25.
_____. "Technological Developments in Europe: 1100-1400." Pp. 168-219 in G. S. Métraux and F. Crouzet, eds., *The Evolution of Science; Readings from the History of Mankind.* New York: New American Library, 1963.
_____. *Engineers of the Renaissance.* Cambridge, Mass.: M.I.T. Press, 1966. Pp. 256. A reconstruction of Renaissance technology based on the author's extensive original research in this field.

GINESTIER, PAUL. *The Poet and the Machine.* Translated by M. B. Friedman. Chapel Hill: Univ. North Carolina Press, 1961. Pp. 183. Reviewed in *Technology and Culture* 3, no. 2 (Spring 1962): 230-34.

GLOAG, JOHN. *Artifex; or, The Future of Craftsmanship.* New York: Dutton, 1926. Pp. 96.

GOBLOT, H. "L'interaction des techniques dans leur genèse," *Revue philosophique de la France et .de l'étranger* 90, no. 2 (April-June 1965): 207-16.

HACKER, BARTON C. "Greek Catapults and Catapult Technology: Science and Technology, and War in the Ancient World," *Technology and Culture* 9, no. 1 (January 1968): 34-54.

HALL, A. RUPERT. "The Scholar and the Craftsman in the Scientific Revolution." Pp. 3-23 in M. Clagett, ed., *Critical Problems in the History of Science.* Madison: Univ. Wisconsin Press, 1959.
_____. "The Historical Relations of Science and Technology," *Imperial College of Science and Technology* (November 19, 1963), pp. 119-29. Also published as a separate ten-page pamphlet, London: Univ. London, 1963.

HALL, EVERETT W. *Modern Science and Human Values: a Study in the History of Ideas.* Princeton, N.J.: Princeton Univ. Press, 1957. Reprint, New York: Delta, 1966. Pp. 483. "Each of the two parts of the ensuing study is chiefly directed to the portrayal of a major revolution in men's attitudes toward facts and values, and the methods of investigating or assaying them." This is traced from medieval backgrounds to the present age. Part 1, "Attainment of the Method of Modern Sciences"; Part 2, "Toward an Independent Investigation of Values."

"The Hand: a *Lithopinion* Survey of the Craftsmen's Position in Today's World," *Lithopinion* (the Graphic Arts and Public Affairs Journal of Local One, Amalgamated Lithographers of America) 2, no. 1 (first quarter 1967), issue no. 5: 66–71. "Somehow, we have to close the gap between the hand and the brain. Our security, the quality of our lives, our comforts, and the forward movement of the country depend as much on the one as the other. . . . It is a much deeper question of remaining a healthy nation by not drifting away from the human heritage of the hand which, with its incredible skills, masters the physical world for us. It is a question of being whole men and whole women. . . ." See also in this issue E. Swayduck's "A 'Quantum Jump' in Craftsmanship," and J. J. McFadden's "Automation Needs Skilled Hands."

HANKINS, THOMAS L. *Jean d'Alembert: Science and the Enlightenment.* Oxford: Clarendon, 1970. Pp. 260.

HATFIELD, H. STAFFORD. *The Inventor and His World.* New York: Dutton, 1933. Reprinted, London: Pelican, 1948. "The distinction between applied science and invention, to my knowledge, was made by only one writer, the most important writer on technology, perhaps; I am referring to H. S. Hatfield and his *The Inventor and His World.* Hatfield does not draw the distinction explicitly, but he uses it clearly and systematically enough. Applied science, according to his view, is an exercise in deduction, whereas invention is finding a needle in a haystack." — Joseph Agassi, "The Confusion between Science and Technology in the Standard Philosophies of Science," *Technology and Culture,* vol. 7, no. 3 (Summer 1966).

HAUSMAN, CARL R. "Mechanism or Teleology in the Creative Process," *Journal of Philosophy* 58, no. 20 (September 28, 1961): 577–84.

HEILBRONER, ROBERT L. "The Impact of Technology: the Historic Debate." Pp. 7–25 in J. T. Dunlop, ed., *Automation and Technological Change.* Englewood Cliffs, N.J.: Prentice-Hall, 1962.

———. "Do Machines Make History?" *Technology and Culture* 8, no. 3 (July 1967): 335–45. Examines the role of technology in determining the nature of the socioeconomic order. See also Heilbroner's "Technological Determinism," in *Between Capitalism and Socialism* (New York: Vintage, 1970), pp. 147–64.

HEYDE, JOHANNES ERICH. "Zur Geschichte des Wortes 'Technik' " [On the history of the word 'Technik'], *Humanismus und Technik* 9, no. 1 (December 19, 1963): 25–43. Detailed hisorico-etymological study with philosophical implications. Cf. also a similar study in W. Schadewaldt, *Natur Technik, Kunst* (Göttingen: Musterschmidt, 1960).

HIERS, JOHN T. "Robert Frost's Quarrel with Science and Technology," *Georgia Review* 25, no. 2 (Summer 1971): 182–205. Frost denies the primacy of science-technology in human affairs without completely rejecting them. Instead he advocates "a pluralistic attitude toward all humanistic endeavor." Science, religion, and art each has its own peculiar validity.

HILLEGAS, MARK R. *The Future as Nightmare: H. G. Wells and the Anti-Utopians.* New York: Oxford Univ. Press, 1967. Pp. 200. A reassessment of the central importance of H. G. Wells in the development of literary reactions to modern science. Besides Wells's *A Modern Utopia* (1905) — a work which should not be read without some acquaintance with Wells's later, less optimistic thought, for example, *The Outlook for Homo Sapiens,* (1942) — there are a number of other 20th-century utopias and anti-utopias which deal especially with the ethical and political problems of modern science and technology. Among these are Karl Capek's *R.U.R.* (1923); Eugene Zamyatin's *We* (1924); Aldous Huxley's *Brave New World* (1932) and

Island (1960); C. S. Lewis's trilogy, *Out of the Silent Planet* (1938), *Perelandra* (1944), and *That Hideous Strength* (1946); George Orwell's *1984* (1949); and Ernst Jünger's *Heliopolis* (1950).

"The Historical Relations of Science and Technology." Symposium in *Technology and Culture* 6, no. 4 (Fall 1965): 547–95. Contains J. J. Beer's "Introduction," D. J. de Solla Price's "Is Technology Historically Independent of Science? A Study in Statistical Historiography," R. P. Multhauf's "Sal Ammoniac: a Case History in Industrialization," C. W. Condit's "Comment: Stages in the Relations between Science and Technology," and R. E. Schofield's "Comment: On the Equilibrium of a Heterogeneous Social System."

HOLTON, GERALD, ed. *Science and Culture.* Boston: Beacon, 1967. Pp. 348. Essays first published in *Daedalus* (Winter 1965). Divided into three parts. Part 1, "Definitions," includes: H. Levin's "Semantics of Culture," J. S. Ackerman's "On Scientia," E. R. Leach's "Culture and Social Cohesion: an Anthropologist's View," T. Parsons's "Unity and Diversity in the Modern Intellectual Disiciplines: the Role of the Social Sciences." Part 2, "On Coherences and Transformations," includes: H. Brooks's "Scientific Concepts and Cultural Change," G. Holton's "The Thematic Imagination in Science," D. K. Price's "The Established Dissenters," G. Kepes's "The Visual Arts and Sciences: a Proposal for Collaboration," and M. Mead's "The Future as the Basis for Establishing a Shared Culture." Part 3, "On Disjunction and Alienation," includes: O. Handlin's "Science and Technology in Popular Culture," E. Weil's "Science in Modern Culture," H. Marcuse's "Remarks on a Redefinition of Culture," D. Bells' "The Disjunction of Culture and Social Structure: Some Notes on the Meaning of Social Reality," R. Dubos's "Science and Man's Nature," and R. S. Morison's "Toward a Common Scale of Measurement."

HUGHES, THOMAS PARK, ed. *The Development of Western Technology since 1500. Main Themes in European History Series.* New York: Macmillan, 1964. Pp. 149.

IGGERS, GEORG G. "The Idea of Progress: a Critical Reassessment," *American Historical Review* 71 (1965–66): 1–17.

JENKINS, IEUAN MILES LEWIS. *Science and Technology.* London: Hamilton, 1965. Pp. 127.

JEWKES, JOHN, DAVID SAWERS, and RICHARD STILLERMAN. *The Sources of Invention.* 2d revised ed. New York: Norton, 1969. Pp. 372. 1st ed., 1958. A documented study of case histories which concludes that "there is nothing in the history of technology in the past century and a half to suggest that infallible methods of invention have been discovered or are, in fact, discoverable." Moreover, "the theory that technical innovation arises directly out of . . . advance in pure science does not provide a full and faithful story of modern invention," and there is nothing to "support the view that inventions can be predicted or that forecasts of their consequences can provide secure grounds for anticipatory social action."

JOLIVET, R. "Création et poésie," *Giornale di metafisica*, vol. 22, nos. 2–3 (March–June 1967).

JONES, HOWARD MUMFORD. "Ideas, History, Technology," *Technology and Culture* 1, no. 1 (Winter 1959): 20–27.

JONES, RICHARD FOSTER. *Ancients and Moderns: a Study of the Rise of the Scientific Movement in Seventeenth Century England.* 2d ed. Berkeley: Univ. California Press, 1965. Pp. 354. A valuable historico-philosophical study.

KAHN, ARTHUR D. " 'Every Art Possessed by Man Comes from Prometheus': The Greek Tragedians and Science and Technology," *Technology and Culture* 11, no. 2 (April 1970): 133–62.

KÄMF, HELLMUT. " 'Ars imitatio naturae,' " *Geschichte in Wissenschaft und Unterricht* 10 (1959): 84–94.

KAPOOR, A. *The Role of Science and Technology in the Development of Man through the Ages.* Delhi: Rainbow, 1965. Pp. 523.

KENNER, HUGH. "From Technology, a Style," *National Review* 21, no. 31 (August 21, 1969): 809-10. A brief review suggesting that the international style of architecture is a function of technology. "Immense machines that make the building workable, stacked on the acreage that will make it profitable, so dictate its structure as to leave the architect little more than a cosmetic function."

KERÉNYI, CARL. "Myth and Technique," *Diogenes* 49 (Spring 1956): 24-39.

KIRK, G. S. "Greek Science," *Philosophy Today* 5, no. 2 (Summer 1961): 108-13. First published in *Listener* 55 (February 23, 1961): 345 ff. Good summary from a typically modern point of view, which explains the absence of technology in the ancient period as a result of an emphasis on theory. Cf. also Nicholas Lobkowicz, *Theory and Practice: History of a Concept from Aristotle to Marx* (Notre Dame, Ind.: Univ. Notre Dame Press, 1967).

KLEMM, FRIEDRICH. *A History of Western Technology.* Translated by D. W. Singer. New York: Scribner's 1959. Pp. 401. From *Technik; eine Geschichte ihrer Probleme* (Freiburg: Alber, 1954). A collection of historical documents from technologists, theologians, naturalists, poets, economists, statesmen, etc., with a running commentary. Focuses on how historical circumstances altered technical development, and how intellectual forces interacted with technical progress. Reprinted, Cambridge, Mass.: M.I.T. Press, 1968.

KORFMACHER, W. C. 'Mechanization and Aesthetic Achievement," *Classical Bulletin* 38 (January 1962): 40.

KRAFFT, FRITZ. "Bemerkungen zur mechanischen Technik und ihrer Darstellung in der klassischen Antike" [Remarks on mechanical technology and its conception in classical antiquity], *Technik-Geschichte* 33, no. 2 (1966): 121-59.

———. "Die Stellung der Technik zur Naturwissenschaft in Antike und Neuzeit" [The place of technology and natural science in ancient and modern times], *Technik-Geschichte* 37, no. 3 (1970): 189-209.

KRANZBERG, MELVIN, and CARROLL W.

PURSELL, JR., eds. *Technology in Western Civilization.* 2 vols. New York: Oxford Univ. Press, 1967. A basic work. Attempts to show the effects of technological achievement upon the cultural, economic, and social patterns of human life. Vol. 1, *The Emergence of Modern Industrial Society — Earliest Times to 1900* (pp. 816); vol. 2, *Technology in the Twentieth Century* (pp. 800). Articles by a number of major figures in the field. Among them: R. J. Forbes, L. White jr., A. R. Hall, D. J. de S. Price, E. S. Ferguson, R. Burlingame, R. P. Multhauf, P. F. Drucker, C. R. Walker, R. Theobald, J. B. Rae, D. N. Michael, K. E. Boulding, etc. Valuable bibliography.

KRETZSCHMER, FRITZ. "Rätsel der antiken Technik" [The riddle of ancient technology], *VDI-Zeitschrift* 100, no. 24 (1958): 1169-71.

KUHNS, R. "Art and Machine," *Journal of Aesthetics and Art Criticism* 25, no. 3 (Spring 1967): 259-66.

LANDES, DAVID S. *The Unbound Prometheus: Technological Change and Industrial Development in Western Europe from 1750 to the Present.* Cambridge: Cambridge Univ. Press, 1969. Pp. 566. An extended version of the author's "Technological Change and Development in Western Europe, 1750-1914," in *Cambridge Economic History of Europe*, vol. 4, Habakkuk and Postan, eds., *The Industrial Revolutions and After* (London: Cambridge Univ. Press, 1965), part 1, pp. 274-603.

LASLETT, PETER. *The World We Have Lost.* London: Methuen, 1965. Pp. 280. A sympathetic account of the structure of English society before the Industrial Revolution. Detailed and documented.

LAYTON, EDWIN T. *Revolt of the Engineers; Social Responsibilities and the American Engineering Profession.* Cleveland: Case Western Reserve Univ. Press, 1971. Pp. 286. Basically a historical study of the abortive revolt of "engineering progressives" against business rationality in the years before and after World War I.

LEDERER, EMIL. "Technology." Pp. 553-59 in Edwin R. A. Seligman, ed., *En-*

cyclopedia of the Social Sciences. New York: Macmillan. 1934. Contains a good bibliography to 1934. See Lederer's *Technischer Fortschritt und Arbeitslosigkeit* (Tubingen: Mohr, 1931). Cf. also the *Encyclopedia* articles on "Machines and Tools," "Engineering," "Invention," "Industrialism," etc.

LENIHAN, JOHN M. A. "The Triumph of Technology," *Philosophical Journal: Transactions of the Royal Philosophical Society of Glasgow* (Edinburgh/London) 6, no. 1 (1969): 12–18. The eighth Lord Kelvin Lecture to the Royal Philosophical Society of Glasgow. "My purpose tonight is to examine another aspect of our debt to Lord Kelvin by considering how his life and work illuminate the relationship between science and technology. In particular I shall try to explain the importance of technology as the inspiration of science."

LENOBLE, ROBERT. *Mersenne ou la naissance du mecanisme.* Paris: J. Vrin, 1943. Pp. 633.

LEROI-GOURHAN, ANDRÉ. *Evolution et techniques.* 2 vols. Paris: Michel, 1943–45. Vol. 1, *L'homme et la matière;* vol. 2, *Milieu et techniques.* An anthropologist's study. Vol. 2 contains an interesting taxonomic classification of different kinds of machines.

LEWIS, ARTHUR O., JR., ed. *Of Men and Machines.* New York: Dutton, 1963. Pp. 349. A collection documenting the general cultural reaction to technology since the Industrial Revolution. Includes short selections by Capek, Mumford, Emerson, Bacon, Dickenson, Whitman, Butler, Frost, Poe, Forster, Snow, Bradbury, etc.

LILLEY, SAMUEL. *Men, Machines, and History.* New York: International Publishers, 1966. Pp. 352. A Marxist survey of the history of technology and its social effects from the beginning of agriculture to the space age, stressing "the tremendous scope and potential of the technological revolution."

LOBKOWICZ, NICHOLAS. *Theory and Practice: History of a Concept from Aristotle to Marx.* Notre Dame, Ind.: Univ. Notre Dame Press, 1967. Pp. 442.

LOWITH, KARL. "The Fate of Progress."

Pp. 145–61 in Arnold Levison, ed., *Nature, History and Existentialism.* Evanston, Ill.: Northwestern Univ. Press, 1966. Distinguishes progress, development, and change, then argues that the modern notion of progressive time is dependent on a rejection of eternal or universal time.

McMULLIN, ERNAN. "Medieval and Modern Science: Continuity or .Discontinuity?" *International Philosophical Quarterly* 5, no. 1 (February 1965): 103–29. A slightly different version of this paper first appeared in G. F. McLean, ed., *Philosophy in a Technological Culture* (Washington, D.C.: Catholic Univ. America Press, 1964), pp. 55–87. Argues for a basic continuity.

MAHR, OTTO. "Technikgeschichte: zur Autonomie der Technikgeschichte" [History of technology; on the autonomy of the history of technology], *Sudhoffs Archiv zur Geschichte der Medizin und der Naturwissenschaften* 42 (1958): 46–56.

MAIER, CHARLES S. "Between Taylorism and Technocracy: European Ideologies and the Vision of Industrial Productivity in the 1920s," *Journal of Contemporary History* 5 (1970): 27–61.

"The Making of Modern Science: Organization of Science and Technology." Part 7, pp. 661–753 in A. C. Crombie, ed., *Scientific Change.* New York: Basic, 1963. Contains articles by D. S. L. Cardwell, C. F. Carter, A. T. Grigoryan and B. G. Kuznetsov, and N. A. Figurovsky. Commentary and discussion by E. Ashby, R. H. Shryock, H. J. Habakkuk, A. R. Ubbelohde, W. A. Smeaton, D. W. Singer, S. Galin, R. Taton, B. Glass, M. Daumas, V. P. Zubov, D. S. L. Cardwell, C. F. Carter, A. T. Grigoryan, and N. A. Figurovsky.

The Man-made World: a Course on the Theories and Techniques That Contribute to Our Technological Civilization. Engineering Concepts Curriculum Project. 3 vols. New York: McGraw-Hill, 1968. A good introduction to the engineering way of thinking and doing.

MANUEL, FRANK EDWARD, ed. *Utopias and Utopian Thought.* New York: Houghton Mifflin, 1966. Pp. 321. First published in *Daedalus* (Spring 1965). Contains L. Mumford's "Utopia, the City and the Machine," J. R. Pierce's "Communications Technology and the Future," and B. de Jouvenel's "Utopia for Practical Purposes."

MARX, LEO. *The Machine in the Garden: Technology and the Pastoral Ideal in America.* New York: Oxford Univ. Press, 1964. Pp. 392. Reviewed by Harold D. Woodman in *Technology and Culture* 4, no. 4 (Fall 1965): 661-64.

———. "American Institutions and Ecological Ideals," *Science* 170 (November 27, 1970): 945-52. What are the prospects for fulfilling the ecological ideal of a healthy, life-enhancing interaction between man and the environment? Characterizes key social institutions from an ecological perspective and suggests the striking convergence of the scientific and the literary criticism of our national life-style.

MAYR, OTTO. *The Origins of Feedback Control.* Cambridge, Mass.: M.I.T. Press, 1970. From *Zur Frühgeschichte der technischen Regelungen* (Munich: Oldenbourg, 1969). A good history of automatic controls in mechanical systems.

———. "Adam Smith and the Concept of the Feedback System: Economic Thought and Technology in 18th-Century Britain," *Technology and Culture* 12, no. 1 (January 1971): 1-22.

MEIER, HUGO A. "Technology and Democracy: 1800-1860," *Mississippi Valley Historical Review* 43, no. 4 (March 1957): 618-40. Good account of the rise of the American optimistic attitude toward industrialization. Cf. also "American Technology and the Nineteenth-Century World," *American Quarterly* 10 (Summer 1958): 116-30.

MERTON, ROBERT K. *Science, Technology and Society in Seventeenth Century England.* New York: Fertig, 1970. Pp. 279. First published, 1938. Reprinted, New York: Harper Torchbooks, 1972.

MORISON, ELTING E. *Men, Machines, and Modern Times.* Cambridge, Mass.: M.I.T. Press, 1966. Pp. 235. "Made up of seven papers and a reflective summary, Morison's book is a valuable potpourri of naval history, reflections on bureaucracy, computer lore, steel making in nineteenth-century America, and an example of systematic social engineering. The unifying factor is Morison's attempt to examine attitudes and actions involving social change." — review essay by B. Barber, *Technology and Culture* 8, no. 4 (October 1967): 525-33.

MULTHAUF, ROBERT P. "The Scientist and the 'Improver' of Technology," *Technology and Culture* 1, no. 1 (Winter 1959): 38-47.

MUSSON, A. E., and ERIC ROBINSON. *Science and Technology in the Industrial Revolution.* Toronto: Univ. Toronto Press, 1969. Pp. 534. Argues that, contrary to received historical opinion, the Industrial Revolution was not a triumph of empiricism or illiterate practical craftsmen working independently of scientific knowledge. Early modern technology was already intimately bound up with modern science. Focuses on the early development of engineering and chemical industries.

NASH, RODERICK. *Wilderness and the American Mind.* New Haven, Conn.: Yale Univ. Press, 1967. Pp. 256. Contains a good bibliographic essay on the problem of the relation between man and nature in the West.

"Natural Science and Technology in Relation to Cultural Institutions and Social Practice." Part 2, pp. 135-222, in Charles A. Moore, ed., *Philosophy and Culture East and West.* Honolulu: Univ. Hawaii Press, 1962. Includes: W. H. Werkmeister's "Scientism and the Problem of Man," S. Kramrisch's "Natural Science and Technology in Relation to Cultural Patterns and Social Practices in India," N. Bammate's "The Status of Science and Technique in Islamic Civilization," H. Yukawa's "Modern Trend of Western Civilization and Cultural Peculiarities of Japan," and H. Shih's "The Scientific Spirit and Method in Chinese Philosophy."

NEEDHAM, JOSEPH. *Science and Civilization in China.* Cambridge: Cambridge Univ. Press, 1954-71. A monumental work. See especially vol. 4, part 2, "Mechanical Engineering."

_____. *Clerks and Craftsmen in China and the West: Lectures and Addresses on the History of Science and Technology.* In collaboration with Wang Ling, Lu Gwei-djen, and Ho Ping-yu. Cambridge: Cambridge Univ. Press, 1970. Pp. 375. "Of the projected seven 'Volumes' of *Science and Civilization in China,* four have thus far appeared in six large tomes. Meanwhile the project has produced a flood of by-products. Two monographs have been published: *The Development of Iron and Steel Technology in China* and *Heavenly Clockwork.* The present collection reprints nineteen lectures and papers, revised to avoid repetition, having to do with scientific and technical matters; at least fourteen others were omitted from this selection. Two other collections of papers are to appear, more sociological and philosophical in content, *Within the Four Seas—the Dialogue of East and West* and *The Grand Titration—Science and Society in East and West.*"—J. K. Fairbank, *Technology and Culture* 12, no. 2 (April 1971): 328. Numerous articles by Needham are also valuable. See, e.g., "Science and Society in East and West," *Centaurus* 10, no. 3 (1964): 174–97, a philosophical analysis of why Chinese civilization was more efficient than the West in applying natural knowledge to practical human needs between the 1st and 15th centuries.

NEF, JOHN U. *War and Human Progress; an essay on the Rise of Industrial Civilization.* Cambridge, Mass.: Harvard Univ. Press, 1950. Pp. 464. General argument: Peace rather than war has provided the impetus to real technological progress. At the same time, modern technology, which was supposed to eliminate war progressively by supplying men all those things they traditionally went to war to get, has wound up extending the nature of war itself because of the way it has undermined traditional religious convictions. Reprinted as a Harper Torchbook (1963), under the title *Western Civilization since the Renais-*

sance; *Peace, War, and the Arts.* Another paperback reprint by Norton (1968) under the original title.

_____. "The Genesis of Industry and the Origins of Modern Science." Pp. 200–269 and 288–92 in *Essays in Honor of Conyers Read* (Chicago: Univ. Chicago Press, 1953).

_____. *Cultural Foundations of Industrial Civilization.* Cambridge: Cambridge Univ. Press, 1958. Pp. 163. An economic historian goes beyond his specialization to explain "the rise of industrialism from the vantage point of general history." Especially good on politics of the Enlightenment. For brief statements of some main themes from this book, see "The Genesis of Industrialism and of Modern Science" in N. Downs, ed., *Essays in Honor of Conyers Read* (Chicago: Univ. Chicago Press, 1953), pp. 258–63, and the pamphlet *Civilization, Industrial Society, and Love* (Santa Barbara, Calif.: Center for the Study of Democratic Institutions, 1961).

_____. *The Conquest of the Material World.* Chicago: Univ. Chicago Press, 1964. Pp. 408. Reviewed by D. F. Dowd in *Technology and Culture* 6, no. 4 (Fall 1965): 651–53.

NELSON, B. "Scholastic Rationales of Conscience; Early Modern Crises of Credibility, and the Scientific-Technocultural Revolutions of the 17th and 20th Centuries," *Journal of the Scientific Study of Religion* 7 (Fall 1968): 170–77.

NISBET, ROBERT A. *Man and Technics.* University of Arizona Bulletin Series. Tucson: Univ. Arizona Press, 1956. Pp. 24. The importance of technics for modern man is not founded on the linear development of machines and technical insights but, rather, upon a world view which makes control of the environment central. A discussion of some factors responsible: development of the modern state, decline of medieval social organizations, rise of the doctrine of progress, etc.

_____. *Social Change and History: Aspects of the Western Theory of Development.* New York: Oxford Univ. Press, 1970. Pp. 335. Historically and analytically this book sets forth the essential sources and contexts of the Western idea of social development.

O'BRIEN, ROBERT, and the editors of *Life. Machines.* New York: Time, Inc., 1964. Pp. 200. A good pictorial introduction.

ONG, WALTER J., S.J., ed. *Knowledge and the Future of Man: an International Symposium.* New York: Simon & Schuster, 1968. Pp. 276. Contains a number of interesting essays, among the more important being: G. Kepes's "Art for a Changing Scale," J. Macquarrie's "The Doctrine of Creation and Human Responsibility," J. R. Zacharias's "The Spirit of Science and Moral Involvement," J. T. Noonan's "From Social Engineering to Creative Charity," and K. Rahner's "Christianity and the New Earth."

–––––– *Rhetoric, Romance, and Technology: Studies in the Interaction of Expression and Culture.* Ithaca, N.Y.: Cornell Univ. Press, 1971. Pp. 352.

ONIGA, TEODORO. "Cybernétique et création," *Cybernetica* 10, no. 3 (1967): 173-93.

PAPARELLA, BENEDICT A. "Progress and Modern Man," *Thomist* 25 (July 1962): 419-43. A generally superficial contrast of modern with Greek and medieval man, followed by the assertion (buttressed with many quotations) that human nature does not change. Concludes with a watery argument for the Christian direction of technology so that technological change equals progress.

PARENTE, M. ISNARDI. "Platone e la prima Accademia di fronte al problema delle idee degli 'artefacta,' " *Rivista critica di storia della filosofia,* vol. 19, no. 2 (April-June 1964).

PARKINSON, G. H. R. "The Cybernetic Approach to Aesthetics," *Philosophy* 36 (January 1961): 49-61.

PFLAUM, WALTER. "Leonardo da Vinci als Ingenieur" [Leonardo da Vinci as engineer], *Humanismus und Technik* 2, no. 1 (April 15, 1954): 24-38.

POLLACK, NORMAN. *The Populist Response to Industrial America.* Cambridge, Mass.: Harvard Univ. Press, 1962. Pp. 166. Argues that Populism was not a retreat from industrialism but a class

movement for social reform. According to a review by Theodore Saloutos in *Technology and Culture* 5, no. 3 (Summer 1964): 455-57, the thesis is probably true but unsupported by the evidence offered.

PRICE, DEREK J. DE SOLLA. *Little Science, Big Science.* New York: Columbia Univ. Press, 1963. Pp. 119. See also Price's *Science since Babylon* (New Haven, Conn.: Yale Univ. Press, 1961).

–––––– . "Automata and the Origins of Mechanism and Mechanistic Philosophy," *Technology and Culture* 5, no. 1 (Winter 1964): 10-23. See also Price's "An Ancient Greek Computer," *Scientific American* 200, no. 6 (June 1959): 60-67; "On the Origin of Clockwork, Perpetual Motion Devices, and the Compass," in *Contributions from the Museum of History and Technology, United States National Museum Bulletin* 218 (Washington, D.C.: Smithsonian Institution, 1959): 81-112; and "Gods in Black Boxes," in Edmund A. Bowles, ed., *Computers in Humanistic Research* (Englewood Cliffs, N.J.: Prentice-Hall, 1967), pp. 3-7.

PURSELL, CARROLL, W., JR., ed. *Readings in Technology and American Life.* New York: Oxford Univ. Press, 1969. Pp. 470. Although primarily historical, a valuable collection. Contains, for instance, Timothy Walker's "Defense of Mechanical Philosophy," a reply to Thomas Carlyle's attack on the mechanical age in "Signs of the Times" in the *Edinburgh Review* of 1829.

PURVER, MARGERY. *The Royal Society: Concept and Creation.* Cambridge, Mass.: M.I.T. Press, 1967. Pp. 246. See also Sir Harold Hartley, ed., *The Royal Society: Its Origins and Founders* (London: Royal Society, 1960).

RAE, JOHN B. "The 'Know-How' Tradition: Technology in American History," *Technology and Culture* 1, no. 2 (Spring 1960): 139-50.

REINGOLD, NATHAN. "Alexander Dallas Bache: Science and Technology in the American Idiom," *Technology and Culture* 11, no. 2 (April 1970): 163-77.

RIDER, K. J. *History of Science and Technology: a Select Bibliography for Students.*

2d ed. London: Library Association, 1970. Pp. 75. Short annotations. Favorably reviewed by E. S. Ferguson in *Technology and Culture*, vol. 12, no. 4 (October 1971).

RIVERS, J. P. W. "Technology and Literature," *New Scientist* 35 (1967): 355–57.

ROLLER, D. H. D., ed. *Perspectives in the History of Science and Technology*. Norman: Univ. Oklahoma Press, 1971.

ROSENBLOOM, RICHARD S. "Men and Machines: Some 19th-Century Analyses of Mechanization," *Technology and Culture* 5, no. 4 (Fall 1964): 489–511. An examination of 19th-century theories about the nature of factory mechanization and its effects on the working population. A discussion of Charles Babbage, Andrew Ure, Karl Marx, David A. Wells, Carroll Wright, and R. W. Cooke-Taylor.

ROSENBROCK, H. H. "Control: Past, Present and Future," *Radio and Electronic Engineer*, vol. 37, no. 1 (January 1969).

ROSENFIELD, LEONORA. *From Beast-Machine to Man-Machine: Animal Soul in French Letters from Descartes to La Mettrie*. New York: Oxford Univ. Press, 1941. Pp. 353.

ROSSI, PAOLO. *Philosophy, Technology and the Arts in the Early Modern Era*. Translated by S. Attanasio. New York: Harper Torchbooks, 1970. Pp. 194. From *I filosofi e le macchine (1400–1700)* (Milan: Feltrinelli, 1962). The Italian version was reviewed by M. Bunge in *Technology and Culture* 7, no. 1 (Winter 1966): 73–74: "This is a scholarly yet highly readable account of the revolution of the arts and crafts and the birth of modern technology during the Renaissance and the first century of the modern era."

RÜSEN, JÖRN. "Geschichte der Technik in philosophischer Perspective: Kritische Bemerkungen zu zwei Arbeiten über 'Philosophie der Technik' " [History of technology in philosophic perspective: critical remarks on two works concerning "philosophy of technology"], *Technik-Geschichte* 38, no. 1 (1971): 1–16. A critical commentary on H. Beck's *Philosophie der Technik; Perspectiven zu Technik, Men-*

schheit, Zukunft, and K. Schilling's *Philosophie der Technik; die geistige Entwicklung der Menschheit von den Anfängen bis zur Gegenwart.*

RUYER, R. "Les limites du progrès humain," *Revue de métaphysique et de morale* 4 (October–December 1958): 412–23.

SANZO, E. "William Blake and the Technological Age," *Thought* 46 (Winter 1971): 577–91.

SARTON, GEORGE. *The History of Science and the New Humanism*. Cambridge, Mass.: Harvard Univ. Press, 1937. Reprint, Indiana Univ. Press, 1962. Pp. 196. After outlining the "Faith of a Humanist," the author analyzes "The History of Science and the History of Civilization," discusses the progress of scientific thought since ancient times in "East and West," and proposes a solution for the educational and cultural crisis of our time in "The New Humanism" and "The History of Science and the Problems of Today." Cf. also Sarton's *A History of Science: Ancient Science through the Golden Age of Greece* (New York: Wiley, 1964).

SCHMITT, PETER J. *Back to Nature: the Arcadian Myth in Urban America*. New York: Oxford Univ. Press, 1969. Pp. 230.

SCHWEITZER, ALBERT. *The Philosophy of Civilization*. Translated by C. T. Campion. New York: Macmillan, 1949. The sections on the origins of modernity briefly discuss Francis Bacon and the rise of an ethical will to transform the world.

SEIBICKE, WILFRIED. *Versuch einer Geschichte der Wortfamilie um* τέχνη *in Deutschland vom 16. Jahrhundert bis etwa 1830* [Inquiry into the history of the word-family τέχνη in Germany from the 16th century to about 1830]. Technikgeschichte in Einzeldarstellungen, no. 10, Düsseldorf: VDI-Verlag, 1968. Pp. 392. Photomechanical reprint of a doctoral dissertation, and a work of much broader scope than its title indicates. Contains the most extensive etymology of the words τεχνη, *Technik,* and *Technologie* available. Extensive notes and good bibliography. Cf. also J. E.

Heyde, "Zur Geschichte des Wortes 'Technik,'" *Humanismus und Technik* 9, no. 1 (1963): 25–43; and W. Schadewaldt, "Die Begriffe 'Natur' und 'Technik' bei den Griechen," in *Natur, Technik, Kunst* (Gottingen: Musterschmidt, 1960).

"Selected Texts," *Impact of Science on Society,* vol. 2, nos. 3–4 (July–December 1951). Includes selected texts on science from Descartes, Rosseau, d'Alembert, Condorcet, Laplace, Saint-Simon, Valéry, Lionais, Rostand, followed by an annotated bibliography of work in French (1900–1950) on science and society.

SHEPARD, PAUL. *Man in the Landscape: a Historic View of the Esthetics of Nature.* New York: Knopf, 1967. Pp. 290.

SIMON, WILHELM. "Geologie: Naturwissenschaft zwischen Technik und Humanismus" [Geology: natural science between technology and humanism], *Humanismus und Technik* 5, no. 2 (April 7, 1960): 105–29.

SINGER, CHARLES. *Technology and History.* L. T. Hobhouse Memorial Trust Lecture, no. 21. London: Oxford Univ. Press, 1952. Pp. 19. Argues against the conception of history as a narrative of public events and a study of the formation of nations. Divides history of technology into six periods and argues that in the modern period (beginning about 1450–1650) "techniques have become dependent both for their practice and their development on science." The major characteristic of modern technology is its identification with science; "self-conscious science began to determine the main direction of technology." "The instrumental extension of the senses . . . has now assumed such proportions that every man of science has become a technician." There is also a brief account of the importance of Diderot, Fremont, and Dickinson to the history of technology.

SINGER, CHARLES, E. J. HOLMYARD, A. R. HALL, and TREVOR I. WILLIAMS, eds. *A History of Technology.* 5 vols. New York: Oxford Univ. Press, 1955–58. A basic history. *Technology and Culture,* vol. 1, no. 4 (Fall 1960), is entirely devoted to reviews of this work. Cf. also the one-volume summary and condensation of this massive work by T. K. Derry and Trevor I. Williams, *A Short History of Technology: From the Earliest Times to A.D. 1900* (New York: Oxford Univ. Press, 1961).

SKLAIR, LESLIE. *The Sociology of Progress.* New York: Humanities Press, 1970. Pp. 272. Surveys the history of the idea of progress stressing the correlation between this idea and the growth of science and technology. As an alternative to previous theories, this book develops a sociological meaning for progress based on the influence of scientific discoveries and technological innovations on society.

―――. "Moral Progress Revisited," *Philosophy and Phenomenological Research* 31 (March 1971): 433–39.

SMITH, CYRIL STANLEY. "Matter versus Materials: a Historical View," *Science* 162 (November 8, 1968): 637–44. Demonstrates the aesthetic interest of early craftsmen in their materials.

―――. "Art, Technology, and Science: Notes on Their Historical Interaction," *Technology and Culture* 11, no. 4 (Autumn 1970): 493–549.

SOLMSEN, FRIEDRICH. "Nature as Craftsman in Greek Thought," *Journal of the History of Ideas* 24, no. 4 (October–December 1963): 473–96.

SPIEGEL, H. W. "Theories of Economic Development: History and Classification," *Journal of the History of Ideas* 16, no. 4 (October 1955): 518–39.

SPIER, ROBERT F. *From the Hand of Man; Primitive and Preindustrial Technologies.* Boston: Houghton Mifflin, 1970. Pp. 159. "Modern technology is so complex that there is real value in examining primitive and preindustrial technologies, for which this work provides a description. . . . However, scholars who want to pursue the subject beyond the simplest introduction have better books to consult, including volume 1 of Singer's *History of Technology.* There is also available in French a similar but far more detailed treatment in the two-volume work by André Leroi-Gourhan (*L'homme et la matière* and *Milieu et techniques . . .). . . .* The book by Spier includes fifty-eight drawings of simple technological devices, that by Leroi-Gourhan includes 1,199 such drawings."—C. W. Meighan, *Technology and Culture* 13, no. 1 (January 1972): 66.

SUSSMAN, HERBERT L. *Victorians and the Machine; the Literary Response to Technology.* Cambridge, Mass.: Harvard Univ. Press, 1968. Pp. 261. Argues that the literary modes developed and the emotional attitudes held by Victorian writers have largely determined the modern responses to the machine.

SYPHER, WYLIE. *Literature and Technology: the Alien Vision.* New York: Random House, 1968. Pp. 257.

"Technology." Pp. 576–98 in David L. Sills, ed., *International Encyclopedia of the Social Sciences.* Vol. 15. New York: Macmillan and Free Press, 1968. Contains two articles: Robert S. Merrill's "The Study of Technology" and Warner Schilling's "Technology and International Relations." Each has a good basic bibliography. Cf. also the articles on "Automation," "Cybernetion," etc.

"Technology as Cause in History," Symposium in *Technology and Culture* 2, no. 3 (Summer 1961): 219–39. This symposium consists of two articles: R. Burlingame's "Technology: Neglected Clue to Historical Change" and L. Mumford's "History: Neglected Clue to Technological Change," with H. S. Hughes's "Commentary: Technology and the History of Ideas."

TEDDER, O. "Vom Wesen der englischen Technik" [On the essence of English technology], *Schweizerische Technische Zeitschrift* 53, nos. 32–33 (1956): 688–90.

TEGGART, FREDERICK J. *The Idea of Progress: a Collection of Readings.* Berkeley: Univ. California Press, 1927. Pp. 367. A good selection of materials from ancient to modern.

THOMAS, WILLIAM L., JR., ed., in collaboration with CARL O. SAUER, MARSTON BATES, and LEWIS MUMFORD. *Man's Role in Changing the Face of the Earth.* Chicago: Univ. Chicago Press, 1956. Vol. 1, pp. 488. Vol. 2, pp. 760. A massive interdisciplinary evaluation.

This is an updating of George Perkins Marsh's classic *The Earth as Modified by Human Action* (1875); reprinted as *Man and Nature* (Cambridge, Mass.: Harvard Univ. Press, 1965). Along this same line see Philip Wagner's more concise *The Human Use of the Earth* (New York: Free Press, 1964).

THOMPSON, EDWARD P. *The Making of the English Working Class.* New York: Vintage, 1963. Pp. 848. Three great influences molded the English working class in the period 1792–1832: the political counterrevolution, the increase in population, and the "technological aspects of the Industrial Revolution." See also the author's "Time, Work-Discipline, and Industrial Capitalism," *Past and Present* 38 (December 1967): 56–97.

THORNDIKE, LYNN. *A History of Magic and Experimental Science.* 8 vols. Vols. 1–2, New York: Macmillan, 1923; vols. 3–8, New York: Columbia Univ. Press, 1934–58.

TIMM, ALBRECHT. "Technologie und Technik im Übergang zwischen Mittelalter und Neuzeit" [Technology and technique in transition from the Middle Ages to modern times], *Vierteljahrsschrift für Sozial- und Wirtschaftsgeschichte* 46 (1959): 350–60.

―――. "Technologie und Polytechnika—ein Blick in die Geschichte" [Technology and polytechnics—a glimpse into history], *Humanismus und Technik* 8, no. 3 (April 22, 1963): 115–26.

―――. "Geschichte der Technik und Technologie: Grundsätzliches vom Standort des Historikers" [History of technique and technology: fundamental issue from the standpoint of the historian], *Technik-Geschichte* 35, no. 1 (1968): 1–13.

TOULMIN, STEPHEN, et al. *Seventeenth Century Science and the Arts.* Princeton, N.J.: Princeton Univ. Press, 1961. Pp. 137.

TOY, F. C. "Technology and the Literature and Philosophy," *Memoirs and Proceedings of the Manchester Literary and Philosophical Society* 99 (1957–58): 5–19.

TREUE, WILHELM. "Technikgeschichte

und Technik in der Geschichte" [History of technology and technology in history], *Technik-Geschichte* 32, no. 1 (1965): 3-18.

————. "Technischer Fortschritt," *Die deutsche Berufs- und Fachschule* 63, no. 7 (1967): 481-89.

TROITZSCH, ULRICH. *Ansätze technologischen Denkens bei die Kameralisten des 17. und 18. Jahrhunderts* [Beginnings of technological thought in the political economists of the 17th and 18th centuries]. Schriften zur Wirtschafts und Socialgeschichte, vol. 5. Berlin: Duncker & Humbolt, 1966. Pp. 193. Analyzes the literary background against which Johann Beckmann founded technology as a university subject with his *Anleitung zur Technologie* (1777).

TROST FRIEDRICH. "Die Technik in der Frühzeit des Menschen" [Technology in the early period of man], *Humanismus und Technik* 5, no. 3 (August 15, 1958): 186-201.

UBBELOHDE, ALFRED RENÉ. Edwardian Science and Technology; Their Interactions," *Nature* (London) 194 (1962): 1110-13.

————. *Man and Energy*. London: Hutchinson, 1954; New York: Braziller, 1955. Revised ed., Baltimore: Penguin Books, 1963. Pp. 247. Historical survey of the growth of human power over energy from the prehistoric domestication of slaves and animals to the modern control of atomic energy. Some description is also given to growth of knowledge about energy, with special reference to the laws of thermodynamics. "One of the most deep-seated instincts of mankind—to control and dominate its environment—is responsible for the developments of modern technology."

USHER, ABBOTT PAYSON. *A History of Mechanical Inventions*. Boston: Beacon, 1929. Pp. 411. A sketch of the history of mechanical invention intended for students of economic history who want to understand the historical development of the production and application of power. Influential work which raises a number of fundamental issues.

VARTANIAN, A. *La Mettrie's l'homme machine; a Study in the Origins of an Idea.* Princeton, N.J.: Princeton Univ. Press, 1960.

VEACH, HENRY B. *Two Logics: the Conflict between Classical and Neo-analytic Philosophy.* Evanston, Ill.: Northwestern Univ. Press, 1969. Pp. 280. Although only marginally relevant, the author suggests that modern logic is inherently technological or manipulative, whereas classical Aristotelian logic is not.

VERGEZ, A. "Technique et morale chez Platon," *Revue philosophique de la France et de l'etranger,* vol. 81, no. 1 (January-March 1956).

VERNANT, J. P. "Remarques sur les formes et les limites de la pensée technique chez les Grecs," *Revue d'histoire des sciences et de leurs applications,* vol. 10, no. 3 (July-September 1957).

VIERECK, PETER. "Poet in the Machine Age," *Journal of the History of Ideas* 10 (January 1949): 88-103.

WALKER, P. G. "The Origins of the Machine Age," *History Today* 16 (September 1966): 591-600.

WEST, THOMAS REED. *Flesh of Steel: Literature and the Machine in American Culture.* Nashville, Tenn.: Vanderbilt Univ. Press, 1967. Pp. 155. Two of the most distinctive characteristics of machines are strict discipline and massive energy. This study examines the works of Sherwood Anderson, Waldo Frank, John Dos Passos, Sinclair Lewis, and Carl Sandburg and the ways they deal with the machine in terms of these two characteristics which form "a unity as well as a polarity."

WHITE, LYNN, JR. "Technology and Invention in the Middle Ages," *Speculum* 15 (1940): 141-59. Reprinted in Hermann Ausubel, ed., *The Making of Modern Europe,* Vol. 1: *The Middle Ages to Waterloo* (New York: Dryden, 1951), pp. 47-58.

————. "Tibet, India and Malaya as Sources of Western Medieval Technology," *American Historical Review* 63, no. 3 (April 1960): 515-26.

————. *Medieval Technology and Social Change.* New York: Oxford Univ. Press, 1962. Pp. 194. Three studies of technology and social change in the European Middle Ages. Chap. 3, "The Medieval Exploration of Mechanical Power and Devices," is especially valuable as an argument for the continuity of medieval and modern technology. See also "Technology in the Middle Ages," in M. Kranzberg and C. W. Pursell, eds., *Technology in Western Civilization*, vol. 1 (New York: Oxford Univ. Press, 1967).

―――. "The Medieval Roots of Modern Technology and Science." Pp. 19–34 in K. F. Drew and F. S. Lear, eds., *Perspectives in Medieval History*. Chicago: Univ. Chicago Press (for Rice University in Houston, Texas), 1963. The technological dominance of Western culture which is characteristic of the modern world begins to be evident in the early Middle Ages and is clear by the later Middle Ages.

―――. "The Act of Invention: Causes, Contexts, Communities, and Consequences." Pp. 102–16. in C. F. Stover, ed., *The Technological Order*. Detroit: Wayne State Univ. Press, 1963.

―――. "What Accelerated Technological Progress in the Western Middle Ages?" Pp. 284–88 in A. C. Crombie, ed., *Scientific Change*. New York: Basic, 1963. German translation in *Technik-Geschichte* 32, no. 3 (1965): 214–19.

―――. *Machina ex Deo: Essays in the Dynamism of Western Culture*. Cambridge, Mass.: M.I.T. Press, 1968. Pp. 186. Reprinted as *Dynamo and Virgin Reconsidered; Machina ex Deo* (Cambridge, Mass.: M.I.T. Press, 1971). Collects a number of important essays; among them: "The Changing Canons of Our Culture," "Dynamo and Virgin Reconsidered," "The Historical Roots of our Ecologic Crisis," and "Engineers and the Making of a New Humanism."

―――. "The Iconography of Temperantia and the Virtuousness of Technology." Pp. 197–219 in T. K. Rabb and J. E. Sergel, eds., *Action and Conviction in Early Modern Europe; Essays in Memory of E. H. Harbison*. Princeton, N.J.: Princeton Univ. Press, 1969. Traces, first, the increased emphasis on temperance as the highest virtue

in 13th- and 14th-century thought and, second, the emergence of the mechanical clock as the image of the disciplined life. Reveals some of the historical foundations of Puritan asceticism.

WHITEHEAD, ALFRED NORTH. *Science and the Modern World*. New York: Macmillan, 1925. Numerous reprints. The first half of this book is a pioneering study of the social and political implications of the rise of the modern scientific and technological attitude toward the world; the second half addresses itself to the cosmological implications of the discovery of relativity and quantum mechanics. As Whitehead's treatment indicates, at some point the philosophy of technology becomes involved with problems of the philosophy of nature. Other works in this same area which might be consulted are: E. A. Burtt, *Metaphysical Foundations of Modern Physical Science*, revised ed. (New York: Doubleday Anchor, 1954); R. G. Collingwood, *The Idea of Nature* (Oxford: Clarendon, 1945); E. J. Dijksterhuis, *The Mechanization of the World Picture* (Oxford: Clarendon, 1961); Alexandre Koyré, *From the Closed World to the Infinite Universe* (Baltimore: Johns Hopkins Press, 1957); Thomas Kuhn, *The Structure of Scientific Revolutions* (Chicago: Univ. Chicago Press, 1964); and the following works by Anneliese Maier: *An der Grenze von Scholastik und Naturwissenschaft; die Struktur der materiellen Substanz, das Problem der Gravitation, die Mathematik der Formalatituden* (Rome: Edizioni di storia e letteratura, 1952); *Ausgehendes Mittelalter; gesammelte Aufsatze zur Geistesgeschichte des 14. Jahrunderts*, 2 vols. (Rome: Edizioni di storia e letteratura, 1964–67); *Metaphysische Hintergrunde de spatscholastischen Naturphilosophie* (Rome: Edizioni di storia e letteratura, 1955); and "Philosophy of Nature at the End of the Middle Ages," *Philosophy Today* 5, no. 2 (Summer 1961): 92–107.

WIENER, PHILIP P., and AARON NOLAND, eds. *Roots of Scientific Thought: a Cultural Perspective*. New York: Basic, 1957. Pp. 677. A collection of essays from *Journal of the History of Ideas*. L. Edelstein's "Recent Trends in the Interpretation of Ancient Science" has an extended discussion of ancient science and technology arguing that contrary to common assumptions the Greeks were not hostile to technology. P. Wiener's "Leibniz's Project of a Public Exhibition of Scientific Inventions" is also relevant.

WOLF, ABRAHAM. *A History of Science. Technology and Philosophy in the Sixteenth and Seventeenth Centuries.* With the cooperation of F. Dannemann and A. Armitage. 2d revised ed. by Douglas McKie. London: Allen & Unwin, 1950. Pp. 692. First published, 1935.

———. *A History of Science, Technology, and Philosophy in the Eighteenth Century.* 2d revised ed. by Douglas McKie. London: Allen & Unwin, 1952. Pp. 814. First published, 1939.

ZASTRAU, ALFRED. "Technik und Zivilisation im Blickfeld Goethes" [Technology and civilization from Goethe's perspective], *Humanismus und Technik* 4, no. 3 (June 15, 1957): 134-56. See also the author's "Gedanken über die Menschlichkeit in Zeugnissen Goethes," *Humanismus und Technik* 6, no. 2 (March 8, 1959): 63-80.

ZILSEL, EDGAR. "The Sociological Roots of Science," *American Journal of Sociology* 47, no. 4 (January 1942): 544-62.

———. "The Genesis of the Concept of Scientific Progress," *Journal of the History of Ideas* 6, no. 3 (June 1945): 325-49. Argues that the idea of progress originated in the technological writings of superior artisans. But cf. A. C. Keller, "Zilsel, the Artisans, and the Idea of Progress in the Renaissance," *Journal of the History of Ideas* 11, no. 2 (April 1950): 235-40. Both the Zilsel and the Keller essays are reprinted in P. Wiener and A. Noland, eds. *Roots of Scientific Thought* (New York: Basic, 1967).

ZÜRCHER, RICHARD. "Imagination und Technik," *Reformatio* 9 (1960): 415-21, 487-90.